Water Trading and Global Water Scarcity

Water scarcity is an increasing problem in many parts of the world, yet conventional supply-side economics and management are insufficient to deal with it. In this book the role of water trading as an instrument of integrated water resources management is explored in depth. It is also shown to be an instrument for conflict resolution, where it may be necessary to reallocate water in the context of increasing scarcity.

Recent experiences of implementation in different river basins have shown their potential as instruments for improving allocation. These experiences, however, also show that there are implementation challenges and some limitations to trading that need to be considered. This book explores the various types of water trading formulas through the experience of using them in different parts of the world. The final result is varied because, in most cases, trading is conditioned by the legal and institutional framework in which the transactions are carried out. The role of government and the definition of water rights and licenses are critical for the success of water trading.

The book studies the institutional framework and how transactions have been undertaken, drawing some lessons on how trading can improve. It also analyses whether trading has really been a positive instrument to manage scarcity and improve water ecosystems and pollution emission problems in those parts of the world which are most affected. The book concludes by making policy proposals to improve the implementation of water trading.

Josefina Maestu is Coordinator of the United Nations Office to Support the International Decade for Action "Water for Life" 2005–15, based in Zaragoza, Spain. For 12 years she has been a lecturer on economics of natural resources management at the University of Alcala, Spain.

About Resources for the Future and RFF Press

Resources for the Future (RFF) improves environmental and natural resource policy-making worldwide through independent social science research of the highest caliber. Founded in 1952, RFF pioneered the application of economics as a tool for developing more effective policy about the use and conservation of natural resources. Its scholars continue to employ social science methods to analyze critical issues concerning pollution control, energy policy, land and water use, hazardous waste, climate change, biodiversity, and the environmental challenges of developing countries.

RFF Press supports the mission of RFF by publishing book-length works that present a broad range of approaches to the study of natural resources and the environment. Its authors and editors include RFF staff, researchers from the larger academic and policy communities, and journalists. Audiences for publications by RFF Press include all of the participants in the policy-making process – scholars, the media, advocacy groups, NGOs, professionals in business and government, and the public.

Water Trading and Global Water Scarcity

International Experiences

Edited by Josefina Maestu

LONDON AND NEW YORK

First published 2013
by RFF Press

2 Park Square, Milton Park, Abingdon, Oxfordshire OX14 4RN
711 Third Avenue, New York, NY 10017

Routledge is an imprint of the Taylor & Francis Group, an informa business

First issued in paperback 2017

British Library Cataloguing in Publication Data
A catalogue record for this book is available from the British Library

Library of Congress Cataloging-in-Publication Data
Water trading and global water scarcity : international experiences / edited by Josefina Maestu. – 1st ed.
p. cm.
Includes bibliographical references and index.
1. Water transfer. 2. Water-supply–International cooperation. I. Maestu, Josefina.
HD1691.W3655 2013
333.33'9–dc23
2012021794

ISBN13: 978-0-415-63821-0 (hbk)
ISBN13: 978-1-138-57310-9 (pbk)

Typeset in Baskerville
by FiSH Books Ltd, Enfield

Contents

Illustrations

Boxes

Figures

Tables

Preface

From Rosegrant and Dinar's *Markets for Water: Potential and Performance* (1998) to Anderson, Scarborough and Watson's *Tapping Water Markets* (2012), there has been increasing interest in water trading. Not surprisingly, as there is growing evidence of water scarcity in the world, changing and growing economies will need to consider water reallocation as an option.

Our book has been a long time in the making. For me it all started when, from 1990 to 1995, there was a serious drought in Spain. There had been no precedent for decades and it provided painful evidence of the limitation of existing water supply policies as the only answer. José Manuel Naredo, the initiator of water economics in Spain, and a mentor to so many of us, asked me then to reflect on what happened and make proposals for making more flexible the water licensing/water use rights system in Spain. This was included in a 1997 publication on *The Water Economy in Spain.*

As coordinator of Thematic Week Seven of the so-called Water Tribune, the EXPO Zaragoza 2008 gave me an opportunity to bring together an impressive group of some of the best water economists in the world, to present and discuss water trading. Since then much has happened. The different chapters of this book are based on the original papers of the conference, and they have been updated, and many redrafted, to reflect what has changed.

I hope you will find this book useful. The key contribution is to play out an updated discussion of the state of the art, as reflected in real experiences in water trading – how they have helped us so far to deal with growing water allocation issues, not in theory but in practice, where they have been applied. These experiences show that there are implementation challenges and some limitations to trading that need to be considered. Most of this book is therefore about experiences, practical policy implementation, and how to improve it.

Water trading has been controversial, with strong detractors and defenders. This book does not hide this. This is why it brings different perspectives and, importantly, deals explicitly with some of the major concerns on water trading, such as privatization of the commons, environmental impacts, legal reforms, monopolistic and speculative practices, and price formation. The book does not advocate that water trading can be the single overall response to water scarcity as opposed to others such as infrastructures (supply-side solutions) or government regulation/intervention. The intention is to show how water trading can complement these and be one more instrument of integrated water resources management. It is part of the solution, and one that needs to be applied wisely.

I owe my thanks to a number of people and institutions:

- First and foremost to the EXPO Zaragoza, which in 2008 brought together an impressive group of water economists from all over the world, for discussions in the Water Tribune. This book has been possible because of the vision and the herculean efforts of many in the EXPO, and especially those of Roque Gistau, Jerónimo Blasco and Eduardo Mestre. Thanks to them and the present managers for their interest in disseminating the results (the legacy) of the work during thematic week seven and for giving us the opportunity to do so through this book.
- To my team of economic analysis of the Ministry of Environment from 2003 to 2009, and especially to those that supported the organization of the thematic week: Almudena Gómez-Ramos and Sici Sanchez. This book needs to be a tribute to them and the other members of my team (Alberto del Villar, Carlos Mario Gómez, Carlos Gutierrez, Daniel Cabello and Lorenzo Domingo), as well as to the Heads of Planning of the River Basin Authorities in Spain and its coordinators Teodoro Estrela and Federico Estrada, who took to water economics and economic instruments during those seven magic years in Spain.
- To the authors who have contributed to the book. Their enthusiasm has provided the necessary energy for me to continue. In spite of the potential complexity of a book with so many authors, their professionalism has made it mostly a real pleasure. I am honored for the trust they have shown toward me.
- To my present colleagues in the United Nations Department of Economic and Social Affairs (UN-DESA), and to the United Nations for allowing me to publish this book. A disclaimer is in place since this book does not reflect in any way the opinion of the UN. It was prepared before I joined the UN.
- Last, but not least, to Roger, Diana and Robin for putting up with my often "absent mind" and for their encouragement, and to my parents (Ceferino and Josefina) for providing always such an example of ethical behaviour and hard work. To all in my numerous family, for being always such a challenge, and for providing me always love and the necessary security to take on such difficult tasks as this one.

This book is intended for practitioners, for researchers, academics and students. It may also be of interest to industry and advocacy groups.

Josefina Maestu

Contributors

Editor

Josefina Maestu is (since September 2009) Coordinator of the United Nations Office to Support the International Decade for Action: "Water for Life" 2005–2015. With an academic background in economics, planning, and organizational studies, she has an extended professional career in the field of water, in academia, in civil servant positions, and in international consultancy. She has served as senior water adviser in the Ministry of Environment, Rural and Marine Affairs, where she coordinated the preparation of the economic analysis of River Basin Management Plans for the implementation of the Water Framework Directive in Spain. This included the use of economic instruments (including trading, pricing, taxes) as part of the measures of River Basin Plans. She was earlier a consultant for the European Commission and the World Bank doing evaluation of projects and participatory water management. She has been a lecturer on economics of natural resources management at the University of Alcala in Spain, for 12 years. She has worked, researched and published extensively on applied water economics and management.

Authors

Pedro Arrojo is Emeritus Professor in the Department of Economic Analysis of the University of Zaragoza (Spain), in which he served for almost his entire career. He chaired the Expert Commission for the Spanish Ministry of Environment to draft a report on "Exchange of water use rights in Spain – an update review." He has edited, written, or co-authored chapters in 74 books.

Carl J. Bauer is Associate Professor, School of Geography and Development, University of Arizona, where he teaches and does research about comparative and international water law and policy. He previously worked at Resources for the Future in Washington, DC. He has published two books (in English and Spanish) about water rights and water markets in Chile.

Henning Bjornlund is a Research Chair in Water Policy and Management at University of Lethbridge, Alberta, Canada and a Professor at the University of South Australia. He has researched water management and policy issues in Australia and in Canada. His work concentrates on water markets and policy reforms in Australia, their design and impact on irrigators and rural communities. Recently he has focused on community acceptability of policies to share water resources.

Javier Calatrava is Associate Professor of Agricultural Economics and Policy at the Technical University of Cartagena, Spain (UPCT). His work focuses on economics and policy of water resources. He has participated in 12 public research projects, and authored 40 academic references, including 12 articles in JCR-referred journals and 10 chapters in international books.

Francisco Cubillo is the Deputy Director of Research, Development & Innovation in the Madrid water Supply company, Canal de Isabel II. He has chaired the Group on Efficient Urban Water Management of the International Water Association between 2001 and 2011, and currently Chairs the Committee for Research, Development and Innovation of the Spanish water and wastewater suppliers association. He teaches in different specialized courses, and has published 14 books and more than 100 technical papers.

Joseph W. Dellapenna is a Professor of Law at the Villanova University School of Law, in Villanova, Pennsylvania, USA. He has taught water law for 40 years at universities on four continents and has worked as a lawyer on water problems on three continents. He is co-editor of *The Evolution of the Law and Politics of Water* (a global history of water law) and *Waters and Water Rights*, as well as the author of numerous authors and chapters in several books.

Guillermo Donoso is a Professor at the Department of Agricultural Economics, Pontificia Universidad Católica de Chile. He was also the Dean of the College of Agricultural and Forestry Engineering at theis University and Director of the Department of Agricultural Economics. His publications include *Water Markets and Institutions in Bolivia* and *Mercados (de Derechos) de Agua: Experiencias y Propuestas en América del Sur* (CEPAL, 2004).

Antonio Embid Irujo is Professor of Administrative Law, University from Zaragoza. He is Doctor Honoris Causa of two universities in Argentina and he has been given the "Gran Cruz de la Orden de Alfonso X el Sabio." He was President of the Regional Parliament of Aragon. He has extensive academic and professional experience as an administrative and water law advisor in Spain and Latin America, where he has been developing their water law legislation. He is the sole author of 23 books and has co-authored 38. He has collaborated in 68 publications and has written more than 80 articles.

Teodoro Estrela is Deputy Technical Director at the Júcar River Basin Authority and a Professor of the Technical University of Valencia. He has been Head of planning of the Jucar River Basin and Deputy Water Director on Water Planning of the Spanish Ministry of Environment and Rural and Marine Affairs from 2005 to 2009. He was in charge of the preparation of the river basin plans, implementing the Water Framework Directive in Spain. He has numerous publications on water issues in journals and books.

Franklin M. Fisher, Ph.D., is the Jane Berkowitz Carlton and Dennis William Carlton Professor of Microeconomics, Emeritus at the Massachusetts Institute of Technology, where he taught for 44 years. He is the sole or senior author of 18 books, most notably *Liquid Assets: An Economic Approach for Water Management and Conflict Resolution in the Middle East and Beyond*, the major book of the Water Economics Project – an international project of which he has served as the Chair since 1982.

Alberto Garrido is Professor of Agricultural and Resource Economics, Technical University of Madrid; Director of Research Centre for the Management of Agricultural and Environmental Risks, and Deputy Director of the Water Observatory of the Botin Foundation. His work focuses on natural resource and water economics, and policy and risk management in agriculture. He is the author of 130 academic publications, including *Water Footprint and Virtual Water Trade in Spain: Policy Implications and Water for Food in a Changing World.*

Carlos Mario Gómez is Professor of Economics at the University of Alcala de Henares and associated Researcher at IMDEA in Madrid. He is specialized in applied water economics. He has worked on this for the Ministry of the Environment, and the river basin authorities in Spain. He is involved in different Spanish and European research and development projects. He has authored several book chapters and peer-reviewed papers.

Almudena Gómez-Ramos is Associate Professor of Agricultural Economics of University of Valladolid (Spain). Her research is focused on agricultural economics and management of water resources. She has been involved in the implementation of European Water Framework Directive in Spain through collaborations with the Ministry of Environment, and River Basin Authorities. She has more than 20 publications related with water economics and risk managements in drought contexts.

Stefan Görlitz is a junior environmental expert working at InterSus Sustainability Services, an environmental policy consulting company

(SME) based in Berlin, Germany. InterSus is specialized in environmental economics and policy, focusing on water management issues. Mr. Görlitz is a Geographer, and holds an M.A. in Environmental Management.

Ronald C. Griffin is professor of water resource economics at Texas A&M University. He is the author of *Water Resource Economics: The Analysis of Scarcity, Policies, and Projects* (MIT Press, 2006), the editor of *Water Policy in Texas* (RFF Press, 2011), and a co-editor of the journal *Water Resources Research.* He specializes in water studies pertaining to demand, pricing, policy, marketing, and cost–benefit analysis.

Shreekant Gupta is Associate Professor at the Delhi School of Economics, University of Delhi and Adjunct Associate Professor at LKY School of Public Policy, National University of Singapore. He has been Director, National Institute of Urban Affairs, New Delhi. His publications include two books: *India and Global Climate Change: Perspectives on Economics and Policy from a Developing Country* and *India: Mainstreaming Environment for Sustainable Development.*

Ellen Hanak is a senior policy fellow at the Public Policy Institute of California (PPIC), where she specializes in the economics of water and natural resource management. She has held research positions with the French agricultural development research system, the World Bank, and the Council of Economic Advisors. Recent co-authored books include *Managing California's Water: From Conflict to Reconciliation* and *Comparing Futures for the Sacramento–San Joaquin Delta.*

Kristiana Hansen is an Assistant Professor in the Department of Agricultural and Applied Economics at the University of Wyoming. Her current research focuses on water markets, payments for ecosystem services, and reclamation after energy extraction.

Richard Howitt is a Professor of Agricultural and Resource Economics at the University of California at Davis. His research interests are in the following areas: disaggregated economic modeling methods, testing market mechanisms for the allocation of natural resources, experimental economics, and implementing empirical dynamic stochastic methods. Recently Howitt has co-authored two books on the Sacramento Delta and a third book on the future of California water management.

Eduard Interwies is the founder and director of InterSus Sustainability Services. He has extensive knowledge and experience in water and environmental economics and policy through participating in (and in a number of cases leading) more than 100 policy consultancy and research projects in the national and international context. He is the author of various book chapters, reports and articles.

David Katz is a lecturer in the Department of Geography and Environmental Studies at Haifa University in Israel. His research focus is on environmental and resource economics and policy, especially concerning water conservation and transboundary resource management. David was formerly the Director of the Akirov Institute for Business and Environment at Tel Aviv University.

Cody L. Knutson is a Research Associate Professor with the National Drought Mitigation Center, School of Natural Resources, University of Nebraska–Lincoln. He specializes in water scarcity, drought, and climate change. He has co-authored the UN International Strategy for Disaster Reduction's *Drought Risk Reduction Framework and Practices*, FAO's *Near East Drought Planning Manual*, and UNDP's *Community Water Initiative: Fostering Water Security and Climate Change Adaptation and Mitigation*.

Jay R. Lund is Director of the Center for Watershed Sciences and Ray B. Krone Professor of Environmental Engineering at the University of California–Davis. He specializes in the application of systems analysis, optimization, and economics to water and environmental problems. He has published many papers and several books, including *Managing California's Water: From Conflict to Reconciliation* (2011).

Jennifer McKay is the Director Centre for Comparative Water Policies and Laws, University of South Australia and is a part-time Commissioner of the South Adelaide Environment, Resources and Development Court. She has researched, taught, and consulted on water resource management and law issues throughout Australia, India and the USA. She has a B.A. Hons (Melbourne), LLB Adelaide, Ph.D. (Melbourne) and Diploma in Human Rights Law from an American University in 2009.

Nirmal Mohanty is Vice President at India's National Stock Exchange. His experience has been in financial and infrastructure sector reforms with the Government of India, Infrastructure Development Finance Company (IDFC), IMF, and State Bank of India. He led the team of the report of the Committee on Infrastructure Financing and the team preparing the India Infrastructure Report 2009. He has contributed to the book *India's Water Economy: Bracing for a Turbulent Future*.

Dannele E. Peck is an associate professor of Agricultural and Applied Economics at the University of Wyoming, Laramie, Wyoming, USA. She is an associate editor of the journal Natural Resources Policy Research. Her research interests include water resources, decision-making under uncertainty, and management issues at the livestock–wildlife interface.

Robert J. Rose is a policy analyst for US EPA in the Office of Water (since 2007). His professional experience includes designing and implementing a national standard for energy efficiency of commercial buildings and on market-based approaches to address water quality. He has served as the EPA's expert on water quality trading for Chesapeake Bay, and was responsible for increasing the states' and stakeholders' knowledge and awareness of policy choices.

Peter Rossini is a Senior Lecturer in Property and Researcher in the Centre for Regulation and Market Analysis at the University of South Australia. His research includes the analysis of water market prices in Australia.

Ricardo Segura is a Civil Engineer (1968) and has a Master's in Hydrology (1976). He became a civil servant in 1973 when he joined the Institute for Agrarian Development. Since 1990 he has worked in the Water Directorate of the Spanish Government, where he has taken different responsibilities including water planning and regulation of water use rights (concessions) and water trading initiatives. He has participated in different international cooperation missions/projects in Ecuador, Brazil, Nicaragua and Paraguay.

Antonio Serrano Rodríguez is a Professor of Spatial development in the Civil Engineering Faculty of the University of Valencia and Director of the Master and postgraduate course of spatial development planning of FUNDICOT. He has been in high public responsibility positions in Spain: Director of Spatial Planning and Urbanism (1991–4) and Secretary General for Biodiversity and Territory (2004–8) when he undertook major legal and policy reforms, including water reallocation for biodiversity protection. He has written and published extensively on spatial development and has been the editor of several publications.

Miguel Solanes is a Senior Researcher with IMDEA Water. He has been Water Law Advisor at the UN, first in New York, and then at ECLAC, Chile, for 24 years. His experience is in water law, water rights, and water markets. He has done extensive research on the regulation of public utilities.

Sarah Wheeler is a Senior Research Fellow at University of South Australia. She has worked as an economist in a variety of organizations within Australia and internationally, and has published extensively on water markets and farmer behavior.

Jeffrey Williams is Professor of Agricultural and Resource Economics and holder of the Daniel Barton DeLoach Endowed Chair at the University of California at Davis. His research is on price relationships among markets and on the institutional features of markets. He is interested in institutions that deal with variability in commodity supplies, such as swap markets and futures markets.

Jesús Yagüe is a Civil Engineer (1970). He worked for 14 years as a Consultant, 5 years in Venezuela in the hydroelectricity development of "Uribante Caparo". In 1985 he became a civil servant in Spain working in dam security, water planning and regulation. As deputy director of "water public domain" (regulator) for 15 years he was in charge of water use rights and trading initiatives.

Mike Young is Professor of Water and Environmental Policy at the University of Adelaide and was the founding Executive Director of its Environment Institute, specializing in the development of market-based approaches to the resolution of environmental challenges. He holds an Honorary Professorship with University College London, and has been appointed to the Gough Whitlam and Malcolm Fraser Chair in Australian Studies at Harvard University for 2013–14.

1 Introduction

Myths, principles and issues in water trading

Ronald C. Griffin, Dannele E. Peck and Josefina Maestu

There is growing evidence of increasing water scarcity in the world. This general scarcity is often exacerbated by the more frequent occurrence of extreme weather events. Cities and agriculture are struggling to make ends meet with their current water endowments. Increasing competition often means that there is not enough water for fragile habitats and the natural environment. Two linked features of the problem are rising water scarcity and more expensive infrastructure. Both issues underscore deficiencies within the conventional supply-side alternatives that are commonly suggested.

A key water management option – water trading – appears to be underutilized. Water trading can contribute to the efficiency of water use, and it can decouple economic growth and water use, thus contributing also to environmental flows. This book seeks to illuminate the role of water trading as an instrument of integrated water resources management. Trading is also a potential instrument for conflict resolution, where it may be necessary to reallocate water to address scarcity.

Water trading is based on voluntary agreements to transfer water among different users. Recent experiences of implementation in different river basins have demonstrated the potential of trading as a strategy for improving allocation. These experiences, however, also show that there are implementation challenges and some limitations to be considered.

Mounting experience with water trading is reshaping not just the management of this resource, but the way we envision it. Where there were once substantive barriers to reform because "water is different," institutional change in various parts of the world is demonstrating that water is not so different. At the root of this revelation is water's intersection with people. There's a reliable constancy about people: their respect for a commodity is related to its experienced financial value. By tapping into this feature of human behavior, water trading is making some progress in alleviating excess demand in many regions of the world. Moreover, if economists are right about the adjustments that people are pursuing under trading, then this reduction in excess demand is improving overall human welfare. Whether these claims are true and whether these policies can be further refined will be analyzed for many years to come. For now, water trading represents an intriguing group of policy opportunities to harness human self-interest for improved water management.

Water's peculiarities pose challenges, however, and water trading is not a panacea for scarcity. In spite of these limitations, the hardships of scarcity are spurring many governments to give water trading a try. Policy experimentation is under way, especially in water-scarce parts of the world, as countries seek out better ways to address the unique challenges that come with water. These trials and their results are a boon for water-scarce regions that have yet to embark on reform efforts, for there are valuable lessons to be gained. To aid these opportunities, the objective of this book is to assemble many of these stories and attempt to isolate the emergent keys and hurdles to successful water trading. By placing many perspectives and experiences (not all of which are in agreement) "under one roof," we hope to propel further investigation, experimentation, and reform. By bridging many countries' contemporary water policy transitions, this text shows how ideas have evolved since Easter *et al.*'s 1998 book.

In the remainder of this introductory chapter, we aim to illuminate the water trading debate by clarifying basic terminology, dispelling some of its more divisive myths, highlighting a few foundations for success, and pointing out some persistent stumbling blocks. If these objectives can be achieved, readers should be primed to more systematically comprehend the water trading stories that follow.

Polarizing ideologies

Steadfast ideologies and confused rhetoric have frustrated the advance of water policy. To get on the right path, so that the potential role of water trading can be accurately cast, people should set aside their prior indoctrinations and begin using terminology more accurately and consistently. This is difficult in ordinary water policy settings. Few decision-makers and voters have formal training in water management or economics, and therein lies a major hurdle. Limited education in either area allows dogma and symbolism to drive debate and subsequent policy decisions.

Two polarized "thought sets" are especially problematic. On the one hand is the "water is too different to sell" argument, which rejects water trading on the grounds that it is socially objectionable and doomed to fail before it even begins. On the other hand is the "water is no different from other commodities" perspective, which steadfastly promotes water trading as the best means to improve economic well-being. This pro-market viewpoint maintains that water trading is the natural solution. Although both sides have defensible origins, debate about water trading continues precisely because it neither fails miserably nor works perfectly. Moreover, fervent application of either of these perspectives tends to be harmful in policy design. Juxtaposing them, however, can help delimit the range of debate and pinpoint issues that need to be confronted.

Origins of water trading myths

Water trading opponents usually espouse a forceful, water-is-different view that tends to exaggerate biological requirements for water (i.e. classify non-needs as needs), confuse capital scarcity with water scarcity, and claim a public entitlement to water. They sometimes add other arguments for good measure, such as preserving agrarian economies or protecting impoverished peoples. There is a lot of false imagery underlying rhetoric about water (Kelso 1967), obscuring both the nature of water scarcity issues and the true character of available policy approaches.

A polar ideology is offered by water trading proponents, who hold fast to the belief that a market system is solely capable of extracting the greatest economic benefit from scarce water resources. Mainstream economists often advocate this view, as do pro-business and freedom-from-government factions. Their faith in the omnipotence of markets makes it difficult for them to appreciate why water might be different. Economists even have an indisputable mathematical theorem proving that a market system for all commodities will result in an efficient economy. Students of economics are so well indoctrinated about the power of markets that they sometimes forget the special conditions required for this ideology to hold. The tripping point here is that some of these conditions are unmet when the commodity under consideration is water. This pitfall is well appreciated within certain fields of economics, but it can be too technical or unwanted to influence the market-faithful crowd.

The critique here will be that water is sufficiently different to derail disciplinary ideology about market performance, but not so different that it renders markets unproductive. Water trading is indeed imperfect, but with the right circumstances and design it can increase society's net benefit from scarce water resources. To establish a viable platform for critiquing the water trading stories in this book, we begin by clarifying confusing terminology and dispelling common myths underlying the anti-trading viewpoint.

What is "water trading" and what is not?

Supplying water is a notoriously capital-intensive process, more so than all other utilities (National Research Council 2002: 12). Therefore, the pivotal social problem is often the scarcity of capital for capturing, pumping, treating, and conveying water, or for collecting and treating wastewater. This is not a problem to which the notion of water trading is applicable, and it is a mistake to confuse the privatization of water infrastructure with the assignment and trade of private property rights to water. Owners of privatized water infrastructure need not and commonly do not own marketable water rights. This is not to say that those who privately own infrastructure could not also have private ownership of raw water inputs (i.e. naturally occurring ground or surface water). They are compatible, but they address rather different social problems.

Some regions have ample raw water, but their publicly managed water utilities or districts lack the infrastructure or incentives to process and deliver this water to consumers. In some cases, private capital and private service delivery is introduced as an alternative solution to public capital. Here, privatization refers to ownership of built facilities and the right to profit from operating those facilities. The hope is that the pursuit of profit through the processing and transporting of raw water will result in improved facilities, more efficient operations, or a higher quality or more widely available product.

Other regions establish their facilities and operations successfully, but find themselves short on naturally occurring water to process and distribute. It is in these situations that the introduction of enforceable and transferable water rights, for at least a portion of a jurisdiction's water resources, may make a social contribution. Without any changes to infrastructural capacity or its ownership, water law can be reformed to transform a system of public property in water[1] (including private use rights) to one where a portion of a jurisdiction's water use rights is regarded and enforced as a divisible, transferable, privately managed asset that can be bought, sold, or leased, in whole or in part. While it might be said that the water resource has been *privatized* in these circumstances, this term invites confusion with infrastructure privatization and is not advisable. Separation of these two issues will help clarify whether trading opponents object to privately held water rights or infrastructure privatization.

The public's right to water

The anti-trading viewpoint commonly argues that we should not assign private rights to water because everyone has a right to water. Actually, all people do have this right if a government's political process decides they do, just as a society may decide all people have a right to food, a minimum wage, housing, or health care. Regardless of society's decision, this circular argument sidesteps the real issue at hand: how to manage water resources when they are scarce. Moreover, public rights can be reserved in all transitions to tradable water.

Societies often manage important resources (e.g. food) by creating private property rights to them, thereby establishing incentives for more efficient production and consumption of these scarce things (with appropriate interventions to deal with externalities and other market failures). This strategy has often been successful for non-water resources, such as land. Moreover, it is not uncommon for a government to partition a resource, assigning private use rights to a portion of it while keeping the remainder within the public domain (such as when private land coexists with public parks, forests, or wildlife reserves). Environmental water reservations and water rights for public taps and other publicly provided water or wastewater services can easily coexist with private property in water. Whether a full or partial private property strategy can be usefully applied to water is the question at hand. It is not helpful to reject water trading on the grounds that it conflicts with public rights. The two are not incompatible and may complement each other, as the land example illustrates.

Property rights as a key to success

Water use rights are pivotal for water trading. The effectiveness of private property rights hinges, however, on the extent to which they can be unambiguously defined, well enforced, and readily transferred. Some resources and goods are more conducive to management via private property rights, while others are better suited to common property rights or state ownership. Institutional economics informs that a private property right system has both pros and cons, and may not be appropriate in all circumstances. It is well recognized, for example, that private property rights can quell the race to use water resources before others do (i.e. the tragedy of the commons), but they can also be expensive to establish and enforce. If these costs outweigh the benefits of improved resource management, private property rights might not be justified.

Similarly, private property rights can facilitate the reallocation of water to higher-value uses. However, private water rights can also be costly to transfer when third party impacts (externalities) must be controlled via administrative oversight processes. In this case, trade might not move water to its most valued uses, and the other benefits of private water rights (e.g. preventing over-exploitation) might be underachieved.

Thus, even completely defined and enforceable water rights are not foolproof paths to efficiency. No property rights system can eliminate all inefficiencies or resolve all disputes. It is therefore important to consider relative strengths and weaknesses of each system. The system that maximizes society's net benefit from a particular water resource will depend largely on the characteristics and uses of that resource, some of which create the potential for market failures. These failures certainly reduce the effectiveness of water trading, but they do not imply that trading is outperformed by alternative policies.

Market failures as a stumbling block to success

Water trading achieves a socially optimal allocation of water only if the following conditions hold: the market is highly competitive (i.e. individuals cannot control price); buyers and sellers exclusively enjoy and incur all benefits and costs associated with their decisions (i.e. no externalities, public goods, or missing markets), and the market does not suffer greater transaction costs or information imperfections than other allocation mechanisms.

These strong conditions clarify why water trading cannot be the panacea that protagonists want it to be. Nearly every water resource setting violates at least one, if not several, of these conditions. The question is whether market failures are sufficiently severe to render water trading inferior to alternative approaches. Crucial elements of the answer must attend to the important matters of public goods, externalities, and transaction costs.

Is water a public good?

Public property in water does not imply that water is a public good. The use characteristics and ownership institutions of a resource are two separate matters, yet public goodness (when actually present) does raise suspicions about private property's ability to assist water scarcity problems (Myles 1995: ch. 9). Illumination of this matter is best achieved by relying on strict economic doctrine to assess whether a good is public. By definition, a pure public good is both *non-rival* in its use and *non-exclusive*. Acting in concert, these two properties interfere with a market's ability to produce and allocate a commodity efficiently. Hence, the more a given water use takes on these two public good attributes, the less likely it is that water trading is a beneficial instrument for managing that use.

It is a gross oversimplification to label water as a public good, yet some *uses* of water do constitute public goods. To correctly resolve the matter, one must emphasize the different uses of water as distinct applications and then pose the non-rivalry and non-exclusivity queries. Consider a farmer using water for irrigation. As a result of this irrigation, some of the applied water evaporates, transpires into the air or seeps into the soil and is gone (rivally consumed) relative to alternative uses in the area. A slight portion of the applied water may be embodied in the crop and be removed with the harvest (also rival). The remainder of this irrigation water may return to the originating watercourse as runoff, likely at a downstream location where it is unavailable to alternative users in the irrigator's vicinity. Again, the farm's use of water is rival with others within the immediate area, and is therefore not classifiable as a pure public good. Nor is the farmer's use of water non-exclusive, because well practiced administrative options are available for denying/regulating/pricing water to the farmer (regardless of whether those options are actually practiced). Because the farm's use of water is rival and exclusive, it is not a public good. Notice that both properties are assessed with respect to the technical character of the good without regard to in-place institutions and procedures. This is because the definition of a public good revolves around whether rivalry and excludability are *possible* for that use, not whether they actually occur under the current institution.

The same facts about rivalry and excludability generally hold true for water used by households. A household is rivally using water relative to other households and other users in its service area. Water quality tends to be diminished rivally by most ordinary uses as well. Considered carefully then, there are various degrees of rivalry within and between water-using sectors, depending on relative locations and the nature of their use. The fact that there is return flow from users to the original watercourse or to other watercourses does not establish water as a public good, though it is the root of important externalities that bear upon potential market performance (and potential non-market performance too).

Public good *uses* of water do exist, but even these tend to be non-rival only within their use category. For example, water kept instream for biodiversity or recreational support may be non-rivally and non-exclusively enjoyed by its human beneficiaries. Instream use of water is rival, however, against out-of-stream uses to which this water could be alternatively applied in its locale. In some ways, hydropower water use may be non-rival with other local water uses such as water supply or instream flow preservation. Yet, these same uses do not ordinarily agree on the preferred timing of reservoir releases, thereby injecting a degree of rivalry.

Where public good uses truly exist, the potential performance of water markets becomes suspect. People participate in markets because their rivalry in commodity use and potential exclusion motivate it. Markets are less respectful of public good values and will underprovide these goods if given the opportunity (Bergstrom *et al.* 1986). The fact that some public good uses of water do exist in all notable basins (and for many aquifers too) immediately implies that *trading is not an acceptable, stand-alone instrument for managing all water in a region.* With respect to the public good components of this issue, one attractive approach can be to partition the resource, setting aside a quantitatively defined portion of the water where trading is not allowed to affect allocation (and establishing qualitative and biological objectives that need to be respected). With this strategy, if the water resource is subject to seasonal or weather cycles (as is common), then thoughtful rules are needed to define the boundaries between privately and publicly owned water under all reasonably imaginable circumstances.

Are externalities fatal?

While public good issues are not as ubiquitous for water as is commonly claimed (because of the confusion addressed above), any reallocation of water between two parties has the potential to affect third parties. Potential externalities can be listed for both surface water and groundwater scenarios, but such lists are too long for inclusion here (Griffin 2006: §§4.3, 7.5, 7.6, 7.11). The fundamental problem – and the key reason why water is different – is that the stability of water molecules (i.e. they are not destroyed in use), and subsequent hydrologic processes acting upon these molecules, guarantee third-party influences. Such externalities are not automatically corrected by either market or non-market management systems.

The market-oriented solution for externalities is to conduct water trading in an open forum in which (a) proposed water trades are publicly and formally announced, (b) potentially harmed third parties[2] are given the opportunity to lodge objections or are otherwise represented, and (c) a government agency then rules on the acceptability of each trade by applying pre-established guidelines that hopefully pay attention to streamflow effects.[3] This is an imperfect process that increases transaction costs, but such costs are inevitable for non-market systems of reallocation as well.

What about transaction costs and imperfect information?

Beyond public goods and externalities, several other characteristics of water reduce the net benefit society is able to derive from the resource, but do not necessarily constitute market failures. High transaction/information costs discourage trades that would otherwise be beneficial. However, these costs are often high for any reallocation process, market or non-market, due to the complex effects of one person's water use on others. Transaction costs reduce the net gains to reallocation, but they do so regardless of the means by which reallocation is achieved.

Transaction/information costs only constitute a market failure if they cause water trading to achieve a lower social net benefit than other approaches. This may be true in some cases and not in others. One unique effect of transaction costs on water trading, for example, is the potential for buyers or sellers in "thin" water markets (i.e. those with few participants) to exert market power. Market power can cause private and social optimums to differ and therefore reduce social net benefit. Non-market approaches would not suffer this particular consequence of transactions costs, which implies that transaction costs, in this case, might cause a market failure.

Some forms of imperfect information, such as asymmetric information, are clearly capable of causing market failure. An agricultural producer who knows more about the reliability of their water right than a potential buyer, for example, could withhold or distort information to secure a higher price. Water markets may be more vulnerable to asymmetric information than other approaches, such as command-and-control (if the social planner has access to the agricultural producer's information). This form of imperfect information could therefore constitute market failure.

Market failure versus non-market failure

In light of the many complications that arise in water management, it is tempting to adopt a head-in-the-sand posture and prohibit any modifications to a region's historic allocation of water. Unfortunately, growing populations, advancing technologies, evolving economies, and a changing climate place mounting costs on such a strategy. A no-change approach constitutes only a temporary "solution," one whose costs will rise to intolerable levels over time as tension over scarce water intensifies and expensive structural options are eventually proposed.

A sobering question voiced by trading proponents is "Even if there are market failures present, why will non-market institutions do a better job of managing water?" Their argument is that it is illogical to compare water market performance against idealized outcomes that are not achievable by any available management system. This raises the question, "What can non-market policies achieve?"

Non-market systems of water allocation, such as participatory, bureaucratic, judicial, administrative, executive, or voting processes, also cause externalities in reallocation. Political approaches may not consider everyone's benefits and costs; hence, third parties and externalities arise in these systems just as they do for water trading. Furthermore, the costs of addressing these complexities in a political approach can easily be as large as those encountered in water trading.[4] Non-market mechanisms also are capable of generating distributional inequity, where people without political power are disadvantaged, while "well-organized constituencies" are treated preferentially (Wolf 1979). This, in turn, induces people to spend effort and money on enhancing their group status, rather than investing in direct solutions to a water management issue. Rent seeking, corruption, and side payments are potentially worse in political approaches to water allocation (Whitford and Clark 2007).

Resistance to change is another documented attribute of water managers in the absence of market incentives, even in the face of elevated scarcity (Lach *et al.* 2005). Wolf (1979) points out that among the roots of non-market failures are instances in which symbolism is valued more highly than effectiveness, possibly resulting in conflicting stated objectives, high costs of operation, and deficient accomplishments. Symbolic platitudes are rampantly expressed in political dialogue concerning water, so the door is wide open for non-market failures of these types.

"Derived externalities," or unintended consequences of government action, also emerge as possible problems (Wolf 1979). Even judicial approaches to assuaging water conflict, such as Riparian law in the eastern United States, become tedious and expensive for participants, particularly as water scarcity raises the stakes and litigation becomes more frequent. Although attorneys and expert witnesses may view this outcome positively, it detracts from the net benefits that users receive from water.

We should strive to be realistic in appraising non-market approaches and comparing them to water trading. For good or bad, non-market approaches install their own unique problems. They tend to substitute political power for the economic power that drives markets, not that political and economic power are independent. Thus, water tends to be less available to groups with diminished access to the political process. Although public goods, externalities, transaction/information costs, and imperfect information pose significant challenges to water trading, non-market approaches are subjected to the same challenges, and are not necessarily more efficient than imperfect trading.

This book

This book explores opportunities and challenges for water trading through a variety of experiences in different parts of the world. Each experience is unique because trading is conditioned by the legal and institutional framework in which the transactions are carried out. The institutional framework, the definition of water rights, and the means by which transactions are undertaken in various countries provides lessons on how trading may be improved. Contributors to the book investigate whether trading has been a positive instrument for managing water scarcity, improving aquatic ecosystems, and reducing water pollution in regions where such problems are most serious. The book concludes with proposals for improving the implementation of water trading.

This introductory chapter has set the scene for objective analysis of alternative water trading experiences by clarifying the difference between water as a private versus public good, and highlighting the relative strengths and weaknesses of market versus non-market approaches. Water trading has the potential to help address one of the key challenges of the twenty-first century – the global water crisis – and facilitate an on-going trend toward greening the global economy.

Part I of the book presents lessons from water trading experiences in the western US, Australia, Chile, India, and Spain. Each country's experience, and institutional model of water trading, is analyzed by different authors whose perspectives vary. Introducing this section of the book, Josefina Maestu presents an overview of experiences and implementation challenges, as well as some initial reflections on water trading. In Chapter 2, Cody Knutson discusses water stress and drought risk in different regions of the world, physical conditions that make trading a potentially valuable water management tool. Drought events occur frequently and cause important socio-economic and environmental impacts. These recurrent situations necessitate more effective approaches to understanding, anticipating, preventing, and mitigating drought effects. The author discusses the role of trading among the policy measures to improve drought planning and management in water scarce countries.

The California experience, as discussed in Chapters 3 (by Ellen Hanak) and 4 (Kristiana Hansen, Richard Howitt, and Jeffrey Williams), illustrates the potential for demand management (through water trading) as an alternative means to address water shortages, rather than traditional supply-side approaches. California's institutional context and definition of property rights have facilitated water trading as a viable drought management instrument. Emergency water banks, based on a system of purchase-options, were set up during the drought of the 1990s and continue to be used in anticipation of future shortages. This tool and other non-traditional water sources, such as wastewater re-use, have helped California water managers cope with demand growth, droughts, and increasing environmental water requirements. These tools will become increasingly relevant under a changing climate. The authors see a need, however, for new innovations to help resolve third-party effects, cope with growing uncertainty, and reconcile the implications of water's quasi-public characteristics for reallocation, risk reduction, and equity.

Australia has been using water markets for over ten years. These markets have generated significant benefits for water management, but have also generated high environmental costs in some cases. Henning Bjornlund and co-authors (Chapter 5) and Mike Young (Chapter 6) assess a major water reform program in Australia that established a more competitive and market-orientated approach to water management. The authors discuss lessons learned during the development of Australia's water markets, difficulties in making water trading cost-effective, upward trends in market participation over the last 10 years, the role of water brokers in ensuring market transparency and competitiveness, and the extent to which trading has reduced the economic and community impacts of water scarcity.

Chile provides an interesting case study of water markets that have little institutional intervention at the basin level, a system with both strong defenders and detractors. Authors Guillermo Donoso (Chapter 7) and Carl J. Bauer (Chapter 8) analyze water markets in Chile, including recent reforms, from an economic and institutional perspective. They argue that favorable governance, and organization of water rights and the water sector in general are essential for efficient allocation of scarce resources between consumptive users and the environment. Various aspects of the Chilean experience provide useful lessons about the reform of water laws and policies to improve implementation of water trading.

India's water markets provide an interesting contrast to other markets discussed in Part I. Although some of India's water transactions take place in regulated markets, many occur in informal or non-regulated markets. These informal markets are typified by water rights that are legally unclear and by participants who have no legal sanction to trade them. In Chapter 9, Nirmal Mohanty and Shreekant Gupta explore current water management challenges in India, and the potential benefits and costs of establishing formal water markets to facilitate transfers between economic sectors.

Finally (for Part I), the experience of Spain is discussed. Spain is a clear example of a country that requires new policies to meet increasing demand for water and decreasing supplies. Traditionally, water has been traded in some regions of Spain through non-regulated water auctions and intermediaries. Today, Spain has opted to regulate trade through public trading centres and bilateral contracts, which better account for negative environmental or third-party effects. Antonio Serrano Rodríguez (Chapter 10) discusses how water competition and conflict in Spain has made it necessary to introduce trading, and how trading can satisfy regional governments' demand for more decentralized water management. In Chapter 11, Alberto Garrido and co-authors analyze water trading in Spain from 2004 to 2008 and discuss how it mitigated water supply problems. Their assessment considers prices, environmental externalities, economic impacts, and market access. Existing instruments for managing water scarcity and drought in Spain are compared with trading instruments.

Part II of the book explores potential limitations and shortcomings of water trading for managing scarcity and quality. Carlos Mario Gómez, Eduard Interwies and Stefan Görlitz provide an introduction, followed by an assessment by Joseph Dellapenna (Chapter 12) that raises serious questions about the utility of markets as a tool for addressing the global water crisis. Drawing upon legal and economic theory, Dellapenna considers whether markets are appropriate for water resources management, and discusses potential consequences of privatization and commodification of water.

Although water trading may have negative environmental impacts in some situations, water trading may also help preserve environmental flows and contribute to other environmental objectives. Authors David Katz (Chapter 13) and Robert Rose (Chapter 14) assess whether water markets, coupled with transferable emission rights, could enhance minimum flows and achieve water pollution standards with minimal political opposition and cost. Experiences with water quality trading are reviewed. New policies are explored that would augment traditional water quality permit trading programs for point-source pollution with alternative market structures that address non-point pollution sources, such as agriculture.

Part III of the book highlights various regulatory reforms that have been implemented to overcome legal barriers to water trading, and reviews economic and institutional considerations necessary for implementing water trading. Carlos Mario Gómez provides an introduction to this part of the book, discussing the need for water use rights to be carefully defined. Miguel Solanes (Chapter 15) explores institutional, regulatory, and economic conditions that need to be considered in water trading to account for indirect socio-economic effects. Jennifer McKay (Chapter 16) focuses on legal reforms in Australia that support more flexible and efficient mechanisms of water allocation. She argues that coordinated and unified mechanisms are needed to implement water trading in a manner that explicitly considers environmental costs. McKay observes that Australia's legal and institutional frameworks are moving away from decentralized and non-regulated models to more centralized and regulated approaches.

In Chapter 17, Antonio Embid Irujo describes different types of water trading instruments incorporated in Spanish water law since 1999, and other legal reforms implemented to overcome legal barriers to trading. He proposes additional reforms that would harmonize trading and traditional administrative measures to manage water during times of drought. He concludes that new trading instruments may be necessary to deal with increasing uncertainty about future water supplies. Almudena Gómez-Ramos concludes Part III of the book by discussing water management under risk and uncertainty. She analyses the potential for option contracts to facilitate water trading and achieve efficient risk sharing during cyclical water shortages.

Part IV of the book explores incentives and pricing considerations in water trading, and the instruments which can help regulators and policy-makers value water more accurately, and design effective economic incentives for more efficient use. Carlos Mario Gómez opens Part IV by providing an overview on the importance of considering incentives in water trading. In Chapters 19 and 20 Franklin Fisher and Jay Lund present economic models that estimate the marginal value of water. Fisher's model maximizes net benefits from available water, given demand characteristics and supply infrastructure, by identifying optimal water allocations among alternative uses. He applies the model to mountain aquifers located between three countries in conflict: Israel, Jordan, and Palestine. Results reveal that reallocation of water among the countries is less costly, and more eco-friendly, than desalinating seawater, an option that international bodies in this conflict zone are currently pursuing. From the author's perspective, there are clear advantages to treating water as any other scarce economic good; water trading is a win–win solution that can help avoid conflict. Lund applies a hydro-economic optimization model to California to inform integrated water resource management decisions there. The model allows for joint management of ground and surface water, can simulate alternative water allocations among competing economic uses, and provides insights about the economic value of water in alternative uses.

The book's closing chapter provides main conclusions and recommendations from a synthesis of the contributed chapters. Clear messages emerge about the strengths, weaknesses, and appropriate uses of water trading, as compared with other integrated water management instruments. Opportunities and challenges for water trading vary across a diverse and complex set of physical, institutional, and socio-economic settings.

Notes

1. We lump both state property and common property forms in the term "public property."
2. An interesting asymmetry and market failure here is the absent regard for positive externalities. Whereas institutionalized procedures exhibit a desire to mitigate potential negative externalities in water trade, to the point of rejecting the proposal if necessary, the notion of enhancing trades that promise positive externalities has not yet emerged. For example, a surface water trade from an upstream party to someone downstream will increase streamflow on the

portion of the watercourse lying between them. If the environmental value of this gain is not factored into the trade, then the trade will involve an inefficiently low amount of water.

3. The physical settings of some trading situations can yield assurances that third party effects are only remotely possible, or are slight in value. Examples include situations in which traders are drawing their water from a common pool or are utilizing the same infrastructure. In these cases the transfer rules may not require approval by an oversight agency.
4. Note, however, that the goal is not necessarily to select the policy with the lowest information costs.

References

Bergstrom, T., Blume, L., and Varian, H. (1986) "On the Private Provision of Public Goods." *Journal of Public Economics* 29: 25–49.

Easter, K., Rosegrant, M., and Dinar, A. (1998) *Markets for Water: Potential and Performance.* Boston, MA: Kluwer Academic Publishers.

Griffin, R. (2006) *Water Resource Economics: The Analysis of Scarcity, Policies, and Projects.* Cambridge, MA: The MIT Press.

Kelso, M. (1967) "The Water-is-Different Syndrome, or What is Wrong with the Water Industry?" In *Proceedings of the Third Annual American Water Resources Conference,* 176–83. Available from: http://ron-griffin.tamu.edu/reprints/kelso1967.pdf.

Lach, D., Ingram, H., and Rayner, S. (2005) "Maintaining the Status Quo: How Institutional Norms and Practices Create Conservative Water Organizations." *Texas Law Review* 83(June): 2027–53.

Myles, G. (1995) *Public Economics.* Cambridge, UK: Cambridge University Press.

National Research Council. (2002) *Privatization of Water Services in the United States.* Washington, DC: National Academy Press.

Whitford, A. and Clark, B. (2007) "Designing Property Rights for Water: Mediating Market, Government, and Corporation Failures." *Policy Science* 40: 335–51 (doi:10.1007/s11077-007-9048-5).

Wolf, C. (1979) "A Theory of Nonmarket Failure: Framework for Implementation Analysis." *Journal of Law and Economics* 22: 107–39.

Part I

Water trading experiences

Introduction: how we are implementing water trading – learning from experience

Josefina Maestu

Water trading is a reality in western USA, in Australia, in Chile, in India, in South Africa, and in Spain, where increasing scarcity of water has led to an increase in the need to reallocate supplies. There are important lessons to be drawn from this experience.

Increasing water use and pollution emissions are resulting in increasing competition (and, in some cases, conflict and rivalries between uses), and compromising environmental protection. This situation is exacerbated by climate change. Secure and adequate supply of quality water in a context of increasing pressures and uncertainties associated with climate change is the major challenge of the twenty-first century.

Experience shows that water stands apart from other natural resources, because it is context-specific. Where there has been trading, there has been a need to consider the local and basin character of the externalities that can emerge from changes in water use that are potentially harmful for both human uses and the environment. Time, geographical, contamination-base, and cost externalities are common in surface water and groundwater. Experience shows that although trading plays a critical role in balancing supply and demand, trading generally takes place only for a small fraction of the available resources in a given basin.

Part I shows the lessons from implementing trading in some of parts of world. Cody L. Knutson (Chapter 2) discusses water-related problems and how they take different forms in each country and region because of differences in economy, culture, government institutions, and environment. Solutions to water problems need to consider fostering better coordination between water, drought, and climate change planning processes, and the identification, evaluation, and implementation of alternative risk management strategies. Under new climate scenarios, the importance of planning for water scarcity and drought will increase because long-term planning adds resilience to human settlements. As a response to scarcity and uncertainty, prospective analysis is needed to have scenarios and strategic planning on integrated water resources management.

In the western United States, both Ellen Hannack and Richard Howitt *et al.* analyze water trading in rapidly growing regions where water managers face difficult challenges in mobilizing new water supplies to meet new demands. Hannack shows how environmental concerns have curtailed the scope for large new surface storage projects, and basin overdraft limits groundwater's potential as a source of expansion. Drawing on the California experience, she explores modern water planning approaches, which focus on a portfolio of options including non-traditional sources (recycling, underground storage) and more efficient use of existing supplies (conservation and water marketing). It reviews the advantages and drawbacks of the elements of the portfolio and provides examples of innovative planning approaches. California has made considerable progress in expanding non-traditional sources, particularly since the early 1990s, when a prolonged drought and a series of environmental rulings constrained traditional supply sources. Many of the current challenges are institutional, rather than technical. For water marketing, continued progress is needed to resolve concerns of third parties in the source regions. For underground storage, further development of aquifer management protocols is a precondition. Progress in urban conservation will require more aggressive use of tiered water rates and shifts in public perceptions regarding landscaping. Changing public perceptions is also key to the successful expansion of recycled wastewater.

Richard Howitt and co-authors discuss how, depending on the relative importance of water supply uncertainty and impediments to water transfers, markets are forming differently across the western United States. In many locations, trades take the form of short-term leases of water, where the underlying property rights remains unaffected. In other regions, transfers of permanent water rights predominate. Econometric analysis of 3,696 transactions reported in the *Water Strategist* over 1990–2005 supports the conclusion that water property rights have influenced not only whether water trades occur, but also whether trades are permanent transfers of rights or short-term leases. The paper shows that for 14 western states during the years 1999–2002, short-term leases outnumbered permanent-rights sales by an average ratio of eight to one. In states with the largest volume of water trades (Oregon, Texas, Idaho, Arizona, and California) the ratio of leases to sales was nearly eleven to one. Some of the emerging trends in western water markets are reviewed and examined for ways to reconcile the quasi-public characteristics of water with the reallocation, risk reduction, and equity roles that influence water markets. The different systems in the US and Australia for defining transferable water rights and minimizing externalities from water trades are compared.

In Chapters 5 and 6, Henning Bjornlund and colleagues and Mike Young present different views of the successful and evolving experience of implementing trading in Australia. Henning Bjornlund *et al.* discuss how water markets have been vigorously pursued in Australia as instruments to minimize the socio-economic impact of increased scarcity and to provide water for environmental purposes. Initially markets were hesitantly adopted by irrigators, but over the last 10 years market participation has increased significantly both in the entitlement and allocation market, but most predominantly in the allocation market. There is evidence in Australia that both markets have facilitated the anticipated reallocation of water with the associated socioeconomic benefits. However, there is also evidence of declining rural communities as a result of drought and policy-induced scarcity, or, as some argue, as a result of the operations of water markets. Yet there is no real evidence as to whether this decline has been caused by scarcity or markets; or whether, in fact, markets have reduced the socioeconomic and community impact of scarcity. With the latest generation of legislative changes, including statutory-based water planning processes defining the consumptive pool of water and provisions for environmental and other public benefits, more secure water entitlements and water registers and an unbundling of the rights embedded in the traditional water entitlement, the scene seems to be set for the introduction of more sophisticated water market instruments and water products.

Mike Young discussed the types of water licenses that exist in Australia, including "high-security" and "general-security" licenses. In times of scarcity, allocations are made first to high-security licenses, and then any water that is left over is allocated to general-security licenses. He argues that from a trading perspective, one of the most important reforms was a decision to convert most licenses from licenses to irrigate an area of land to volumetric entitlements. In the Murray–Darling Basin this happened many years ago – mostly in the 1960s and 1970s. Without metering and the allocation of volumes of water, many of the water trading arrangements in place today would not be possible.

In Chapters 7 and 8, Guillermo Donoso and Carl J. Bauer present different views of an important case in question: the implementation of water markets in Chile. Donoso argues that since the establishment of the water allocation mechanism based on a market of water use rights in Chile, a series of empirical and theoretical studies have been carried out to determine the existence of several elements: a water use rights market and the number of transactions; water use rights market efficiency; bargaining, cooperation, and strategic behaviors of market participants; and the marginal gains from trade. These studies indicate that the allocation framework based on a market allocation system established by the Water Code in 1981 has been efficient from an investment point of view, mainly due to the water use rights security granted by the legislation. This is evidenced by significant investments that have been undertaken by several economic sectors to improve water use efficiency and to increase the availability of groundwater through exploration. Likewise, the free transaction of water use rights – even though in many areas water use rights markets have not been very active – constitutes an efficient reallocation mechanism, which has facilitated the reallocation of granted rights.

Bauer argues, however, that since Chile's free-market Water Code was passed in 1981, Chile has become known as the world's leading example of the free-market approach to water law and economics: the textbook case of treating water rights not merely as private property but also as a fully marketable commodity. This approach is often referred to as the "Chilean model." In this chapter the author summarizes key aspects of the Chilean experience, including the weak legal reform of 2005, in order to draw broader lessons for international debates about how to reform water laws, policies, and economics. The fundamental conclusion is that the Chilean approach has had some important benefits as well as problems, but overall it is not compatible with the long-term goals of integrated water resources management or sustainable development.

In India, Nirmal Mohanty and Shreekant Gupta (Chapter 9) present how informal water markets have existed for decades in states such as Gujarat and Tamil Nadu. Though these markets have generally been localized and confined to trades between irrigators, they have led to efficiency gains and have also helped resource-poor farmers to access irrigation. The gains, however, have been limited, since inter-sectoral transfers have not taken place. The current challenge in India, therefore, is to establish formal water markets that can expand the scope of trading and make inter-sectoral transfer of water possible. Further, since formal water markets have a legal basis, regulations can be designed to address concerns of ecological sustainability. Markets in water are very important for the urban sector, which faces acute shortages, but has not been able to access informal water markets. Formal water markets can augment urban supply at low cost and relatively quickly. The National Commission on Water Resources estimated that a reallocation of mere 5 percent of irrigation water annually to the urban sector could meet the latter's requirements over the next 50 years. It is, however, important to note that water markets can only supplement and not substitute tariff rationalization and other reforms at the distribution end in urban areas.

In Spain, Antonio Serrano Rodríguez argues in Chapter 10 that in a context of climate change new solutions are necessary. He describes the geography of water conflicts in Spain and discusses the potential of water trading to resolve them. The author examines this potential in relation to other tools, in the context of decentralization, where regional governments are demanding new responsibilities of water management. He argues that trading may have potential to contribute to conflict resolution, but it may also have limitations.

In this context, Garrido *et al.* (Chapter 11) analyze the experience of water trading since 2005. Trading has been possible because of the reform of the Water Law of 1999, which authorized a form of voluntary agreements with compensation. They argue that the experience with water trading in Spain has received moderately positive assessments. They discuss its strengths and weaknesses, and make some recommendations to improve them. They argue that there is great potential for water trading in Spain, in view of the somber projected impacts of climate change in the Iberian basins.

2 Managing water stress, drought and climate change in the twenty-first century

Water trading as part of integrated approaches in water management

Cody L. Knutson

Introduction

Chronic water shortages are a common problem around the world and promise to be an on-going concern in many regions during the twenty-first century. Water shortage problems are expected to continue since growing populations, increasing demands, water pollution, and governance problems continue to strain water supply systems in several regions of the world.

Drought places additional strain on water supply systems. In general, drought originates from a deficiency of precipitation from "normal" over an extended period of time, resulting in a water shortage for some activity, group, or environmental sector. Recent drought events around the world, such as those affecting the Horn of Africa, Europe, and the United States, demonstrate the problems associated with managing water resources during these extreme events.

Climate change adds even more uncertainty about the availability and reliability of water resources. Many studies show that climates are quickly changing around the world and will continue to do so in the future, often at alarming rates. Changing temperatures are altering the timing, intensity, and form of precipitation events and ultimately, the hydrological cycles that humans and environment rely on to meet their basic needs. The uncertainty associated with these changes adds to current water problems and the difficulty of water resources planning.

Specific water-related problems take different forms in each country and region because of differences in economy, culture, governmental institutions, and environment, however, the need for adequate supplies of quality water is a global concern. Increasing pressures placed on our water resources, recurrent drought, and the uncertainties associated with climate change will ensure that water resources management will be one of the major global issues of the twenty-first century. Identifying and implementing effective strategies to address these challenges is critically important for the sustainable development of our finite water resources.

Water stress and scarcity

Freshwater withdrawals around the world increased seven-fold during the last century, causing water scarcity and stress in many regions (UNDP 2006), and these withdrawals are expected to increase (Cosgrove *et al.* 2012). It has been estimated that the number of people at risk from water stress is likely to reach 1.7 billion before 2030 (before 2020 at the earliest), and 2.0 billion by the beginning of the 2030s (Cosgrove *et al.* 2012). Much of this increase is expected to occur in the developing world, in places such as India, China, and north Africa (UNDP 2006). However, there are also regions of the developed world that are also facing the overexploitation of water resources, which is causing problems with water availability and the maintenance of essential ecosystems (e.g. the southwestern United States, Australia's Murray–Darling Basin, and local regions throughout Europe).

Although the physical availability of water is a factor in many locales, current water supply problems are often as much about governance as about the physical environment (UNDP 2006). Management policies have often created biased development frameworks and unsustainable conditions by providing subsidies and incentives for the overexploitation of water resources. However, regardless of the cause, the result is that many people and activities face a shrinking supply of water resources to meet their needs. This has the potential for increasing conflicts between water users and countries that share aquifers, streams, and reservoirs. As demand increases, people are increasingly struggling to secure their water resource allocation.

The occurrence of drought

Drought exacerbates water supply problems by making a difficult situation even worse. Drought affects a wide range of social, economic, and environmental sectors, but will more easily turn into a disaster when water supplies are overburdened and families, communities, and governments lack the capacity to effectively plan for and respond to periods of drought. For example, a combination of multiple years of drought during the 2000s and increasing demands placed on the Colorado River in the western United States reduced water levels in its instream reservoirs (i.e. Lake Mead and Lake Powell) to record lows, and forced the creation of new interstate water allocation guidelines between the riparian states that rely on the river (Reclamation 2007). The rapidly growing city of Atlanta (Georgia, USA) also garnered national attention for coming dangerously close to running out of water during especially severe drought in 2007 (Seager *et al.* 2009).

Similar issues have arisen around the globe where increasing water demands and/or drought are straining water systems and placing livelihoods at risk. For example, in early 2012, it was reported that extended drought conditions and upstream water development had reduced Poyang

Lake (China's largest freshwater lake) to 5 percent of its former area, affecting the fishing industry and water availability for nearby cities (Qian and Zhu 2012; Thibault 2012). A prolonged drought in Australia from 1997 to 2011 (Australia's worst drought in a century) created ecological and economic crises in the Murray–Darling river system, which supports most of the nation's farms (McGuirk 2008; McGrath *et al.* 2012). Similarly, over-abstraction of water and prolonged drought during recent years have caused a wide variety of water-related problem in many parts of Europe (e.g. reduced river flows, lowered lake and ground water levels, and drying up of wetlands), which has increased the focus on enhanced management efforts for dealing with water scarcity and drought (EEA 2009).

Climate change adds even more uncertainty to water management and has the potential to magnify drought and water stress. As discussed by van Lanen (2012), since the impacts of drought usually exacerbated water scarcity, it necessary to have a comprehensive understanding of future risk to identify promising management options.

Climate change scenarios

In 2007, the International Panel on Climate Change (IPCC) released its Fourth Assessment Report on climate change (IPCC 2007). The inclusion of many new studies in the report resulted in the use of stronger cautionary language and substantiation that humans are affecting our global climate. The report concluded that the release of greenhouse gases (e.g. carbon dioxide, methane, nitrous oxides, etc.) is altering the Earth's heat budget and causing climatic changes. These changes include significant alterations of the water cycle and other water-related parameters.

For example, the report states that global temperature has already increased 0.74°C from 1906 to 2005, affecting the regional timing, amount, and distribution of precipitation, as well as mountain glaciers and snow cover that many regions depend on to maintain river flows throughout the year. In general, precipitation has decreased during the last century in the Sahel, the Mediterranean, southern Africa, and parts of southern Asia, and globally the area affected by drought has likely increased during the last three decades. Mountain glaciers and snow cover have also declined around the world, causing increased runoff and early spring peak discharge in many areas. The increased early season runoff may supplement short-term water supplies downstream, but changes the river dynamics and calls into question the long-term viability of glacier- and snow-fed basins.

In terms of future projections, these trends are expected to continue as rising temperatures continue to intensify, accelerate or enhance the global hydrologic cycle (Bates *et al.* 2008; UN 2010). According to the special Technical Paper of the IPCC on climate change and water (Bates *et al.* 2008), models for the twenty-first century are consistent in projecting precipitation increases in high latitudes and parts of the tropics, and decreases in some sub-tropical and lower mid-latitude regions. By the middle of the twenty-first century, in general, annual average river runoff is expected to decrease in some dry regions in the mid-latitudes and tropics, and many semi-arid regions will suffer a decrease in water resources (e.g. Mediterranean basin, western United States, southern Africa, and northeastern Brazil).

Increased precipitation intensity and variability are also projected to increase the risks of flooding and drought in many areas. In terms of drought and water availability, although the frequency of heavy precipitation events is expected to increase for most areas, there may also be longer times between rainfall events. In general, the proportion of land surface in extreme drought at any one time is projected to increase. Exacerbating the situation, water stored in glaciers and snow cover is expected to continue to decline and will not be available to supply as much water during seasonal warm and dry periods to regions supplied by melt water from major mountain ranges. Globally, the negative impacts of future climate change on freshwater systems are expected to outweigh the benefits.

As described by the IPCC (2007), the result of these climatic changes could include the increased exposure of hundreds of millions of people to water stress and heat waves, an increase in endemic morbidity and mortality because of water-related diseases, and greater risk of malnutrition and famine. Additional impacts are also expected to include agricultural yield reductions and changes in vegetation and animal species survival and distribution, increased wildfire occurrence, reduced tourism and recreation opportunities, and significant hydropower losses.

In general, climate change will likely cause additional uncertainty for people trying to maintain their livelihoods, and for water planners and institutions responsible for providing depending amounts of quality water supplies to meet the needs of citizens, industries, and the environment. However, climate change is only one driving force of change for water resources management, together with demographic, economic, environmental, social, and technological forces (UN 2010; Balaji *et al.* 2012). The realization of impacts associated with climate change (including those associated with water stress and drought), can be moderated through the adaptation of appropriate management strategies by a broad range of activities. For example, incorporating climate change information and perspectives into water planning and developing plans to reduce the effects of climate extremes (such as drought) can help develop robust systems that are better able to deal with climatic changes.

The role of water policy and institutions in dealing with drought risk

As stated by Milly *et al.* (2008: 573):

> systems for management of water throughout the developed world have been designed and operated under the assumption of stationarity. Stationarity – the idea that natural systems fluctuate within an unchanging envelope of variability – is a foundational concept that permeates training and practice in water-resource engineering.

The authors assert that recent evidence points to the fact that hydrologic systems are exiting the envelope of contemporary natural variability. Therefore, basing decisions on past statistics and management strategies may no longer be appropriate. They assert that the concept of stationarity is dead, and water planners must seek new ways of incorporating uncertainty and risk management into their planning. The selection of the most appropriate mixture of risk management measures for a particular activity are being identified through a variety of water, drought, and climate change planning activities at a wide range of administrative levels.

In terms of drought, drought planning has received increased attention in recent years by natural hazard and water planners around the world, although much more work is needed (Sivakumar 2011). Drought planning revolves around efforts to better understand the nature of drought and its effects on society and the environment, and using that knowledge to implement appropriate and effective measures to minimize the likelihood of experiencing harm from future drought. Drought planning provides an opportunity for decision makers to identify sectors that are vulnerable to drought and investigate management options before a crisis occurs. With this information, decision makers can identify and implement the most appropriate and cost-effective strategies available in a strategic and systematic manner. This will help foster a more informed decision-making process and the development of efficient drought management programs, which are often outlined in a drought plan. The development of a drought plan requires at least three components:

- monitoring and early warning;
- risk assessment; and
- risk management options.

Drought monitoring and early warning

During the last decade, drought monitoring has made great strides with monitoring and early warning systems being established in several countries and regions. For example, in Africa, regional centers such as the IGAD Climate Prediction and Applications Centre (ICPAC) and the Drought Monitoring Centre (DMC) in Harare, supported by the World Meteorological Organization and Economic Commissions and the Sahara and Sahelian Observatory, provide current data, develop climate outlooks, and issue warnings to national meteorological and hydrological services (UNISDR 2009).

Similar systems have also been, or are in the process of being, developed in several other regions and countries around the world (e.g. Australia, Canada, central Asia, China, Mexico, South Africa, southeastern Europe, and the United States). For example, in southeastern Europe, Slovenia has been selected by the eleven-country region to host the Drought Management Centre for Southeastern Europe (DMCSEE) with support from the UNCCD Secretariat in cooperation with the World Meteorological Organization (www.dmcsee.org). One of the primary goals of the DMCSEE is to establish a regional drought monitoring and early warning system. Drawing on the experience of the DMCSEE, the UNCCD Secretariat and the Organization for Security and Co-operation in Europe (OSCE) are also supporting the creation of a "Drought Management Centre in Central Asia in the context of the UNCCD (DMCCA)," in cooperation with the World Meteorological Organization (UNCCD 2008). Building on such successes, international scientists have also been advocating and investigating the development of a virtual and comprehensive global drought early warning system for the last decade (Svoboda 2012).

The development of a drought monitoring and early warning system is critical for assessing drought conditions, communicating threats, and triggering actions in a systematic and efficient manner as drought conditions intensify. Although progress is being made, much more work is required to develop effective systems across the globe and to provide information at a scale that is meaningful for drought planning at the local level by water managers and other decision makers. Specifically, additional work is needed to:

- enhance monitoring networks with additions stations and more accurate record keeping;
- foster data sharing between organizations;
- create more reliable and sector-specific forecasts;
- develop more robust monitoring systems that include a wide variety of physical and social impact indicators; and
- develop user-friendly early warning delivery systems.

(WMO 2006)

Drought risk assessment

A drought monitoring system will help to identify the risk of experiencing the physical onset of drought (i.e. frequency, severity, and spatial extent). However, vulnerability must also be assessed to fully understand the likelihood of being affected by drought conditions. Many people perceive drought to be largely a natural or physical event. In reality, drought, like other natural hazards, has both a natural and social component. The risk associated with drought for any region is a product of both the region's exposure to the event and the vulnerability of society. That is, the threat of harm from drought is based on a combination of the frequency, duration, and severity of drought events experienced and the susceptibility of people or activities to the negative effects associated with drought. Some regions, people, and activities are more likely to be at risk for a variety of reasons.

Exposure to drought varies regionally and there is little, if anything, we can do to reduce the recurrence, frequency, or incidence of precipitation shortfalls. However, measures can be taken to reduce our vulnerability to drought events. A risk assessment will help policy and decision makers understand the likelihood of experiencing a water shortage, as well as *who* and *what* are vulnerable to drought and *why*, in order to identify appropriate and effective drought risk management options.

As with drought monitoring, progress has also been made in developing risk assessment methodologies and undertaking risk assessments. For example, the National Drought Mitigation Center, USA, has developed the *How to Reduce Drought Risk* guide (Knutson *et al.* 1998), which has been utilized by planners in many communities, provinces, and nations to assess drought risk. The guide has also been tailored to specific regions, as described in the *Near East Drought Planning Manual* (FAO/NDMC 2008). The Mediterranean Agronomic Institute of Zargoza and Unversidad Politécnica de Madrid, Spain, also coordinated the creation of the MEDROPLAN drought management guidelines to assist Mediterranean countries in drought planning, which includes detailed strategies and case studies for conducting risk assessments (Iglesias *et al.* 2007). In addition, UN-Habitat has created the Comprehensive Disaster Risk Assessment Portal (www.disasterassessment.org). The portal provides a forum for members of the disaster management community to exchange tools and case studies related to disaster risk assessment.

Nonetheless, in practice, more research has been undertaken to understand the physical risk of experiencing drought (e.g. frequency, severity, intensity, exposure) than the underlying social vulnerabilities. However, there are some projects that have examined drought vulnerability for a particular region. For example, the United Nations Economic and Social Commission for Western Asia (ESCWA) undertook recent research to better understand the vulnerability of the region to drought (ESCWA 2005). Seeing a need for more research and information on water development and drought, the ESCWA secretariat initiated a series of development reports focused on water resources in the region from 2004 to 2005. The study focused on examining the various components of socio-economic drought and identifying indicators and mapping socio-economic drought vulnerability in the ESCWA region. Similarly, in response to the 2000–1 drought in Central Asia and the Caucasus, the Canadian International Development Agency (under the Canada Climate Change Development Fund) supported research and analysis by the World Bank to better understand the impacts of drought and needs and capabilities in its management and mitigation (see http://go.worldbank.org/8J56VSUEZ0). The effort resulted in two reports that outlined drought occurrence, impacts, vulnerabilities, climate change scenarios, and potential drought management strategies in several countries within the Near East region, including: Azerbaijan, Kazakhstan, Kyrgyzstan, Tajikistan, Turkmenistan, and Uzbekistan (World Bank 2005, 2006).

Despite such gains, the United Nations International Strategy for Disaster Reduction (UNISDR 2009) outlines several recommendations to enhance risk assessment methodology and applications. They stress that risk assessment methodologies, standards, and mapping applications should continue to be tested and modified to meet the needs of stakeholders, such as water managers. They should also be required as part of national and local planning efforts to help ensure they're carried out as administrations and initiatives change over time. Researchers and planning entities should also encourage the development of common methodologies and indicators (e.g. appropriate hazard and vulnerability metrics for water management) for defining and assessing risk to encourage the adoption of standardized international practices. In addition, researchers and practitioners should develop, update periodically, and disseminate risk maps, scenarios, and related information, with a special emphasis on those populations and activities most at risk. Institutions must also cooperate locally and internationally, as appropriate, to monitor and assess regional and transboundary hazards and vulnerabilities and exchange relevant information.

Drought risk management options

Better understanding drought risk allows for more informed decision making and the selection of more appropriate and effective management options. Although they can be described in a variety of ways, these management options typically include a range of drought preparedness, mitigation, response, and recovery strategies (see Box 2.1 for a list of definitions). In the past, drought management efforts have been biased towards implementing drought response measures. Although often beneficial, they fail to reduce the long-term risk of drought when implemented in isolation from other measures. The new paradigm of drought risk management focuses on increasing the implementation of drought preparedness and mitigation measures, so that the need for response and recovery measures will be minimized.

Table 2.1 shows a matrix of the types of drought risk reduction measures and the relevant stakeholders that may be responsible for implementing the actions. Planners should consider how to develop a balanced distribution of drought risk reduction measures and responsibilities across the range of themes and actors. This would include determining which actions

Box 2.1 Typology of drought risk management measures

Drought preparedness is defined as established policies and specified plans and activities taken before drought to prepare people and enhance institutional and coping capacities, to forecast or warn of approaching dangers, and to ensure coordinated and effective response in a drought situation (contingency planning).

Drought mitigation refers to any structural/physical measures (e.g. appropriate crops, sand dams, engineering projects) or non-structural measures (e.g. policies, awareness, knowledge development, public commitment, and operating practices) undertaken to limit the adverse impacts of future drought.

Drought response efforts include the provision of assistance or intervention during or immediately after a drought disaster to meet the life preservation and basic subsistence needs of those people affected. It can be of an immediate, short-term, or protracted duration.

Drought recovery measures are decisions and actions taken after a drought with a view to restoring or improving the pre-drought living conditions of the stricken community, while encouraging and facilitating necessary adjustments to reduce drought risk

Source: modified from UNISDR Terminology of Disaster Risk Reduction (www.unisdr.org)

are to be taken before drought to build long-term resilience to drought, during drought situations to respond to immediate needs (e.g. during advisory, alert, and emergency stages), and during the drought recovery process. The selection of the appropriate combination of drought risk reduction measures must be evaluated in the context of numerous constraints: time, financial and personnel resources, geography, feasibility, the level and nature of development and vulnerability, legality, public acceptance, and liability. In general, choices must be realistic, as well as socially and environmentally appropriate. The activities must also take place on a scale that is meaningful to those who must act, whether at the national, regional, or local level.

Table 2.2 provides a classification of some of the drought risk management options that could be utilized in the water management sector. For example, as previously discussed, drought monitoring, early warning systems, contingency planning, and communication can help an organization, community, or region better prepare for drought. Many of the other available mitigation and response options include augmenting water supplies, reducing water demand, and improving water governance or

Table 2.1 Matrix of drought risk reduction measures and responsible stakeholders

Stages	*Actors*					
	Micro		*Meso*		*Macro*	
	Individuals and households	*Community institutions*	*District or provincial government*	*NGOs*	*National government*	*International donors*
Preparedness						
Mitigation						
Response						
Recovery						

Source: modified from Davies (2000)

minimizing the effects of drought on society and the environment. As for augmenting water supplies, many of the most feasible rivers deemed suitable for large water projects have already been dammed, and critics argue that large rivers damage the environment, are expensive, and often fail to benefit local populations (Cooper 2003). Other augmentation strategies, such as desalinization and moving water over large distances are capital-, labor-, and energy-intensive. Although water resources continue to be tapped and redistributed in many locations and new storage options (e.g. rainwater harvesting and ground water banking) are being implemented, the "slack" in our water systems is tightening. Therefore, efforts are increasingly focusing on addressing water demand and governance through improving water use efficiency and investigating alternative management and governance strategies, such as establishing water markets, banks, and sharing agreements, and developing flexible water management options that can account for future uncertainty.

Although a variety of potential risk management strategies can be identified, most have not been analyzed thoroughly in practice to evaluate their effectiveness in reducing drought risk. Do drought preparedness and mitigation activities reduce the effects of drought? If so, which actions, under what conditions, and to what extent? Are water markets best for a certain location? Are drought response measures a better strategy than drought preparedness and mitigation in some circumstances? There are many such questions that could be asked. Therefore, additional research is needed to identify and evaluate potential risk reduction actions, especially in a real-world setting. To do this, adequate technical and financial resources should be made available to carry out pilot studies and planning processes.

In addition, risk reduction case studies need to be identified and shared via collaborative networks of stakeholders within the water, drought, and climate change planning fields. For example, the United Nations Framework Convention on Climate Change created the Database on Local

Table 2.2 Some potential drought risk reduction measures in water management

Drought preparedness	
Monitoring and early warning	Develop a drought monitoring and forecast system Develop an early warning system to communicate threats and trigger response actions
Planning	Create or update a long-term water management and conservation plan Create or update a hazard/drought contingency plan Foster effective communication channels between stakeholders Conduct post-drought audits to identify lessons learned
Education	Foster youth and adult educational opportunities on water and drought risk management Foster a culture of proactive disaster management among decision makers
Drought mitigation	
Water demand reduction	Public information campaign for long-term water saving Permanent water restrictions for some uses (lawn watering, landscaping, etc) Economic incentives for water savings Reducing agricultural water use (e.g. more efficient technologies, management techniques, and alternative or non-irrigated crops) Dual distribution network for urban use More water-efficient industrial production systems and water recycling Water recycling in industry and businesses
Water supply increase	Control of seepage, evaporation losses, and other unaccounted-for-water Inter-basin and within-basin water transfers Conveyance networks for bi-directional exchanges Re-use of treated water Construction of new reservoirs or increase of storage volume of existing reservoirs Construction of small catchment ponds Desalination of brackish or saline water Develop equitable water rights or priorities Develop conjunctive use systems Foster exchanges among water users (water banking) Long-term changes in reservoir release rules under drought conditions Institutional and legal changes Operational coordination between water systems
Drought response	
Water demand reduction	Public information campaign for water management and saving Water restriction in some urban water uses (i.e., car washing, gardening, etc.) Restriction of irrigation of annual crops Pricing Mandatory rationing

Table 2.2 Continued

Water supply increase	Improvement of existing water systems efficiency (leak detection programs, operations) Use of additional sources of low quality or high exploitation cost water Additional use of aquifers or use of ground water reserves Increased diversion by relaxing ecological or recreational use constraints
Water governance and public relief (i.e. impact minimization)	Temporary re-allocation of water resources Water hauling and distribution Public aids to compensate income loss Efforts to protect fisheries and riparian areas Maintain navigation/recreational infrastructure (e.g. dredging, boat ramp extension)
Drought recovery	
Water governance, public relief, and assistance	Public information campaign to stress cautionary drought recovery Conservative water use to restore hydrological systems Reallocation of water resources as conditions improve Recovery assistance for businesses and industries Public aids to compensate income loss Reintroduction of fisheries and maintenance of riparian areas Shift back into preparedness and mitigation planning for the next drought

Source: modified from Rossi *et al.* (2007)

Coping Strategies to facilitate the transfer of coping strategies/mechanisms, knowledge and experience from communities that have had to adapt to specific hazards or climatic conditions (including drought) to communities that may just be starting to experience such conditions, as a result of climate change (http://maindb.unfccc.int/public/adaptation). Similarly, in the United States, the National Drought Mitigation Center has begun the creation of a Drought Risk Reduction Database, which will facilitate the sharing of risk reduction strategies that have been implemented within the country. With increased understanding, best management practices can be identified and implemented through appropriate policies, agreements, and institutional governance.

A need to integrate drought, water stress, and climate change risk reduction efforts

A wide variety of people and entities, from individual agricultural producers to regional groups of countries, are increasingly identifying appropriate drought risk reduction measures through the development of drought plans and drought planning networks. The UNISDR (2009) discusses many of the drought planning and networking activities occurring around the world, including local level, community, tribal, national, and international drought planning initiatives. At the international level, these drought-related activities include the development of a proposed global drought risk reduction and preparedness network. This network would consist of an interconnected system of regional drought networks linked together to facilitate the exchange of drought-related information and experiences around the world. The network would provide a forum for nations and regions to share experiences and lessons learned to foster their capacity to reduce drought risks. For example, recognizing the need for additional information sharing and networking, the United Nations Development Programme Drylands Development Centre launched the Africa–Asia Drought Risk Management (DRM) Peer Assistance Network, with financial support from the Government of Japan. As reported by the UNDP (2011), this Network was formulated as part of the three-year Africa–Asia DRM Peer Assistance Project (from 2010 to 2013). The objective of the project is to mitigate the risks of drought and improve human livelihoods in Africa and Asia by creating an enabling environment for inter-regional knowledge sharing among drought-prone countries and facilitating the up-scaling of proven DRM best practice in the two regions. The goal of a global drought risk reduction and preparedness network would be to foster and link such networks around the world as they continue to be developed.

There are also a host of other planning and risk reduction efforts around the world that are related to drought, water stress, and climate change. At the local level, for example, most public and private water utilities have a water development plan that outlines a strategy for meeting customer's water needs in the near future. Communities are also increasingly developing hazard mitigation and response plans to increase their long-term resilience and to prepare a strategy for dealing with specific hazards, such as drought. Some cities are also expanding the scope of their planning exercises to include the issue of climate change. For example, 10 of the United States' largest water providers (i.e. Central Arizona Project, Denver Water, Metropolitan Water District of Southern California, New York City Department of Environmental Protection, Portland Water Bureau, San Diego Water Authority, San Francisco Public Utilities Commission, Seattle Public Utilities, Southern Nevada Water Authority, and Tampa Bay Water) have formed the Water Utility Climate Alliance (www.wucaonline.org/html/about_us.html). The goal of the Alliance is to work together to improve research on the effect of climate change on water utilities and develop strategies for adapting to climate change and future water shortages (Denver Water 2008).

State, provinces, and national governments are also following the trend with an increased focus on water, drought, and climate change planning. At times, these efforts are synchronized but, in many cases, they are conducted in parallel or in succession. Some of the planning efforts are initiated by regional or national initiatives and others have been fostered by international programs and activities. For example, the United Nations Framework Convention on Climate Change (UNFCCC) was adopted in 1992 and entered into force in 1994. The Convention states that each signatory party take precautionary measures to anticipate, prevent, or minimize the causes of climate change and mitigate its adverse effects (http://unfccc.int). Similarly, the United Nations Convention to Combat Desertification (UNCCD) was adopted in 1994 and its implementation began in 1996. The UNCCD and its signatory countries are obligated to carry out activities to combat desertification and mitigate the effects of drought (www.unccd.int). These efforts have resulted in the creation of national action plans and programs to address water stress, drought, and climate change, and the establishment of regional networks to share information on strategies and successes.

Furthermore, in 2000, the Millennium Development Goals were established by the United Nations and the international water community to set goals for human development over the next several decades (www.un.org/millenniumgoals). One of the goals is to halve, by 2015, the proportion of people without sustainable access to safe drinking water and basic sanitation. More recently, the landmark Hyogo Framework for Action 2005–2015 was adopted by United Nations Member States in 2005. The Framework outlines priorities to enhance the resilience of nations and communities to natural hazards, including drought and climate change (UNISDR 2009). Such efforts continue to bring attention to the need for additional work in the water arena and provide a unifying theme around which collaborative efforts can be structured.

Although there are numerous organizations, networks, and international agreements and frameworks, the efforts are often fragmented, under-funded, and lack the means and the political will to carry out many of the proposed recommendations. There are also controversies over which entities are appropriate to manage local water resources. Governments traditionally have had responsibility for providing water but that task has increasingly been turned over to private companies (Cooper 2003; Marin 2009). Privatization is often argued as an efficient method for providing water that requires less public investment. However, there are competing concerns over private companies' willingness to serve those in remote areas that are less economical and the poor who are unable to afford the cost of water. In these arguments, strong governmental oversight is advocated to ensure equal access to water, which is viewed as a basic human right.

Additional cross-disciplinary dialogue and collaboration between natural hazard, drought, water, and climate change professionals are necessary to identify appropriate risk reduction measures and coordinate and streamline the development and implementation of risk reduction strategies and initiatives. Dialogue also needs to take place in regards to the appropriate roles to be carried out by relevant stakeholders and at different administrative levels (e.g. individual, private company, non-governmental organization, community, province, national, international). These types of discussions and collaborations have the potential to develop more unified and holistic strategies for addressing the related issues of water stress, drought, and climate change.

Conclusions

During the twentieth century, water use increased more than seven-fold, and the world water situation may reach alarming dimensions in future decades if water availability forecast projections are realized. Increasing water stress and recent droughts, along with unfavorable climate change scenarios, have brought greater awareness of the need to more effectively plan for future water shortages around the world. In the end, tackling the inter-related issues of water stress, drought, and climate change will require a combination of local action, national direction, and regional and global coordination. There is also a place for private investment, as well as public oversight and funding. A balance must also be struck between local autonomy and general standards for meeting basic human rights.

International frameworks and programs, such as the UNCCD and UNFCCC initiatives, the Millennium Development Goals, and the Hyogo Framework for Action, provide important coordinating mechanisms and advocate the importance of risk reduction through more proactive planning and water resources management. However, specific mitigation and preparedness actions will be implemented at the national and local level. Fostering better coordination between water, drought, and climate change planning processes and initiatives at different administrative levels; the identification, evaluation, and implementation of alternative risk management strategies; and providing the funding and political will necessary to carry out these activities is critical for reducing the risks and conflicts associated with water stress, drought, and climate change.

References

Balaji, R., Connor, R., Glennie, P., van der Gun, J., Lloyd, G.J. and Young, G. (2012) "The Water Resource: Variability, Vulnerability, and Uncertainty." In *World Water Development Report 4, Volume 1: Managing Water Under Uncertainty and Risk*, chapter 3. Paris, France: World Water Assessment Programme, UNESCO.

Bates, B. C., Kundzewicz, Z. W., Wu, S. and Palutikof, J. P. (eds) (2008) *Climate Change and Water*. Technical Paper of the Intergovernmental Panel on Climate Change. Geneva, Switzerland: IPCC Secretariat.

Cooper, M. (2003) "Water Shortages: Is there Enough Fresh Water for Everyone?" *The CQ Researcher* 13(27): 649–72.

Cosgrove, C., Cosgrove, W.J., Hassan, E. and Talfre, J. (2012) "Understanding Uncertainty and Risk Associated with Key Drivers." In *World Water Development Report 4, Volume 1: Managing Water Under Uncertainty and Risk*, chapter 9. Paris, France: World Water Assessment Programme, UNESCO.

Davies, S. (2000.) "Effective Drought Mitigation: Linking Micro and Macro Levels". In D. A. Wilhite (ed.), *Drought: A Global Assessment.* New York and London: Routledge.

Denver Water (2008) "Major US Water Agencies Form Climate Alliance." Denver, CO: Denver Water. Available at www.denverwater.org/waterwire/newsrelease/climate_alliance.html (accessed 1 March 2010).

EEA (European Environment Agency) (2009) *Water Resources Across Europe-Confronting Water Scarcity and Drought.* Copenhagen, Denmark: EEA.

ESCWA (Economic and Social Commission for Western Asia) (2005) *ESCWA Water Development Report 1: Vulnerability of the Region to Socio-Economic Drought.* New York: United Nations.

FAO/NDMC (Food and Agriculture Organization of the United Nations/National Drought Mitigation Center) (2008) *The Near East Drought Planning Manual.* Cairo, Egypt: FAO.

Iglesias, A., Cancelliere, A., Gabina, D., Lopez-Francos, A., Moneo, M., and Rossi, G. (eds) (2007) *Drought Management Guidelines.* European Commission – EuropeAid Co-operation office Euro-Mediterranean Region Programme for Local Water Management (MEDA Water) Mediterranean Drought Preparedness and Mitigation Planning (MEDROPLAN). Available at www.iamz.ciheam.org/medroplan/guidelines/archivos/guidelines_english.pdf (accessed 25 July 2012).

IPCC (Intergovernmental Panel on Climate Change) (2007) *Climate Change 2007: Synthesis Report: Summary for Policymakers.* Available at www.ipcc.ch/pdf/assessment-report/ar4/syr/ar4_syr_spm.pdf (accessed 28 March 2012).

Knutson, C., Hayes, M. and Phillips, T. (1998) *How to Reduce Drought Risk.* Available at drought.unl.edu/portals/0/docs/risk.pdf (accessed 11 July 2012).

Marin, P. (2009) *Public–Private Partnerships for Urban Water.* Washington, DC: International Bank for Reconstruction and Development, World Bank.

McGrath, G.S., Sadler, R., Fleming, K., Tregoning, P., Hinz, C. and Veneklaas, E.J. (2012) "Tropical cyclones and the ecohydrology of Australia's recent continental-scale drought." *Geophysical Research Letters* 39: L03404 (doi:10.1029/2011GL050263).

McGuirk, R. (2008) "Australian Minister Says Drought Needs Attention." *USA Today* 18 June. Available at www.usatoday.com/news/world/2008-06-18-2793843786_x.htm (accessed 11 July 2012).

Milly, P. C., Betancourt, D. J., Falkenmark, M., Hirsch, R., Kundzewicz, Z. W., Lettenmaier, D. P. and Stouffer, R. J. (2008) "Stationarity is Dead: With Water Management?" *Science* 319: 573–4.

Qian, W. and Zhu, J. (2012) "Drought Drying Out Poyang Lake." *China Daily* 1 January. Available at www.chinadaily.com.cn/china/2012-01/05/content_14382441.htm (accessed 28 March 2012).

Reclamation (2007) "Colorado River Interim Guidelines for Lower Basin Shortages and Coordinated Operations for Lake Powell and Lake Mead." Available at www.usbr.gov/lc/region/programs/strategies.html (accessed 28 March 2012).

Rossi, G., Castiglione, L. and Bonaccorso, B. (2007) "Guidelines for Planning and Implementing Drought Mitigation Measures." In G. Rossi *et al.* (eds) *Methods and Tools for Drought Analysis and Management*, chapter 16. Dordrecht, The Netherlands: Springer.

Seager, R., Tzanova, A. and Nakamura, J. (2009) "Drought in the Southeastern United States: Causes, Variability over the Last Millennium, and the Potential for Future Hydroclimate Change." *Journal of Climate* 22: 5021–45.

Sivakumar, M. V. L. (2011) "Current Droughts: Context and Need for National Drought Policies." In M. V. L. Sivakumar, R. P. Motha, D. A. Wilhite, and J.J. Qu (eds) *Towards a Compendium on National Drought Policy*. Geneva, Switzerland: World Meteorological Organization.

Svoboda, M. (2012) "Efforting a Global Drought Early Warning System." Proceedings of the 92nd American Meteorological Society Annual Meeting, New Orleans, LA, 24 January.

Thibault, H. (2012) "China's Largest Freshwater Lake Dries Up." *Guardian Weekly* 31 January. Available at www.guardian.co.uk/environment/2012/jan/31/china-freshwater-lake-dries-up (accessed 28 March 2012).

UN (United Nations) (2010) *Climate Change Adaptation: The Pivotal Role of Water*. UN Water Policy Brief. Availabe from www.unwater.org/downloads/unw_ccpol_web.pdf (accessed 28 March 2012).

UNCCD (United Nations Convention to Combat Desertification) (2008) *Dialogue on Establishing a Drought Management Centre in Central Asia in the Context of the United Nations Convention to Combat Desertification*. Available at www.unece.org/fileadmin/DAM/env/SustainableDevelopment/3Session/Informal%20documents/in_session/OSCE_t_concept_paper.pdf (accessed 3 April 2012).

UNDP (United Nations Development Programme) (2006) *Human Development Report 2006. Beyond Scarcity: Power, Poverty and the Global Water Crisis*. New York: Palgrave Macmillan.

UNDP (2011) "UNDP to Scale Up Fight Against Drought in Africa and Asia With Support from Japan." United Nations Development Center Drylands Development Centre. Available at web.undp.org/drylands/docs/drought/AADP_Article.pdf (accessed 3 April 2012).

UNISDR (United Nations International Strategy for Disaster Reduction) (2009) *Drought Risk Reduction Framework and Practices: Contributing to the Implementation of the Hyogo Framework for Action*. Geneva, Switzerland: United Nations secretariat of the International Strategy for Disaster Reduction.

van Lanen, H. A. J. (2012) "Drought, Water Scarcity, and Climate Change." *Geophysical Research Abstracts* 14. EGU2012-2143-2, 9th EGU General Assembly. Availabe from www.geophysical-research-abstracts.net/volumes.html (accessed 11 July 2012).

WMO (World Meteorological Organization) (2006) *Drought Monitoring and Early Warning: Concepts, Progress and Future Challenges*. WMO no. 1006. Geneva, Switzerland: World Meteorological Organization.

World Bank (2005) *Drought: Management and Mitigation Assessment for Central Asia and the Caucasus*. World Bank Europe and Central Asia Region, Environmentally and Socially Sustainable Development Department. Available at https://openknowledge.worldbank.org/handle/10986/8724 (accessed 11 July 2012).

World Bank (2006) *Drought: Management and Mitigation Assessment for Central Asia and the Caucasus: Regional and Country Profiles and Strategies,* World Bank Europe and Central Asia Region, Environmentally and Socially Sustainable Development Department. Available at http://siteresources.worldbank.org/INTECAREGTOPRURDEV/Resources/CentralAsiaCaucasusDroughtProfiles&Strategies-Eng.pdf (accessed 28 March 2012).

3 New frontiers for water management

The California experience

Ellen Hanak

Introduction

California shares the water management challenges of many regions facing rapid population growth and constraints on supply expansion. Environmental concerns have curtailed the scope for large new surface storage projects, and in many cases are resulting in reductions in human uses. Widespread basin overdraft limits the potential of native groundwater as a source of expansion. As a result, the focus of water planning has progressively shifted toward portfolio approaches, which seek to meet urban demand growth by augmenting non-traditional supply sources (such as recycling, underground storage, and desalination) and by improving the efficiency of use of existing supplies (conservation and the market-based reallocation of water rights).

This article explores the advantages and potential problems associated with this new water planning approach in California. It focuses on four of the major "new" sources of water: water markets, groundwater banking, urban conservation, and recycled wastewater. Although each of these sources offers potential advantages, none are entirely straightforward to implement. Underground storage and water marketing are both potentially low-cost alternatives, but each faces significant institutional hurdles. Expansion of recycled water use can require modifications in plumbing systems and changes in public acceptability of re-using treated wastewater. Conservation can be costly in terms of the technological investments needed to enable the savings and the consequences for "quality of life" if it entails restrictions on landscaping, which can account for over half of residential water use.

California is a virtual laboratory for new approaches to water supply planning. The most recent *California Water Plan* (CDWR 2009) projects that a diverse portfolio of non-traditional sources has a far greater potential to augment usable supplies than new surface storage over the next three decades. At the same time, the state's record on implementation provides ample illustrations of the challenges to innovation. Interestingly, many of the innovations that have occurred since the early 1990s were spurred by a combination of natural and legally-generated water shortages: a prolonged drought, from 1988 to 1994, and a succession of environmental rulings requiring cutbacks in human water uses. In the late 2000s, the state was again reeling from these two factors, with three straight years of very low precipitation and new court restrictions on water diversions from a central water conveyance hub to protect endangered fish species. The crisis has led to a new round of policy proposals at the state and local level, spurring a new round of innovation.

The chapter is organized as follows. The next section provides an overview of the basic supply and demand issues facing the state, drawing on the findings of the new *California Water Plan* (CWP, "the Plan"). In the third section, the focus turns to the four key non-traditional sources, with a discussion of advantages and drawbacks and of approaches being used to overcome implementation difficulties.[1] The fourth section summarizes these findings and notes some of the innovation challenges that lie ahead as a result of climate warming.

Supply and demand: the big picture

As in other parts of the western United States, the staples of California's water supply are native groundwater reserves and "developed" surface water – river water harnessed in surface reservoirs and transported through conveyance channels, often across long distances. Surface storage investments were the predominant form of water supply expansion for most of the last century (Reisner 1993; Hundley 2001). Although some of these projects were undertaken locally, federal and state authorities have played a major role. In particular, the federally financed Central Valley Project (CVP), undertaken from the 1930s to the 1950s, serves farmers and cities in this large inland valley. In the 1960s, the State Water Project (SWP) launched investments to deliver water to farmers and cities further south. Southern California is also a prime beneficiary of federal investments along the Colorado River (Figure 3.1).

In normal rainfall years, groundwater provides roughly one-third of all water used by the agricultural and urban sectors combined – and more in dry years – with the balance provided by surface water. For the 1998–2005 period, this combined demand totaled 51.4 billion m^3 of applied water use per year, with 77 percent going to farmland irrigation (Hanak *et al.* 2011).

Every five years or so, the state updates its projections of supply and demand. The most recent *California Water Plan Update*, completed in December 2009, presents several scenarios of demand growth (Table 3.1). The two key sources of growth are urban and environmental uses. Depending on population growth, conservation trends, and land use scenarios, urban demand growth is projected to increase by 1.9 to 12.3 billion m^3 per year (17 to 115%).

Figure 3.1 California surface water projects
Source: CDWR (2009)

Environmental water demands, estimated at over 1 billion m^3 per year since the early 1990s, are projected to double to enhance protections for endangered aquatic wildlife. In contrast, agricultural water demands are expected to decline by 14–17% under the three scenarios as a result of various market forces (shifts to less water intensive crops, development of some agricultural lands) and conservation. On balance, the scenarios imply that overall demand could be relatively stable, increasing by just 5 percent or even declining slightly, if water currently used in farming could

Table 3.1 Water demand growth scenarios and source replacement needs, 2005–50 (m^3 billions)

Projected demand growth	*Scenario 1: slow and strategic growth*	*Scenario 2: current trends*	*Scenario 3: expansive growth*
Urban	1.9	7.4	12.3
Environment	1.9	1.2	0.7
Agriculture	–6.8	–6.2	–5.7
Net change	–3.1	2.5	7.4
Additional potential demand growth (all scenarios)			
Groundwater overdraft replacement		1.2–2.5	
Climate change effects		0.6–3.7	
Net change		1.8–6.2	

Source: CDWR (2009)

Notes: Population in 2005 is estimated at 36.7 million, with growth ranging from 20% (scenario 1) to 62% (scenario 2) to 90% (scenario 3). Conservation in the urban and agricultural sectors ranges from a high of 15% (scenario 1) to a low of 5% (scenario 3). Urban development continues existing trends (scenario 2) or more compact (scenario 1) or more sprawling (scenario 3). Irrigated acreage declines from 2% (scenario 1) to 8% (scenario 2) to 11% (scenario 3)

be transferred to urban or environmental uses and California maintains current trends or experiences slower population growth and more compact land uses and conservation. However, there is a potential for net demands to increase by as much as 14 percent with more rapid, sprawling growth, and by another 12 percent to make up for an estimated 1.2–2.5 billion m^3 of annual groundwater overdraft and to respond to climate change. (The warming associated with climate change could increase water demands in all three sectors by 0.6–3.7 billion m^3, depending on whether the future is drier as well as warmer).

To replace these sources and meet new demands, the Plan explored the potential for mobilizing water from a wide range of sources between now and 2030 (Figure 3.2). The low-end figures show gains based on current path actions; high-end estimates imply stepped up efforts. The prominence of non-traditional sources is striking. The three largest categories, each potentially generating over 1.5 billion m^3 per year, include urban conservation, underground storage, and municipal wastewater recycling. By contrast, new surface storage under state and federal sponsorship is expected to generate at most 1.2 billion m^3 annually. Anticipated gains from agricultural use efficiency are also more limited, with up to 1 billion m^3 per year in net reductions anticipated. A host of other strategies – desalination, cloud seeding, and improvements in conveyance facilities and operations – each have the potential to generate roughly 0.5 billion m^3 per year.

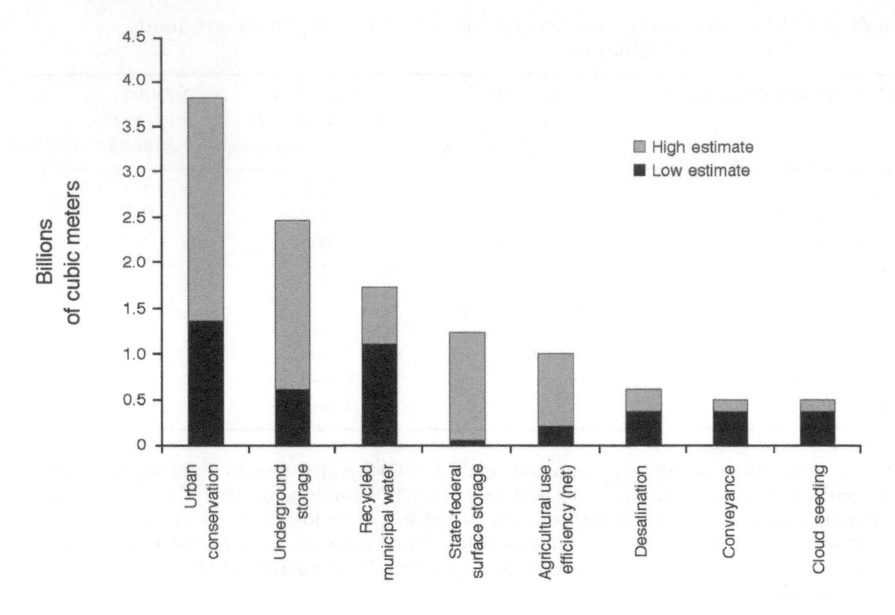

Figure 3.2 Annual production potential from new water sources and conservation, 2000–30
Source: CDWR (2009)

Simply summing these strategies overstates the net supply potential, because some – for instance, surface and groundwater storage – could compete for the same supplies or facilities. But there are also opportunities for synergies among portfolio elements. In particular, the potential for water marketing creates incentives for both agricultural use efficiency and for underground storage. Also, the estimates exclude two options: regional and local surface projects (for which no figures were available) and voluntary reductions in agricultural water use for reasons other than efficiency gains (which proved too contentious a topic for the planning process, as discussed below).

Although there have been some debates on the particulars, the Plan's main message – that California's water supply needs will increasingly be met through a diverse set of options – is now widely accepted. Many of these options are considered more environmentally friendly than traditional surface storage projects, and they are often less costly. Yet although none is entirely new and untested, each presents challenges.

Promises and pitfalls of the non-traditional supplies

The following discussion indicates the relative attractiveness and drawbacks of the four main options for reallocating and augmenting supplies. Cost estimates are for annual deliveries of "raw" water, excluding treatment costs to meet drinking water standards. They include the amortized costs of capital plus operations and maintenance. The discussion begins with the water marketing, which does not explicitly appear in Figure 3.2, but which has significant potential to be both low-cost and beneficial to the environment.

Water transfers

Determining the amount available from future transfers has been a contentious issue for the *CWP* update, because some agricultural interests argue that transfers do not augment supplies. Agricultural water-use efficiency gains, listed in Figure 3.2, do imply transfer activity. But, as noted, agricultural water use is also likely to decline because of various market forces, opening up greater market potential.

State policy began actively promoting water marketing in the early 1980s, by changing the laws to facilitate transfers. One essential step was to make it possible to lease water without losing the water rights. The state played a major role in launching the market during the drought of the late 1980s and early 1990s, initially by arranging for individual drought purchases, and then by running a large-scale drought water bank (Hanak 2003). Since the early 1990s, both federal and state officials have promoted the use of the water market to meet environmental demands as well.

As a result of these actions, the state's water market has grown steadily since the early 1990s drought, with over 1.5 billion m^3 annually in actual flows exchanged by the early 2000s, and another 1 billion m^3 in commitments to transfer water under long-term leases and permanent sales (Figure 3.3). Market growth since the mid-1990s occurred despite favorable precipitation conditions, spurred in part by farmers seeking replacement water to compensate for cutbacks due to new environmental restrictions. Since the early 2000s, several large long-term leases and permanent transfers have been arranged to support development plans in urban areas. Although the growth of these longer-term transactions is a sign that the market is maturing, there are also concerns that new rigidities in the approval process and new regulatory constraints on the use of conveyance infrastructure have hampered the market's ability to respond to drought. Thus, short-term transfers do not appear to have increased during the prolonged drought at the end of the 2000s, and the state-run drought water bank in 2009 purchased less than a quarter of the volumes sought (Hanak *et al.* 2011).

The main obstacles to transfers stem from their potential to harm "third parties" – those other than the buyer and seller (National Research Council 1992; Hanak 2003). In many regions outside of California, one of the key concerns is for third party impacts to the environment, because transfers can alter the water supply conditions upon which wildlife depends. In California, this can also be a concern, but there are legal protections: California law requires transfers to mitigate potential environmental harm,

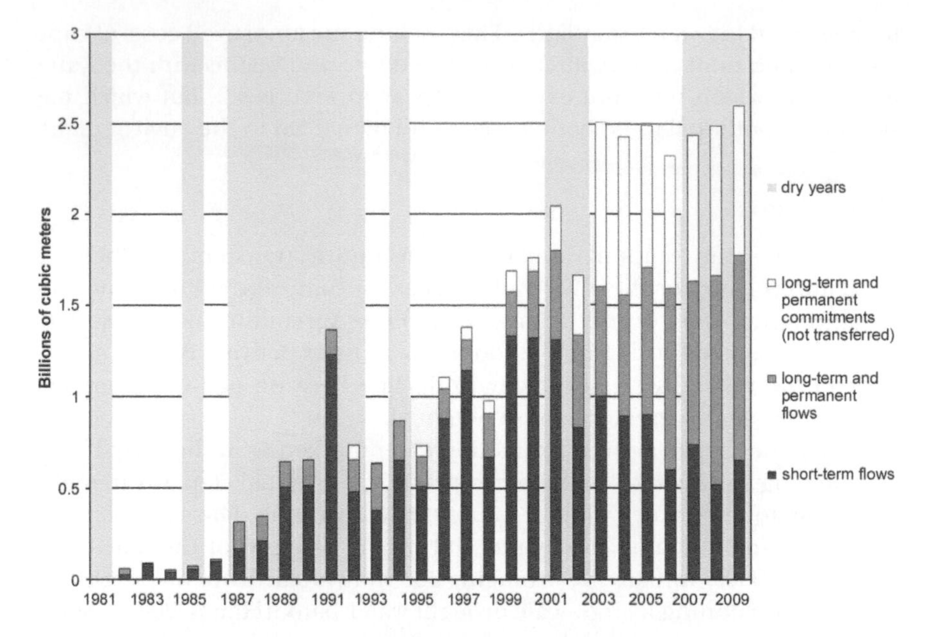

Figure 3.3 California: annual water transfers, 1981–2009

Source: Hanak *et al.* (2011)

Note: The figure shows actual flows under short- and long-term lease contracts, estimated flows under permanent sale contracts, and the additional volumes committed under long-term and permanent contracts that were not transferred in those years. Dry years are years categorized as dry or critical as measured by the Sacramento River 40-30-30 index.

and both government and civil watchdogs may object to proposed transfers on these grounds. As is generally the case in the water law of states in the western United States, California's "no injury" protections also apply to other water users. However, these protections do not generally extend to another key set of third parties – the residents of source communities. If transfers are associated with a decline in farming, such communities may fear the potential for an associated drop in local business activity and tax receipts. As a result, there can be considerable political pressure against transfers (Hanak and Dyckman 2003).

Over this time, buyers and sellers have gained experience in dealing with third party concerns, and more recent deals aim to limit the risk of economic harm to source communities. Such concerns have led to the establishment of mitigation funds for some transfers and to rules limiting the amount of land fallowing in any given area. Both principles have been applied in two large long-term transfers of Colorado River water from agricultural to urban areas (the Imperial Irrigation District to San Diego County, and the Palo Verde Irrigation District to the Metropolitan Water District of Southern California).

Although lead times to meet environmental and community requirements can be substantial, transfers do indeed provide a relatively low-cost water source to urban agencies, with annual prices ranging from under US$100 per 1000 m^3 for local deals within the Central Valley to US$500 per 1000 m^3 or more for deliveries to cities on the Southern Coast or to farmers with high-value crops facing water shortages. The shift toward long-term and permanent transfers has reduced the relative importance of farmers as buyers (whereas they were the primary purchasers during the 1990s, they now account for just a quarter of all contractual commitments and just a third or all purchases). Meanwhile, urban agencies now account for nearly half of all commitments and over a third of flows. Environmental programs have also benefited from the market, with state and federal purchases of up to a third of total volumes for instream flow and wildlife habitat.

Underground storage

Underground storage, or groundwater banking, involves the conjunctive use of surface and groundwater. Conjunctive use exploits the cross-year variability of rainfall, promoting greater use of groundwater in dry years to maximize underground storage of excess surface water in wet years. "Active recharge" programs use spreading ponds or injection wells. The alternative is "in-lieu" recharge, whereby water users substitute pumping with surface water use in wet years to allow the aquifer to replenish faster. Either method generally requires unused space in the aquifer, made available by excess pumping in prior years. In parts of urban Southern California, active recharge programs have existed for decades (Blomquist 1992; CDWR 2003b).

More recently, water users have recognized that groundwater banks can store water not only for those overlying the basin, but also for users elsewhere in the state, in a manner similar to surface water reservoirs. Successful projects of this nature have developed in Kern County, at the southern end of the Central Valley, where irrigation districts are storing water that urban utilities may call on in dry years (Thomas 2001). There has also been some experimentation with using relatively full aquifers – such as those north of Sacramento – for storage. In such cases, the retrieval occurs first, to be followed by recharge. According to the latest *CWP*, artificial recharge has accounted for 1–1.5 billion m^3 in recent normal to wet years, or roughly 6 percent of average annual groundwater use.

Groundwater banking projects can deliver water at a very low cost. A group of projects submitted to California Department of Water Resources (CDWR) for financial support had a weighted average annual cost of US$136 per 1000 m^3. (Not all of these estimates included the costs of acquiring the surface water for storage, which can vary from negligible to several hundred dollars per 1000 m^3, depending on the source and the year.) Reaching Figure 3.2's upper end of 2.5 billion m^3 would also require substantial investments in conveyance and re-operation of surface reservoirs.

Relative to surface storage, groundwater banking is generally considered an environmentally friendly option. However, it has some potential drawbacks from both technical and institutional standpoints. First, both storage and retrieval are slower than with surface storage. When the objective is to capture and store a large volume of flood flow during a relatively short amount of time, recharge capacity may be a limiting factor. Similarly, retrieval from groundwater banks is often limited by pumping capacity. Second, water quality concerns may arise from mixing water from different sources. This presents an obstacle, for instance, to storage of recycled water in the Mojave Basin (Victor Valley Wastewater Reclamation Authority 2004) and to storage of treated drinking water in some Central Valley communities (Cooper 2004). Contamination from overlying land use also raises water-quality issues for conjunctive use in some areas.

Third – and perhaps most importantly – groundwater banking can only be successful when there is a sound basin management system (Thomas 2001; Hanak and Dyckman 2003). Without clear accountability procedures, bankers run the risk of not being able to retrieve the water they store, and their neighbors run the risk of seeing the aquifer depleted from excessive retrieval. These issues are particularly important when outside parties are involved – and they have led to some of the same types of conflicts as those associated with water transfers. Because groundwater management is a local prerogative in California, there is considerable variation in the extent to which protocols and procedures are in place. Most southern California basins are actively managed, and the progress made since the mid-1990s in Kern County has been facilitated by management protocols. Improvements in management are a priority elsewhere in the Central Valley to realize the full potential of this water supply strategy.

Urban conservation

Conservation is a demand-side measure to free up supplies. The numbers reported in the *CWP* have generated considerable discussion because the potential savings appear so high – on the order of 3.8 billion m^3 per year. This estimate may overstate potential savings, however, because it represents the maximum feasible level attainable with today's technology irrespective of costs (California Bay Delta Authority 2005). Estimates that do consider cost-effectiveness are more modest but nevertheless substantial. The California Bay Delta Authority study estimated potential annual water savings of up to 2.6 billion m^3 at an annual cost of US$270 to US$650 per 1000 m^3, and a study by the Pacific Institute (Gleick *et al.* 2003), concluded that 2.8 billion m^3 could be saved for US$740 per 1000 m^3 or less – a threshold the authors deemed relevant for most alternative sources. These various estimates relate to applied water rather than consumptive use. For many urban agencies, this is a relevant metric, because demand growth is measured in terms of applied water use. However, applied use measures overstate net savings to the overall system somewhat, because many return flows from inland areas either recharge the groundwater basin or are reused downstream. (For the more populous coastal areas, most excess applied water is "lost" to the ocean).

Because it makes no additional demands on water resources, conservation is the ultimate environmentally friendly option. It can also be both cost-effective, and – if the savings are durable – reliable. Yet, as Gleick and his colleagues acknowledge, there may be considerable "educational, political, and social barriers" to achieving these savings. California's experience over the past 15 years highlights both the potential and the challenges.

Conservation programs have been promoted actively since the early 1990s drought. Much of the focus has been on non-price tools, including "soft" programs, such as public education, and "hard" programs, such as regulations. There has also been some use of pricing tools, particularly rebates for adopting more water-efficient technologies.

Statewide regulations introduced in the early 1990s included requirements to use low-flow toilets and showers in new construction. Programs to encourage various other measures, including technology retrofits in older homes, have been spearheaded by the California Urban Water Conservation Council (CUWCC), a voluntary association of water utilities formed in 1991. The CUWCC promotes and tracks the adoption of 14 Best Management Practices, and it is nationally recognized as a leading authority on conservation. (For its materials on conservation programs, see www.cuwcc.org).

The programs have sparked some noted successes. Thanks to an aggressive low-flow toilet retrofit program, the City of Los Angeles was able to make up for substantial surface water cutbacks required as part of an environmental mitigation settlement (Los Angeles Department of Water and Power 2001). More generally, the six-county service area of greater Los Angeles, served by the Metropolitan Water District of Southern California (MWDSC), has reduced per capita use by over 10 percent since the late 1980s, saving enough water to accommodate most of the growth the region has experienced since then (MWDSC 2005). Indoor plumbing retrofits have played a central role.

The tougher challenge is outdoor uses, which account for roughly half of residential water use, and even more in hotter inland areas. California's growth patterns are compounding this challenge, with half of the new residents by 2030 expected to settle in inland counties (Johnson 2005). The development footprint in these regions is also less "water-wise." They have a higher share of single-family homes (Hanak and Davis 2006), which use more water than multi-family units (Dziegielewski *et al.* 1990). Single-family lots in inland areas are also larger than those in temperate coastal zones. Given climate and size differences, a typical grass-covered yard in the fast-growing high-desert areas like Palm Springs (Riverside County) or

Lancaster (eastern Los Angeles County) has consumptive water needs nearly three times as high as a yard in coastal Santa Monica (Hanak and Davis 2006).

Technology fixes are also emerging for outdoor water use: "smart" irrigation systems that are sensitive to the weather can reduce watering by 20 percent or more. Utilities are also looking to landscaping solutions, to cut back on the use of turf and other plants more suited to wetter climates. Since the early 2000s, MWDSC and its member agencies have been promoting "California-friendly" gardens and encouraging builders and garden supply chains to participate (www.bewaterwise.com). A few localities are following the regulatory approach of Las Vegas and some Arizona utilities, restricting turf in new homes to just a portion of the total area (e.g. back yard only). And some are considering the Las Vegas model of offering financial incentives for landscaping changes, by paying customers to replace turf with low water-using plants (see www.snwa.com; Hanak and Browne 2006). Many of these policies have been endorsed by a state-sponsored Landscape Task Force, composed of stakeholders from the water and landscaping sectors (CUWCC 2005).

The other side of incentives is the water rate structure. In particular, tiered rates, which charge higher marginal rates for higher levels of water use, can be an important tool for outdoor conservation (Chestnutt *et al.* 1997; Olmstead *et al.* 2007; Mansur and Olmstead 2012). Recent analysis – using multiple regression techniques to control for weather and other factors – finds that California communities using tiered rates had ten percent lower household water use than communities with uniform rates, and 25 percent lower use than communities without volumetric fees (Hanak 2008).

California made progress in tiered rate adoption during the early 1990s drought, but there was little forward movement through the mid-2000s (Figure 3.4). Progress has been slowest in the inland areas where this could make the biggest difference. As of 2003, half of the state's population was subject to a tiered rate structure (Hanak 2005). But in the Central Valley, where summers are hot and lots are larger, this figure dropped to under one-fifth, and nearly half of all homes did not even have water meters.

For a variety of reasons, rate reform has proven extremely contentious in some inland areas. No doubt, residents appreciate the ability to live in the midst of green oases during the hot, dry summer, and they recognize that the introduction of meters or tiered rates could make this more expensive. Such objections have recently arisen in the area around Palm Springs – a desert zone where landscaping accounts for up to 80 percent of residential use – where the local utility has been pushing for the introduction of tiered rates (Desert Sun Wire Service 2008). But fear of the unknown may also be a factor. In professional meetings discussing rate reform, utility officials have noted that some residents assume even low water users will pay higher bills, when in fact the opposite is often true. Similarly, "rate shock" for high

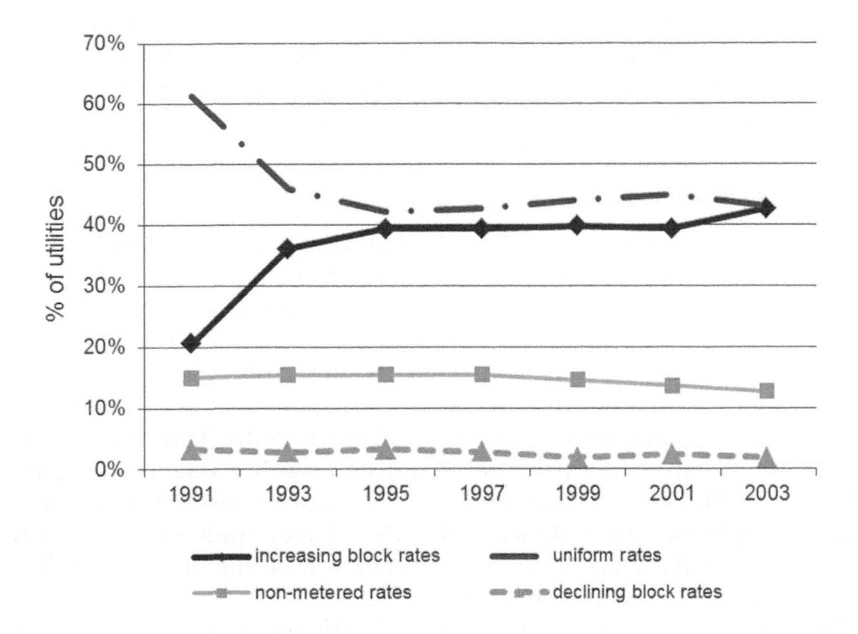

Figure 3.4 Utility rate structures in California, 1991–2003
Source: Hanak (2005)
Note: The chart reports the share of utilities with each rate structure (total = 100 percent), using data from 214 utilities present in the survey in all years

water users can lead to pressure to undo reforms (American Water Works Association 2004). This suggests that public education programs are an essential part of a reform package.

Another barrier to conservation programs is that the savings might make room for growth (Hanak and Browne 2006). When the public holds such views, utility boards, elected by local voters, are often in weak positions to counter them. This is where state regulations may play a useful role. When California's legislature passed a law in 2004 requiring the phase-in of water meters, it took the heat off local officials, who are now free to lobby for earlier introduction of the measures (e.g. Hood 2005). Given the potential savings, requiring utilities to adopt tiered rates – or to justify why they are not necessary – is another potential area for legislative mandates.

The drought of the late 2000s spurred a new round of conservation efforts, as urban utilities adopted both non-price and price-based incentives to reduce water use. These efforts were often very effective, resulting in 10–20 percent reductions in per capita water use. (The concurrent economic downturn also contributed to aggregate demand reductions, particularly in areas with foreclosed and unsold homes). In late 2009, the state adopted new legislation intended to consolidate these savings, requiring urban utilities to meet a target of 20 percent reductions in per capita water use by 2020. This legislation will encourage utilities to maintain the momentum on conservation and prevent the consumption rebound that often occurs after droughts are over.

Recycled municipal water

To some, recycling wastewater is just another form of conservation, because it augments usable supplies from a given water source. But quite different issues are at stake. Because most recycled water is not sufficiently processed or certified to meet drinking water standards, it requires separate plumbing. Incremental processing and redistribution costs can also be high. When limited to outdoor uses, recycled water must be sold at a discount, and it risks being in excess supply in wet winter months. Thus, although it is relatively reliable, recycled water is not necessarily a financial bargain. The potential for cost effectiveness is greater for new construction and new treatment plants. The California Recycled Water Task Force (CDWR 2003a) estimated average unit costs of expansion on the order of US$740 per 1000 m^3, including treatment and delivery.

Recycling is not always as environmentally friendly as it might appear at first glance, either. Recycling typically results in reduced discharges of treated effluent into rivers and streams. If the resulting change in stream flows will have negative effects on wildlife habitat, communities may be required to modify their plans. This occurred in the coastal city of San Luis Obispo, where the recycling plan conflicted with endangered steelhead trout habitat (*Water Reuse News* 2003).

Finally, and perhaps most importantly, the public needs to be convinced of the safety of recycled water. In California, there have been several well-publicized cases of public resistance. In the mid-1990s, the city of Los Angeles launched a project to recharge the groundwater basin with tertiary-treated recycled water and invested in a new treatment plant. In 2000, when the project was about to come on line, bad publicity of what came to be known as the "toilet to tap" program forced the city to abandon the recharge plans and instead try to find irrigation and industrial customers (Sanitation Districts of Los Angeles County 2005).

Using recycled water outdoors has also sparked controversy. In the Bay Area community of Redwood City, officials planned to introduce recycled water for some outdoor uses as a way to accommodate growth (Redwood City 2004). Some residents were concerned about potential health risks of switching to recycled water on lawns and playing fields. Following a year of contentious debates, a modified plan was approved in early 2004. The compromise required that recycled water not be used in areas where children play and that some playing fields switch to artificial turf.

These factors help keep recycled water use to only 310 million m^3, less than one-tenth the volume of water that gets processed by wastewater treatment plants each year. The task force's projections of a three- to four-fold expansion over the coming decades assume that utilities will be able to overcome this resistance through public education and outreach. One promising enterprise is Orange County's "Groundwater Replenishment System," which began recharging the groundwater basin with 86 million m^3 per year of highly purified recycled water in 2007 (www.gwrsystem.com). Cognizant of the pitfalls of bad publicity in neighboring Los Angeles, local officials have devoted a great deal of attention to public education since the early planning stages, and they have used opinion polling to help shape the message. In light of recent water shortages, Los Angeles' mayor has announced that his city will relaunch a recycled water initiative – this time, taking a cue from the more positive experiences in Orange County (Connell 2008). The state is also pursuing regulatory changes to facilitate the use of recycled water, including for direct potable use.

Conclusions and future challenges

California's recent experience offers some interesting insights into the changing world of water planning in regions facing rapid population growth and environmental constraints on supply expansion. Constraints on the continued development of the two traditional mainstays of the water supply portfolio – surface storage and native groundwater reserves – need not spell an impending water crisis. Measures to broaden the portfolio with non-traditional sources (recycling, underground storage) and with more efficient use of existing supplies (conservation and water marketing) can help supplies keep pace with demand. These new sources often come in at lower financial cost and pose fewer risks to the environment than the traditional elements of the portfolio.

However, each of these strategies requires planners to tackle some important institutional and political obstacles. For water transfers, finding ways to mitigate economic harm in the source regions is essential. For groundwater banking, solid institutional rules on the use of the aquifer and good monitoring are preconditions. For urban conservation, much of which now relates to outdoor uses, there is a need to work with the public, builders, and landscape professionals to make new technologies and landscaping alternatives available. There is also a need to engage in rate reform – a sometimes delicate process given the sense of entitlement many large water users have, especially when conservation is used to support new growth. State and federal governments are most effective in a supporting role, with local and regional agencies taking the lead. Financial incentives, technical support, legislation, and regulations all have a role to play in spurring local and regional water agencies to collaborate, innovate, and invest in the water supplies for the twenty-first century.

Looking ahead, California's water planners will also face new challenges as a result of climate warming. Although there is still uncertainty about the impacts on average levels of precipitation, models agree that the state will see a reduction in snowpack and a shift in seasonal runoff as a result of warming, a process already under way (Cayan *et al.* 2006; Knowles *et al.* 2006). Many of the portfolio management tools that California water managers have been developing to cope with demand growth, droughts, and increasing environmental water requirements will also serve them well in adapting to climate change (Hanak and Lund 2012). But a changing climate will also require new types of experimentation and innovation in several areas. First, it will require new thinking about the inter-relationship between water supply and flood management, to make better use of surface and underground reservoirs as runoff patterns shift. Second, it will require continued innovation in water markets to develop tools to cope with greater uncertainty – such as long-term, multi-year options. Third, because warming also has negative implications for habitat of protected aquatic species, it will require new assessments of how to manage environmental flows effectively, while continuing to make water available for human uses. In the future, as now, California will not be alone in needing to meet these challenges, and there is much to be gained by exchanging information and experiences.

Note

1. For a more detailed discussion of the full range of supply options, see Hanak (2007) and Hanak *et al.* (2011).

References

American Water Works Association (2004) *Avoiding Rate Shock: Making the Case for Water Rates.* Denver, CO: American Water Works Association.

Blomquist, William (1992) *Dividing the Waters: Governing Groundwater in Southern California.* San Francisco, CA: Institute for Contemporary Studies Press.

California Bay Delta Authority (2005) *Final Draft Year 4 Comprehensive Evaluation of the CALFED Water Use Efficiency Element.* Sacramento, CA: California Bay Delta Authority.

Cayan, D., Luers, A., Hanemann, M., Franco, G. and Croes, B. (2006) *Scenarios of Climate Change in California: An Overview.* Publication CEC-500-2005-186-SF. San Diego, CA: California Climate Change Center.

CDWR (California Department of Water Resources) (2003a) *Water Recycling 2030: Recommendations of California's Recycled Water Task Force.* Sacramento, CA: California Department of Water Resources.

CDWR (2003b) *California's Groundwater.* Bulletin 118-Update 2003. Sacramento, CA: California Department of Water Resources.

CDWR (2009) *California Water Plan Update.* Bulletin 160-09. Sacramento, CA: California Department of Water Resources.

Chestnutt, Thomas, W., Janice A. Beecher, Patrick C. Mann, Don M. Clark, Michael Hanemann, W., George A. Raftelis, Casey N. McSpadden, David M. Pekelney, John Christianson and Richard Krop (1997) *Designing, Evaluating, and*

Implementing Conservation Rate Structures. July. Sacramento, CA: California Urban Water Conservation Council.

Connell, Rich (2008) "LA Prepares Massive Water-Conservation Plan." *Los Angeles Times* 15 May.

Cooper, Audrey (2004) "Water Storage Plan Hits Snag: Pollution Rules Stall Proposal to Go Underground." *San Joaquin Record* 31 August.

CUWCC (California Urban Water Conservation Council) (2005) *Water Smart Landscapes for California. AB 2717 Landscape Task Force Findings, Recommendations and Actions.* Report to the Governor and the Legislature. Sacramento, CA: California Urban Water Conservation Council.

Desert Sun Wire Service (2008) "Residents to Oppose Water Fee Increases." *Desert Sun* April 21.

Dziegielewski, B., Opitz, E. and Rodrigo, D. (1990) *Seasonal Components of Urban Water Use in Southern California.* Carbondale, IL: Planning and Management Consultants, Ltd.

Gleick, Peter H., Dana Haasz, Christine Henges-Jeck, Veena Srinivasan, Gary Wolff, Katherine Kao Cushing and Amardip Mann (2003) *Waste Not, Want Not: The Potential for Urban Water Conservation in California.* Oakland, CA: The Pacific Institute.

Hanak, Ellen (2003) *Who Should Be Allowed to Sell Water in California? Third Party Issues and the Water Market.* San Francisco, CA: Public Policy Institute of California.

Hanak, Ellen (2005) *Water for Growth: California's New Frontier.* San Francisco, CA: Public Policy Institute of California.

Hanak, Ellen (2007) "Finding Water for Growth: New Sources, New Tools, New Challenges." *Journal of the American Water Resources Association* 43(4): 1024–35.

Hanak, Ellen (2008) "Is Water Policy Limiting Residential Growth? Evidence from California." *Land Economics* 84(1): 31–50.

Hanak, Ellen and Margaret K. Browne (2006) "Linking Housing Growth to Water Supply. New Planning Frontiers in the American West." *Journal of the American Planning Association* 72(2): 154–66.

Hanak, Ellen and Matthew Davis (2006) "Lawns and Water Demand in California." *California Economic Policy* 2(3).

Hanak, Ellen and Caitlin Dyckman (2003) "Counties Wresting Control: Local Responses to California's Statewide Water Market." *University of Denver Water Law Review* 6(2): 490–518.

Hanak, Ellen and Jay R. Lund (2012) "Adapting California's Water Management to Climate Change." *Climatic Change* 111: 17–44 (doi:10.1007/s10548-011-0241-3).

Hanak, Ellen, Jay Lund, Ariel Dinar, Brian Gray, Richard Howitt, Jeffrey Mount, Peter Moyle and Barton "Buzz" Thompson (2011) *Managing California's Water: From Conflict to Reconciliation.* San Francisco, CA: Public Policy Institute of California.

Hood, Jeff (2005) "Lodi Council Wants Water Meters in Sooner." *Stockton Record* 7 December.

Hundley, Norris (2001) *The Great Thirst: Californians and Water, A History.* Revised edition. Berkeley, CA: University of California Press.

Johnson, Hans P. (2005) "California's Population in 2025." In Ellen Hanak and Mark Baldassare (eds.), *California 2025: Taking on the Future.* San Francisco, CA: Public Policy Institute of California.

Knowles, Noah, Michael D. Dettinger and Daniel R. Cayan (2006) "Trends in Snowfall versus Rainfall in the Western United States." *Journal of Climate* 19: 4545–59.

Los Angeles Department of Water and Power (2001) *Year 2000 Urban Water Management Plan.* June Los Angeles, CA: Los Angeles Department of Water and Power.

Mansur, Erin T. and Sheila M. Olmstead (2012) "The Value of Scarce Water: Measuring the Inefficiency of Municipal Regulations." *Journal of Urban Economics* 71: 332–46.

MWDSC (Metropolitan Water District of Southern California) (2005) *The 2005 Regional Urban Water Management Plan.* Los Angeles, CA: Metropolitan Water District of Southern California.

National Research Council (1992) *Water Transfers in the West: Efficiency, Equity, and the Environment.* Washington, DC: National Academy Press.

Olmstead, Sheila M., Michael Hanemann, W. and Robert N. Stavins (2007) "Water Demand Under Alternative Price Structures." *Journal of Environmental Economics and Management* 54: 181–98.

Redwood City (2004) *The Redwood City Recycled Water Task Force Report.* March. Redwood City, CA: Redwood City.

Reisner, Marc (1993) *Cadillac Desert: The American West and Its Disappearing Water.* New York: Penguin Books.

Sanitation Districts of Los Angeles County (2005) *Final Palmdale Water Reclamation Plant 2025 Facilities Plan and Environmental Impact Report.* September. Los Angeles, CA: Sanitation Districts of Los Angeles County.

Thomas, Gregory (2001) *Designing Successful Groundwater Banking Programs in the Central Valley: Lessons from Experience.* August. Berkeley, CA: Natural Heritage Institute.

Victor Valley Wastewater Reclamation Authority (2004) "VVRWA Subregional Facilities Draft Program EIR/EIS." Available at www.trgandassociates.com/vvwra/Documents.RWP.htm (accessed 15 October 2005).

Water Reuse News (2003) "San Luis Obispo Constructing Reclaimed Water System." *Water Reuse News* 5 September.

4 Water trades in the western United States

Risk, speculation and property rights

Kristiana Hansen, Richard Howitt and Jeffrey Williams

Introduction

Population growth and the increased demand for environmental protection in the western United States are placing additional demands on existing water supplies. The increasing environmental and fiscal costs of development limit western states' ability to meet these demands by traditional structural supply augmentation. Excess urban and environmental water demands must therefore largely be met by conserving and reallocating existing supplies. Academics and water agencies alike now widely acknowledge that water trading plays in important role in this process by offering a clear measure of value for conservation and by providing a voluntary, self-compensating mechanism for reallocation.

While several states have had active water-rights markets for many years, until 20 years ago water markets were not marked by large trades whether on an absolute basis or in proportion to state water consumption. For small trades, the costs of defining permanent water rights remained sufficiently low to enable trades to occur. In the early 1990s, droughts, environmental constraints on water development, and the cost of new water supplies pressured large water-using states – such as Oregon, Texas, Idaho, Arizona, and California – to consider alternative water-market structures. These states initially proposed market structures based on those in Montana, New Mexico, Utah, and Wyoming (states with long histories of water sales), which predominantly based their water markets on the sale of permanent water rights between private parties. Water masters in these states, who often had the role of approving the transferable quality of the water rights, registered the sales.

This simple pure-market system has not dominated the development of water markets in the west in recent years. Table 4.1 shows that for 12 western states during the years 1990–2008, short-term leases outnumbered permanent-rights sales by an average ratio of 16 to 1. In states with the largest volume of water trades (California, Arizona, Idaho, and Texas) the ratio of leases to sales was nearly 22 to 1. Several factors contribute to this domination of the water market by leases.

Table 4.1 Transacted volume for reported water transactions, 1990–2008

	Lease		*Sale*		*Total*			
State	*Observations*	*Quantity (m^3 millions)*	*Observations*	*Quantity (m^3 millions)*	*Observations*	*Quantity (m^3 millions)*	*Lease-to-sale ratio*	*Transactions as % of total use*
AZ	59	9,392	106	399	165	9,791	23.54	5.55
CA	614	10,624	77	553	691	11,177	19.22	1.33
CO	116	606	1,361	136	1,477	742	4.47	0.23
ID	117	5,966	19	155	136	6,121	38.59	1.31
MT	41	63	2	–	43	63	–	0.03
NM	90	702	54	44	144	746	16.01	0.96
NV	4	39	276	152	280	191	0.26	0.29
OR	86	1,148	18	29	104	1,177	–	0.61
TX	230	1,447	61	295	291	1,741	4.91	0.31
UT	21	155	44	69	65	224	2.25	0.19
WA	40	162	10	105	50	267	1.54	0.16
WY	52	390	1	–	53	390	–	0.25
Total	1,470	30,695	2,029	1,936	3,499	32,630	15.86	0.96

Source: Transaction data from *the Water Strategist*; total water use from US Geological Survey.
Notes: "Observations" indicates the number of transactions reported by the *Water Strategist* within each state for each contract type. Note that each transaction entry in the *Water Strategist* often represents more than one buyer–seller pair. The entries have been separated into multiple transactions when there is enough information on buyer and seller identity and /or new uses to do so. Otherwise, the number of transactions reported above and in subsequent tables may represent multiple buyer–seller pairs.

Sales reported in the *Water Strategist* for Montana and Wyoming were less than 500,000 m^3, so they are not reported here. Lease-to-sale ratios were not calculated for Oregon, Montana, and Wyoming due to low sales volume.

The two most important factors are:

- the prevalence of usufructory (appropriative) rights, and
- the concern for third-party effects on water exports in exporting regions.

This combination of private and public rights makes the simple, permanent sale of water rights more complex than initially envisaged.

Because states usually retain residual water rights, and state or federal agencies hold contracts for a substantial quantity of the developed surface water, many water users may claim only usufructory rights to water. Such rights do not automatically guarantee rights holders the same volume as transferable rights. Transferable rights need to be defined as a function of usufructory rights. Transferable water rights are proscribed by a wide array of ancillary property rights, instituted to protect third parties and the public interest. Thus, we can describe water as a quasi-public good. Clearly these restrictions on water trades discourage fluid markets, while short-term leases limit the ability of markets to reduce supply-side risk to water buyers. This chapter reviews some of the emerging trends in western water markets and examines ways to reconcile the quasi-public characteristics of water with the reallocation, risk reduction, and equity roles that western US water markets are expected to play.

A quantitative measure of western US water markets

The economic literature on western US water markets usually focuses on actions in a single state. Comparable data across western states is hard to find. We have compiled 3,499 transactions from a monthly trade journal called the *Water Strategist* from 1990 to 2008. We know of no other data set as comprehensive as this one. The *Water Strategist* reports rights transfers and leases (including price, quantity, buyer and seller identification, buyer and seller use, and some additional contract terms) in 16 western states on a monthly basis.

Several other studies have used the *Water Strategist* data. The most recent of these is Brewer *et al.* (2008), who use the data to survey how western water markets have developed in response to the large disparity in value between urban and agricultural water uses. Brewer *et al.* find that the number of agriculture–urban transfers is increasing over time, and that water is increasingly transferred under longer-term leases and sales rather than one-year leases. They also find that urban users pay more to buy and lease water relative to agricultural users, though variance of price across states is much greater than variance across sectors within a state.

Using data compiled from the *Water Strategist*, Brown (2006) estimates the effect of time, precipitation, population, buyer use, groundwater, and transaction size on sale and lease price. Consistent with our observations from the *Water Strategist*, he finds that sale price is increasing over time but is unaffected by drought periods. Lease price is higher during drought periods than during wet periods and for transfers to municipal and environmental uses than to agriculture.

Brookshire *et al.* (2004) also use the *Water Strategist* database (over the years 1990–2001) to estimate a demand function for water rights for three of the most active markets in the western United States, in Arizona, New Mexico, and Colorado. They find that municipal and agricultural buyers pay significantly more for water rights than governmental buyers, who are primarily purchasing for environmental uses mandated by state or federal law. They also emphasize that much of the variation in price and quantity across the three markets is due to institutional differences. For example, rights in the Colorado-Big Thompson Project (CBT) in Colorado are homogeneous and transaction costs are low, whereas in New Mexico, there is a backlog of groundwater rights to be adjudicated by the Office of the State Engineer before transfer can occur.

Table 4.1 indicates the quantities of water transferred by lease and sale for 1990–2008. The four states that transfer the most water in absolute terms are California, Arizona, Idaho, and Texas. These states also report the highest volume transferred through leases. Most leases are short-term; only 15 percent of leases (representing 7% of transaction-year volume) were for durations of longer than one year. Long-term leases are in some respects more similar to rights transfers than to short-term leases. They allow water agencies to avoid repeated negotiation costs associated with multiple transactions and exposure to future price uncertainty, often without the more burdensome regulatory restrictions imposed on rights transfers. The buyer and seller in a number of transactions reported in the *Water Strategist* explicitly identified their short-term leases as a means of acquiring information before entering into longer-term contractual arrangements.

The most striking feature within this table is the variation in state lease-to-sale ratios reported in the second-to-last column of Table 4.1. For example, for every 19 m^3 of water transferred under lease in California during the study period, only 1 m^3 was transferred under sale. For every 4.5 m^3 transferred under lease in Colorado, 1 m^3 was transferred under sale. This stark difference is undoubtedly due to the institutional conditions that prevail in the Central Valley Project (CVP) in California and the CBT in Colorado. The CBT operates as a single water district, encompassing both agricultural and urban areas, which lowers trading costs significantly compared to the more jurisdictionally fragmented CVP. Second, proportional water rights in the CBT make rights transfers easy compared to the CVP, where the priority rationing system requires that water be quantified and potentially adjudicated before it is traded away (Carey and Sunding 2001).

Table 4.1 indicates that although water markets are growing, total volume traded remains small compared to overall consumption. In most states, less than 1 percent of water consumed is transferred through sales or leases each year. When water is transferred under long-term lease or a sale, only the quantity transferred in the first year of the contract is reported in Table 4.1. As these numbers include only transaction-year volume, they under-report the significance of sales and long-term leases relative to short-term leases. When volume is calculated cumulatively (so that the volume transferred under a long-term lease in 2001 is counted in 2001 and each subsequent year of the study period), transferred water is still less than 3 percent of consumptive use in most states.

Table 4.1 also indicates the number of transactions within each state for each contract type. Markets are clearly thin. In approximately half of the 228 state-year cells within this dataset, there are no transactions reported in the *Water Strategist*. When and where transactions do occur, there are few buyers and sellers.

Table 4.2 compares average volume-weighted lease and sale prices for transactions over the study period. "Lease price" is the cost in dollars per $m^3$of acquiring a pre-specified volume of water in each year of the contract. "Sale price" is the total cost of obtaining a right to every m^3 of water each year, in perpetuity. Sale and lease price distributions are both skewed, with small, high-value transactions driving average results. The highest prices observed in the data set are small mining transactions in remote regions. Further, sale transactions on the Front Range of the Rocky Mountains in Colorado are generally small and frequent, due to low transaction costs associated with transferring water within the CBT. To compensate for the skewness of the price data, the prices we present are volume-weighted.

Across states and years there is large variation in lease and sale prices. The final column of Table 4.2 reports the implicit capitalization rate, which is the ratio of annual lease price to total sale price. States' individual implicit capitalization rates vary greatly. The ratios for Oregon and Washington are relatively large, in part because it was common over the study period for irrigators to lease out water at very low prices in wet years to avoid losing rights due to non-use. A variety of state and non-profit institutions have developed in those states to facilitate such transfers. The average implicit capitalization rate for the entire dataset is 5.95 percent (6.78% excluding CBT sale transactions), below the market cost of borrowing money of 8 percent. The low implicit capitalization rates are likely due to the fact that many transfers occur at administratively set prices that do not reflect water value. Most notably, irrigators often pay subsidized rates for leased water. Further, municipal users are willing to pay a premium to acquire water rights. Interestingly, the average capitalization rate for the final four years of the study period is 5.64 percent (6.12% excluding CBT sale transactions), which indicates an increased premium on water rights during the lead-up to the collapse of the housing market in 2008. However, the implicit capitalization rates reported here remain below the market cost of borrowing capital. In the absence of transaction costs, risk and uncertainty, those wanting water would be indifferent between purchasing a right, which yields a flow of water each year in perpetuity, and acquiring water each year from the market in the form of a short-term lease, each year for all time. Several characteristics might prevent parity between the two alternatives. First, buyers may be willing to pay more than the annualized sale cost for an annual lease, because the decision to postpone an investment, such as the purchase of a water right that is very expensive to reverse due to high transaction costs, has value (Dixit and Pindyck 1994). Alternatively, sellers may require a premium to sell their water rights, which in theory should equal the uncertainty cost to the buyer of repeated exposure to spot prices in the lease market. Buyers may be willing to pay such a premium to purchase a right for the same reason (Howitt 1998).

Trends in the new uses of water are striking. Table 4.3 groups transactions within the dataset by buyer use, for lease and sale transactions respectively. There is much greater annual variability in quantity and number of lease transactions than sale transactions within all three new use categories. This is to be expected, as leases are often in response to short-term fluctuations in hydrological and economic conditions.

Table 4.2 Volume-weighted prices (US$, 2008) for reported water transactions, 1990–2008

State	*Lease price ($/m³)*	*Sale/share price ($/m³)*	*Implicit capitalization rate (%)*
AZ	0.06	0.57	9.71
CA	0.10	0.74	12.99
CO	0.04	3.08	1.35
ID	0.05	0.20	23.85
MT	0.01	2.20	0.49
NM	0.58	1.64	35.13
NV	0.01	5.51	0.22
OR	0.17	0.46	38.08
TX	0.05	1.24	4.30
UT	0.08	1.46	5.49
WA	0.06	0.21	29.25
WY	0.02	2.57	0.71
Total	0.08	1.23	6.78

Source: Data from the *Water Strategist*
Note: CBT sales of 58,881 10^6 m³ are omitted from the price calculation. If included, Colorado sale price increases to US$4.74/m³, the overall total sale price increases to US$1.41/m³, and the overall implicit rate of return is 5.95 percent.

Table 4.3 Lease and sale volume and volume-weighted prices (US$, 2008) by new use

Lease		*A*	*E*	*MI*	*Other*
1990–92	Annual average volume (m³ millions)	285	75	504	1,365
	Percentage of total annual average volume	13	3	23	61
	Average price ($/m³)	0.04	0.05	0.12	0.07
2006–8	Annual average volume (m³ millions)	222	755	143	105
	Percentage of total annual average volume	18	62	12	9
	Average price ($/m³)	0.39	0.04	0.13	0.07
All Years	Annual average volume (m³ millions)	389	458	329	440
	Percentage of total annual average volume	24	28	20	27
	Average price ($/m³)	0.07	0.09	0.12	0.06
Sale share		*A*	*E*	*MI*	*Other*
1990–92	Annual average volume (m³ millions)	1	6	73	0
	Percentage of total annual average volume	1	8	91	0
	Average price ($/m³)	1.13	0.18	0.45	2.94
2006–8	Annual average volume (m³ millions)	3	50	26	59
	Percentage of total annual average volume	2	37	19	43
	Average price ($/m³)	1.60	0.27	7.94	0.88
All years	Annual average volume (m³ millions)	7	27	53	15
	Percentage of total annual average volume	7	26	52	15
	Average price ($/m³)	1.41	0.40	1.96	1.25

Source: Data from the *Water Strategist.*
Note: A, E, MI, and Other represent Agricultural, Environmental, Municipal/Industrial, and Other new uses respectively. "Other" new use transactions include transactions that list multiple new uses or do not specify a new use.

Over the study period, sales and leases to environmental use both increased as percentages of total leases and sale volume transferred. Environmental transfers have been facilitated by institutions such as the Environmental Water Account (EWA) in California, under which state and federal fishery managers purchased water in real-time to augment instream flows at critical periods (Hanak 2003).[1] Loomis *et al.* (2003) suggest that public agencies and non-profit organizations are able to purchase water through voluntary transactions for environmental in-stream flows because the environmental value of water now exceeds the value of water used in irrigation in some parts of the West. Environmental buyers tended to acquire water under short-term lease rather than long-term lease or rights transfers. There were only 105 sale transactions to environmental use. Of the 377 leases to environmental use, only 46 were longer than one year.

Water for agricultural use also tended to be purchased under short-term lease rather than long-term lease or rights transfer. Of the 557 transfers to agricultural buyers, 340 were leases. Of these leases, only 36 were transfers of longer than one year. Sales to agricultural buyers decreased over the study period, though lease activity remained relatively constant.

Municipal agencies acquire water rights with far greater frequency and volume than either agricultural or environmental entities. They are first and foremost concerned with reliability. Table 4.3 shows that municipal buyers consistently paid higher prices for rights than environmental and agricultural users over time and throughout the west. Municipal purchases decreased in absolute and percentage terms. Within the CBT, it is not uncommon for municipal water agencies to purchase a water right and lease the water back to the agricultural seller until the water is needed for municipal use. Table 4.3 indicates that municipal agencies also acquire significant quantities of water under lease. When municipal agencies do acquire leases, they tend to be long-term; of the 457 municipal leases reported in Table 4.3, 122 are for periods longer than one year. The time trends for the aggregate water market in the 12 western states are shown for annual leases and sales of water rights in Figures 4.1 and 4.2, respectively. Figure 4.1 shows that while the number of lease transactions has increased, the total volume of water in lease transactions has fallen. In contrast, Figure 4.2 shows that sales of water rights are increasing in both number and volume of water. One likely explanation is that while lease transactions are increasing, the maturing of the market has caused longer term larger transactions to be in the form of water rights purchase, leaving lease transactions to represent lower volume entry level transactions.

In praise of speculation

Water supply is inherently uncertain, but the demand for water is relatively inelastic, and in some cases counter-cyclical. Thus it is reasonable that reliability and supply risk are foremost for many water managers. We also want

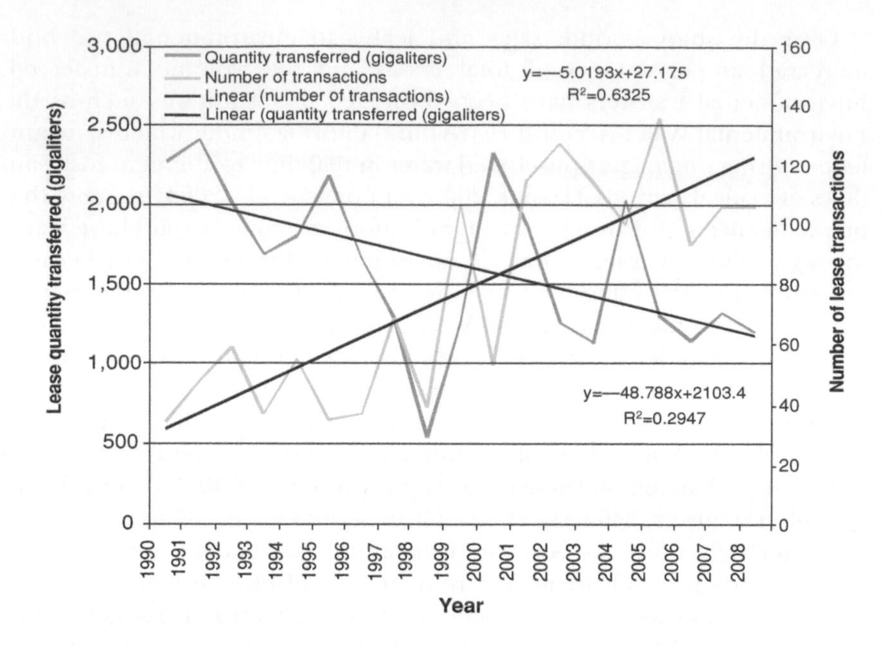

Figure 4.1 Lease transactions and total volume of water traded

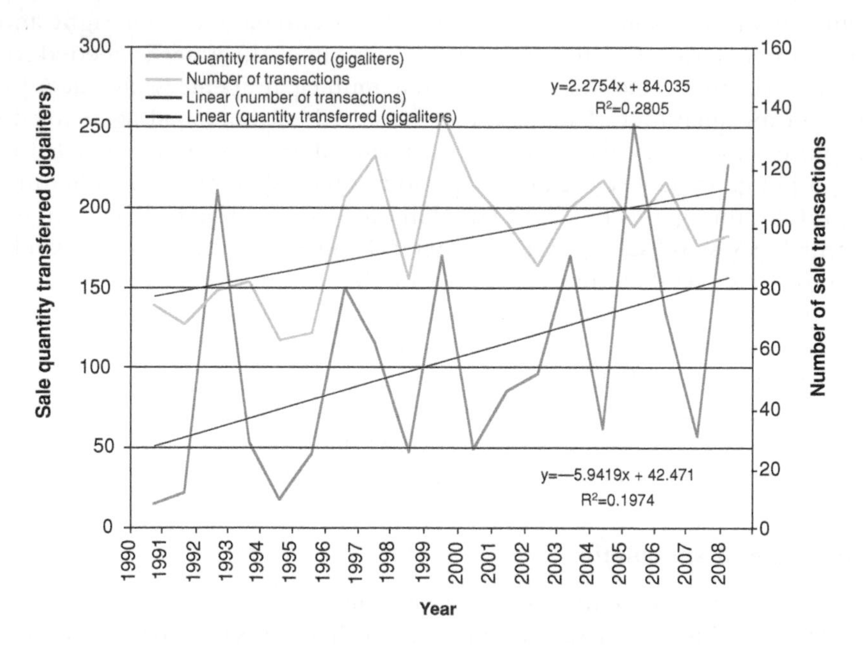

Figure 4.2 Sales of water rights; number and volume of water

to emphasize the difference between speculation and market manipulation. A pure speculator provides a valuable social service by buying risk from producers who want to sell it and concentrate on production. Market manipulation, in contrast, attempts to distort the market for risk or any other product.

The main reason permanent sales of water are preferred to leases is that they automatically solve the problem of supply uncertainty for the purchaser. Paradoxically, however, permanent sales transfer the risk to the seller, who, given the uncertainty about water sales prices, is bound to wonder if he sold too cheaply. Lach *et al.* (2005) shows that supply-risk reduction in the form of water reliability is foremost in the preference function of the majority of water utility managers surveyed. The practices of the CBT project exemplify urban agencies' use of water-rights purchases to reduce risk. Howe *et al.* (1986) report that, in 1982, irrigation farmers used 71 percent of the available water, despite only owning 64 percent of the project's allotments. This outcome resulted from unused urban water rights being leased back to the farmers.

Despite their strong advantages in reducing third-party impacts, short-term leases shift the supply and price risk to the purchaser. Given the relatively inelastic nature of urban water demands and their high use values, the reluctance of urban managers to rely on the undefined potential of annual spot markets seems very rational, despite several successful examples of spot markets. Of the two types of risks in lease markets, it seems that managers are more concerned about supply risk than spot-market price risk.

Option contracts that explicitly separate the cost of risk from the cost of water supply can reduce the cost of both supply risk and price risk. Using option contracts, water purchasers can sell the supply risk to those water rights holders who can absorb the risk at a lower cost. For example, an urban agency with relatively inflexible demands can take an option contract with a farmer who grows annual crops. Clearly, the cost of adjusting crop production to water variability is much lower than an urban rationing or pricing scheme. Like other aspects of water policy, water option markets are more complex than simpler commodities (such as pork bellies). In the first place, a water-options purchaser is only interested in receiving the physical quantity of water when he chooses to exercise the option. In most other option markets, the majority of options are not exercised by the physical delivery of the good in question; options are instead resolved through financial settlements. Uncertainty about when the option will be exercised adds further complication to water market options. Since the holder of the option will only want to exercise it under critical or drought conditions, the option contract must rely on an outside measure of supply – such as a water-supply index – that independently assesses the probability of necessary conditions. Finally, since droughts often occur during a series of years, an option contract with a single exercise option has limited value.

Effectively managing drought risk requires an option contract with several characteristics. First, the contract should run for a period equal to its short-run planning horizon – often seven years or more. Second, the potential to exercise the contract should be available for at least any two years during that period. The contract should also specify the triggering conditions for exercising the option and the strike price. Despite these requirements, urban and agricultural interests have implemented several bilateral option contracts over the past 12 years. Agreements between the Metropolitan Water District of Southern California and several Sacramento Valley irrigation districts are ongoing. Table 4.4 shows the fundamental characteristics of these contracts.

Table 4.4 Option contracts between Sacramento Valley Districts and MWD, 2002–03

	Option price ($/m³)	*Call price ($/m³)*	*Critical year charge ($/m³)*	*Third party cost ($/m³)*	*Quantity (m³ millions)*
Average contract	0.008	0.073	0.020	0.004	Options 206
2003 actual	0.008	0.073		0.004	Calls 148

In 1994–5, the California Department of Water Resources briefly implemented a type of option with a much shorter term (Jerich 1997). The Department had successfully run three emergency drought spot markets in 1991, 1992, and 1994. However, the California hydrologic cycle allows very little time – six weeks – to declare a drought and implement the purchases and contracts needed to run an emergency spot market. Both water sellers and purchasers requested a system that would allow more time to adjust to a drought market. When the water year started with poor rainfall in 1994, the department instituted a one-year option market, which opened in December of that year. The market was aimed at inducing a more responsive supply of water to the water bank, and at creating a price structure that could vary between December and April, as the extent of water supplies became better known. The market took the form of water purchasing options in case the drought year continued. The fixed option price was US$0.003/m³, and the call price of the option was fixed at US$0.030/m³. Early in the water season (December 1994), the projected total demand for the options reached 382 million m³. As the rainfall situation improved, demand for options fell; the department sold only 36 million m³of options (in five separate contracts) during January and February of 1995. Furthermore, the latter part of the season saw substantial precipitation and snow pack, which greatly improved the water situation and reduced the demand for options to zero by mid-April.

Even though the option market had only a small impact in 1995, its operation illustrates that such markets do reduce the risk to buyers and can be operated by public agencies. Given the thinness of such markets and the long periods in which there is no demand for them, it seems very unlikely that they are a viable commercial proposition. However, since they provide a public service through risk reduction, it is appropriate that public water agencies run such markets on a non-profit basis.

Establishing property rights for water trading

From an economic perspective the "beneficial use" doctrine that dominates western water rights is an effective property right for the equitable allocation of a resource that is focused on a single use and in excess supply. These are the conditions that prevailed in the gold mining communities that originated the concept. The modern situation for water users of shifting demands and technologies and ever-present scarcity requires that the voluntary and socially efficient reallocation of water be allowed. A simple redefinition that characterizes socially beneficial water trades as a beneficial use would enable the voluntary reallocation of water to more valuable uses.

Australia and Chile have taken the logical step of decoupling water rights from land. This requires the difficult and expensive legal step of defining water rights on a consumptive basis and also associating priorities with these rights (Young and McColl 2002). This separation of water rights from land rights is more complicated under systems with usufructory rights and the varying time-dependent priorities found in the prior appropriation system prevalent in the western US.

A fragmented and decentralized market requires that the monitoring, enforcement, and information costs are low. It follows that those water rights that are permitted to trade have to be correspondingly simple. California has adopted an unofficial criterion for credible traded water that meets the requirements of simplicity and equity. Essentially the criterion distinguishes between what is colloquially called "wet water" versus "paper water." Paper water refers to the wide range of rights, some of which are either underutilized, or include return flows, or are imprecisely defined. Wet water is defined strictly on the basis of historic consumptive use. While there is a great range of water efficiencies, a simple property right based on average consumptive use provides a low-cost measure that is worth the minor physical inefficiencies. In short, a successful and equitable reallocation must be based on property rights that are a subset of the broader water property rights, but which avoid high transaction costs and third-party impacts.

The state of Colorado has a highly developed code of augmentation requirements that ensure only the net consumptive use is sold out of a given basin. The absence of these provisions in Australian definitions of transferable rights is likely to lead to externalities as water trading develops. In addition, the definition of water rights in Australia seems to give equal priority to nominal water rights that are not in effective use. Giving equal standing in the market to these "sleeper" rights is in sharp contrast to the California practice of ignoring sleeper rights and assigning tradable rights on the basis of historic consumptive use.

In addition to trading water, in many places the only capacity to move water is through a public infrastructure that has de facto rights held by the original contractors. The ability to trade water conveyance capacity is also an important development for efficient water reallocation in the west.

Conclusions

Trading of water in the western US is shown in Figures 4.1 and 4.2 to be active, growing, and consistent with economic principles. Water trades are dominated by short-term leases rather than outright sales of property rights, this is particularly prevalent in states that have stringent environmental regulations, and concerns over third-party impacts. In addition, as water is sold to lower-value uses, leases are more frequently used. An informal case study of examples of corporate attempts to profit from water trading shows that every major corporate foray into this field has met with failure. One can only conclude that the property rights and regulatory restrictions on Western water trades generate high costs for the centralized corporate model. However, the data shows that decentralized small-scale water trades are flourishing. The fear and requirements for anti-speculation legislation seem to be greatly overblown.

In contrast, since supply reliability is of critical importance to many water purveyors, the ability to offset risk by selling options or contingent contracts to willing buyers is a way of increasing the social value of scarce water resources. This will require a legal definition of those water rights that can be traded or held in reserve, to reduce risk.

A comparison of water trading and the definition of tradable water to rights between the western US and Australia shows significant differences in how rights are defined and third-party impacts minimized. Given the similarities of Australia, Spain, and the western US in the distribution of water use, the pressures for changed water use, and the climate, emerging water markets in all three countries can benefit by examination of the good and bad characteristics of the existing water trading systems.

Note

1. The EWA is a fund established by the CALFED Bay-Delta Program that allowed states and federal fishery managers in the San Francisco Bay/Sacramento-San Joaquin Delta to purchase water to increase freshwater inflows and provide better fish protection with a minimum of disruption to water purchasers south

of the Delta. Environmental purchases accounted directly for one-third of traded volume in California in 2001, the first year of EWA operations (Hanak 2003).

References

Brewer, J., Glennon, R., Ker, A. and Libecap, G. D. (2008) "Water Markets in the West: Prices, Trading, and Contractual Forms." *Economic Inquiry* 46: 91–112.

Brookshire, D. S., Colby, B., Ewers, M. and Ganderton, P. T. (2004) "Market Prices for Water in the Semiarid West of the United States." *Water Resources Research* 40: W09S04.

Brown, T.C. (2006) "Trends in Water Market Activity and Price in the Western United States." *Water Resources Research* 42: W09402.

Carey, J. M. and Sunding, D. (2001) "Emerging Markets in Water: A Comparative Institutional Analysis of the Central Valley and Colorado-Big Thomposon Projects." *Natural Resources Journal* 41: 283–328.

Dixit, A. and Pindyck, R. (1994) *Investment Under Uncertainty*. Princeton, NJ: Princeton University Press.

Hanak, E. (2003) *Who Should Be Allowed to Sell Water in California? Third-Party Issues and the Water Market*, San Francisco, CA: Public Policy Institute of California.

Howe, C.W., Schurmeier, D. R. and Shaw, W. D., Jr. (1986) "Innovations in Water Management: Lessons from the Colorado-Big Thompson Project and Northern Colorado Water Conservancy District." In K. C. Frederick and D. C. Gibbons (eds), *Scarce Water and Institutional Change*. Washington, DC: Resources for the Future.

Howitt, R. E. (1998) "Spot Prices, Option Prices and Water Markets." In K. W. Easter, M. W. Rosegrant, and A. Dinar (eds), *Markets for Water: Potential and Performance*, 119–40. Boston, MA: Kluwer Academic Publishers.

Jercich, S. A. (1997) "California's 1995 Water Bank Program: Purchasing Water Supply Options." *Journal of Water Resources Planning and Management* 123: 59–65.

Lach, D., Ingram, H. and Rayner, S. (2005) "Maintaining the Status Quo: How Institutional Norms and Practices Create Conservative Water Organizations." *Texas Law Review* 83: 2027–53.

Loomis, J. B., Quattlebaum, K., Brown, T. C. and Alexander, S. J. (2003) "Expanding Institutional Arrangements for Acquiring Water for Environmental Purposes: Transactions Evidence for the Western United States." *Water Resources Development* 19: 21–8.

Young, M. D. and McColl, J. C. (2002) *Robust Separation*. Canberra, Australia: CSIRO.

5 Water markets and their environmental, social and economic impacts in Australia

Henning Bjornlund, Sarah Wheeler and Peter Rossini

Introduction

Water demand throughout the world is increasing as a consequence of population growth, changing diets, and economic growth. Moreover, since the 1970s environmental awareness and values have gained currency and recreational demand for water has grown. In the future these influences will continue to increase demand, while increases in supply will be limited due to the predicted impact of climate change and limited opportunities for new major water supply infrastructure within sustainable economic, environmental and political constraints. As a consequence water scarcity will increase, especially within semi-arid and arid regions. Failing to manage these growing demands will have significant environmental, social and economic implications, especially within low- to middle-income countries and communities where water is a key factor to reduce poverty and improve public health.

Reflecting these developments and concerns, there has been a major shift in water policy away from supply management; that is, meeting new demand with increased supply, to demand management; that is, meeting new demand by using existing supply more efficiently and reallocating water from existing to new users. This shift has also changed how water is perceived, from being a social good to being an economic good that should be allocated according to market forces. An important element of the new paradigm is the use of market instruments such as full cost recovery prices, private property rights, water markets, and privatization. Many donor agencies, such as the World Bank, have therefore adopted such reforms as conditions of funding for the construction of major water infrastructure in developing countries. Treating water as an economic good is an alien concept to many cultures (Appelgren and Klohn 1999). As a consequence these reforms have been met with considerable, and in some cases violent, opposition in countries such as Bolivia and Sri Lanka (Guntilake and Gopalakrishnan 2002; Hall *et al.* 2005).

This new policy paradigm has been aggressively pursued in Australia since the early 1990s, in response to the serious environmental consequences resulting from the overexploitation of the country's largest water resource, the Murray–Darling Basin (MDB). Market instruments, including water markets, full cost recovery prices and privatization, have been introduced. Water markets have been widely adopted by irrigators to manage scarcity and structural change. Australia therefore represents an interesting case study for countries considering introducing similar changes.

This chapter first discusses how water markets have developed in Australia; it outlines some of the concerns associated with their introduction and the ways in which jurisdictions have tried to deal with these concerns. The second section discusses how market activities and prices have increased over time, how farm businesses have adopted markets, and how important markets are today in determining the allocation of water. The third section provides an overview of the social, environmental and economic impacts of market operations. The final section provides some concluding remarks. The chapter concentrates on developments within the MDB, and most examples are from the Goulburn–Murray Irrigation District (GMID) in Victoria, the country's largest irrigation district.

Development of water markets in Australia

It is important to understand that two different water markets exist: (i) the *entitlement market*, in which the long-term rights to receive water allocations are traded; and (ii) the *allocation market*, in which such allocations are traded. In essence, the entitlement market trades in paper rights to future access to water while the allocation market trades in the actual volumes of water that can be extracted from a source during a specific season and, in some instances, can be carried forward into the next season. In most Australian states there are two kinds of entitlements. Irrigators operating within irrigation districts (district irrigators) have entitlements to have allocations delivered to their farms using the district's infrastructure; while irrigators operating outside districts (private irrigators) have entitlements to extract their allocations directly from the river using their own pumps and infrastructure to transport the water to their fields.

Policy development

Markets for both water allocations and entitlements were first introduced in South Australia (SA) in 1984. Scarcity in the state first emerged as the South Australian Government placed a moratorium on new entitlements and cut existing entitlements according to history of use during the 1970s. When demand from new, efficient and high-value water users (namely horticulture and viticulture) emerged there was an urgent need for a mechanism to reallocate water from existing to new users and, given the nature of these new industries, entitlement trading was perceived necessary to justify the long-term investment.

In the other states allocation trading was first introduced as a result of considerable community concern over the potential impact of entitlement trading. Allocation markets were introduced in New South Wales (NSW) in 1984 and they were piloted in Victoria in 1987. Entitlement markets were introduced in NSW in 1989, but did not take effect in irrigation districts until they were privatized in the mid- to late 1990s. In Victoria, both allocation trading and entitlement trading were formally introduced with the Water Act 1989. However, entitlement trading did not actually commence until late 1991, when the necessary regulations were approved (Bjornlund 1999). Water markets have since been aggressively promoted at the national and basin levels.

Water reforms accelerated in 1994 when the Council of Australian Governments (CoAG) agreed on a new water reform agenda (CoAG 1994). This agenda included a commitment by all jurisdictions to introduce market instruments such as water trading, full cost recovery prices, privatization of the water industry, as well as provide formal entitlements to the environment.

In response to growing environmental concerns an audit of water use within the MDB was initiated in 1995 (MDBMC 1995). It concluded that the level of water use far exceeded what was ecologically sustainable and, worse still, it was likely to continue to increase because previously unused entitlements would be activated as a result of the introduction of water trading. The audit predicted a significant environmental impact if this was allowed to happen. In 1996 it was therefore decided that a cap would be placed on each state's water use. The cap would be set at the volume that would have been used at the 1993–94 levels of development (MDBMC 1996). With total use capped, and water trade activating previously unused water, demand for available resources increased resulting in lower seasonal allocations.

In 2003 CoAG reviewed the 1994 reform agenda and found that good progress was being made, but that the existing water market mechanisms still prevented water markets from achieving their optimal outcomes (CoAG 2003). As a result CoAG introduced the National Water Initiative (NWI) in 2004 (CoAG 2004). Among other things, the NWI aimed to improve the operation of water markets and encourage the emergence of markets in more sophisticated derivative water products by providing:

- better specified and nationally compatible water entitlements, defined as shares of the available resource;
- secure water entitlement registers; and
- national functioning water markets, including the progressive removal of barriers to trade.

Three other important elements of the NWI, encouraging more efficient water markets, were:

- statutory-based water planning processes, defining the consumptive pool of a resource;
- statutory provisions for environmental and other public benefit outcomes, and improved environmental management practices; and
- a commitment to return all currently over allocated or overused systems to environmentally sustainable levels of extraction.

The overarching water trading rules are set out in the national Water Act 2007.

In compliance with the CoAG water reform agenda and the MDB cap, state water policies have undergone considerable changes, with new legislation introduced in SA in 1997 and 2004, in NSW and Queensland in 2000, and in Victoria in 2005. As a part of this process most water authorities have revised their seasonal allocation policies. Traditionally the authorities announced seasonal allocations as a percentage of total entitlement at the beginning of the season, based on water available in reservoirs and predicted seasonal flows based on historical records. These policies provided certainty for irrigators, who could plan their production based on these allocations. Due to a history of exceptionally dry years in some regions of the country, these policies were problematic. Therefore today, most authorities announce the allocation at the beginning of the season based only on what is available in reservoirs and the minimum expected inflows, revising the seasonal allocation on a regular basis as additional water enters the reservoirs. While this represents a more sound environmental policy, it has resulted in irrigators having to plan their production without full knowledge of water availability, therefore increasing the need for efficient markets to help mitigate this increased uncertainty and risk.

These policy changes reduced the volume of water available for consumptive use from most resources. This had two consequences: (i) in many NSW catchments the Water Sharing Plans reduced irrigators' entitlements to water were without compensation; and (ii) mean seasonal allocations declined. This second effect is, however, not only a consequence of policy changes, but also of drought in the basin which was ongoing for more than ten years in the 2000s. This caused increased water scarcity and in 2006 concern over the socio-economic impact of the combination of the drought and achieving environmental flow caused the NSW Government to suspend its water sharing plans.

In effect this action by the NSW Government was a formal deferral of the process of securing water for the environment via the planning processes. Instead, governments and the Murray–Darling Basin Authority (MDBA) are increasingly looking to the market as a key mechanism to secure environmental water. Large-scale programs have been put in place by the Australian Government to obtain water for the environment through market purchases and improved water use efficiency. In 2007, a program was introduced with AU$3.1 billion allocated over ten years to buy back

1,500 Gl of existing MDB water entitlements from willing sellers (Howard 2007). In 2010 the MDBA released its *Guide to the Proposed Basin Plan* ("the Guide"; MDBA 2010), which proposed further reductions in extractive use, and in 2011 it released the proposed Basin Plan which declared a need to achieve a long-term average of 2,750 Gl for environmental water (MDBA 2011). Further reductions would still be sourced from willing sellers, with the federal government expected to fund the additional expenditure.

The drought also put increased pressure on water markets to allow water users to manage this scarcity by moving the available water around between competing users in order to reduce the total socio-economic impact of scarcity. In response to this water exchanges emerged (the first being in 1998) providing fast, efficient, and affordable approvals and transfers.

Another major water management strategy that was implemented in the MDB was carryover. Carryover allows entitlement holders to carry over unused allocations from one water year to the next. Carryover has been available in some systems since the 1990s, but 2007–8 was the first year in which carryover provisions were available in all three southern MDB states. In the GMID, a carryover limit of 30 percent was introduced in 2007–8, with the limit increasing to 50 percent in 2008–9. Spillable water accounts were introduced in 2010–11, allowing irrigators to retain carryover once allocations and carryover are equal to 100 percent of entitlement volume while storage capacity is available. Hence, this enables users to store water above their entitlement volume (NWC 2011).

The economic argument in Australia

The economic arguments for water markets are clear. Water available for extractive use is decreasing. Originally water was allocated in abundance at low or no cost, resulting in excessive water use, rising water tables and increasing soil salinization. Often irrigation was developed on unsuitable and unproductive soils and in locations where the runoff or leaching had a negative impact on river water quality. Further, many of the irrigators were on small lots with inefficient irrigation technology and using water to produce low-value commodities. Many of these farmers did not have the financial or human capacity to improve their productivity. Consequently there was a need to remove water from this land and these users; however, with water attached to land this was not possible. The economic argument is therefore that water entitlements need to be disconnected from land. Markets were introduced to allow water to move to more efficient and higher value users on more suitable land and in more appropriate locations. This will maximize the economic return from the resource and reduce the socio-economic impact of declining access to water for consumptive use while compensating those giving up their water in the process.

During drought allocations are low and there is not enough water to maintain "normal" production. During such periods markets will allow water to move to the highest value users and, more importantly, to users with significant investments in water dependent assets, such as horticultural plantings and dairy herds. These ventures can suffer significant permanent or long-term damage from insufficient irrigation. As a result of insufficient watering, trees and vines can die, or suffer long-term reductions in yields. Dairy farmers might have to reduce their herds if they cannot irrigate their pastures, reducing herds and subsequently increasing them is very costly. Such capital intensive ventures are often associated with high levels of debt that require constant servicing. Reduced annual productions have cash flow implications. It is therefore imperative that these irrigators can gain access to water during periods of drought. On the other hand, most low-value producers are involved in annual cropping that can be reduced or expanded without long-term implications. These farmers are compensated by selling their allocation water to higher value producers during drought as the latter will be willing to pay high prices to protect their long-term investments.

Environmental concern and mechanisms to alleviate concern

The environment has been at the center of much of the debate over water trading in Australia. Concern has been expressed over issues associated with the impact that water use might have in the new location, or the purpose to which the new user might put it. It is envisaged that trade may result in:

- concentrating water use in areas suffering from high water tables;
- moving water into locations where its use might have a negative impact on river water quality;
- moving water use upstream, thereby resulting in reduced river flow from the new point of extraction to the old point of extraction; or
- activating previously unused water leaving less water in the river to support ecosystems.

Concerns have also been expressed about the impact on the exporting land.

To accommodate these concerns each state governments introduced various restrictions on trade, as well as approval processes to ensure that the transfer of water would not cause any environmental damage. For example:

- In SA buyers had to produce an irrigation drainage and management plan to prove that the intended use of the water would not have a negative impact on river water quality.
- In Victoria there were limits on how much water could be traded onto a particular parcel of land. The allowable volume of water depended on the irrigation, drainage infrastructure, and the property's soil type.

- These limits were changed so as to not limit the volume of water attached to a specific parcel of land, but the volume that could be used. This allowed irrigators to purchase extra entitlements so that they could continue irrigation during periods of low allocations.
- The requirements of certain drainage and irrigation infrastructure was later replaced by reference to plans showing best practice irrigation management on suitable soils.
- Spatial restrictions were placed on transfer of water. These restrictions have been constantly modified and made more flexible.
- Low- and high-impact zones were identified with water trading into identified high-impact zones being prohibited or attracting a levy, whereas trading out was encouraged by a government payment in excess of market price.
- Exchange rates were introduced in instances where water is traded between different locations on the river, in order to offset the impact on stream flow and transmission losses.

These trading restrictions proved to significantly increase transaction costs, thereby curtailing the entitlement market in particular. To overcome this problem the right to use the water has been unbundled from the ownership of the water entitlement. SA was the first state to put this into effect with the Water Resources Act 1997. This act introduced two types of licenses, a holding license and a taking license, enabling any person to purchase water entitlements and obtain a holding license. This does, however, not include any right to use the water. To do that a water user must have a taking license, which in turn requires that the user can prove no adverse impact on river water quality. However, the owner of a holding license can sell the water allocations to another user with a taking license. This separates the trading process from the approval process, freeing up the market. Since then, as part of the unbundling process, most states have separated water entitlements from water use rights with a similar effect to the system in SA.

Social and community concerns and mechanisms to alleviate them

Social and community impacts have, together with environmental impacts, been the most important factors causing community concern over the impact of water markets, and especially entitlement markets (Bjornlund *et al.* 2011). These concerns were apparent when markets were first introduced. Indeed, they were a contributing factor to the introduction of allocation markets in most states. At the time it was perceived that the temporary reallocation of water from one user to another would not have permanent impacts, as the long-term entitlement to the water remained in the same location. These concerns were associated with the impact on the welfare of individual farmers, especially those not selling their water, and on the welfare of the wider rural community that depend on irrigation as the engine of economic activity. The irrigators' concerns are mainly associated with the following issues:

- When water is traded out of a supply system, then, given full cost recovery pricing, the cost to remaining irrigators may increase as the maintenance and supply cost are largely fixed.
- If a major part of the water is traded out of a supply channel the water authority might find it unviable to continue to supply water through that channel and close it down, forcing remaining farmers to stop irrigating, leaving both irrigators and authorities with stranded assets.
- Selling water off the poorer soil might result in abandoned farmland with absentee land owners. This could result in unattended land becoming havens for weeds and pests, which will spread to neighboring properties. Additionally, absentee landholders might keep token stock on their property, thereby neglecting to maintain fences, resulting in roaming livestock.
- The separation of land from water rights, and the separation of water entitlements from water use rights, gave speculators an opportunity to buy up water. It was feared that such speculators, often called "Water Barons," will become water hoarders, gaining a monopoly position and restricting supply or pressing up prices to levels in excess of what irrigators can afford.

Traditionally these concerns were overcome by the existence of barriers to trade out of districts. However, under the NWI and the national Water Act 2007 these restrictions have to be removed, giving rise to renewed concern. As part of these water policy reforms a separate supply capacity entitlement has been introduced and the fixed costs of maintaining the supply infrastructure are being charged to this entitlement. Consequently, if irrigators sell their water entitlements, they retain their capacity entitlements and remain responsible to pay these charges. This effectively ensures the viability of supply systems and keeps the cost for remaining irrigators constant. In Victoria the state government has acknowledged that trading has already moved so much water out of some channels that it will ultimately be necessary to close them down. In this respect the government has guaranteed that the remaining irrigators will be compensated for the resulting loss in property value. Finally, to overcome the concern over the "Water Barons," the Victorian Government introduced a limit of 10 percent on the size of the proportion that a single entity can own of a particular water source.

The concerns of the wider community stem from the following key issues:

- If water is traded out of a given area the value of the overall farm produce in this area will decline. Dry land farming is far less productive than irrigated farming and often requires far less farm labor and less need for processing, packaging, services such as accounting and irrigation consulting and farm inputs. A substantial reduction in irrigated activity will therefore result in fewer jobs both on farm and off farm reducing the demand for labor in the local communities and may lead to a reduction in services for the remaining citizens. If both on-farm and off-farm work (as well as local services) decline, there is a fear that people will leave the community, further escalating the depopulation and deservicing of the area. This results in a decline of local social services such as schools, hospitals, medical professionals, local shops, and recreational facilities, such as sporting clubs.

- If irrigation water is traded away from a farm then the land value will decline. If land values decline then the rating base for the local council will decline. A declining rating base can result in two things: either (i) services such as road maintenance, libraries, hospitals and schools are reduced; or (ii) other rate payers have to pay higher rates.
- If people are leaving the affected communities then demand for properties will decline, sending property values down too. This will leave remaining citizens with a declining asset base and can be an incentive for people on low income or transfer income to move into the area to make their transfer income stretch further. This can change the composition of the community.

While both the federal and state governments have traditionally avoided pledging structural adjustment assistance to affected communities, both the NWI and state level policies have indicated a willingness to consider such assistance if the impacts develop to become overly adverse. Indeed, new structural adjustment policies are being planned for release in 2012, in association with the proposed Basin Plan (MDBA 2011). To maintain local rating bases most states are introducing changes to property valuation which will allow them to levy irrigated property at a higher rate than dry land properties. For example, in SA commandable land (i.e. land capable of being irrigated) has a much higher value than dry land does (Bjornlund and O'Callaghan 2004).

Market activities and prices within the Goulburn–Murray Irrigation District

For markets to have a meaningful impact, whether positive or negative, trading needs to be adopted by irrigators so that it influences who owns water entitlements and uses allocation water. This section therefore discusses how water market activities have developed since the early 1990s. Much of this discussion is underpinned by the authors' research within the GMID (Bjornlund and Rossini 2008, 2005, 2007, 2010; Bjornlund 2003a, b, 2006a; Wheeler *et al.* 2009, 2010).

Volume traded in allocation and entitlement markets

Trading in both the allocation and entitlement market within the GMID was initially very subdued as irrigators were unfamiliar with the concept and there were substantial concerns within the community (Figure 5.1). Only 1 to 2 percent of the entitlement base, or less than 30,000 Ml, was traded each year in the allocation market for the first five years. An initial increase took place in 1994–5 as a result of the removal of some trading restrictions. Since 1997, which was the first of many drought years, and the change to allocation rules in 1998, trading has consistently been above 10 percent of the entitlement base, or in excess of 200,000 Ml. To deal with this volume of trade within a reasonable time frame, the Northern Victorian Water Exchange was introduced in 1998. It processed all allocation transfers within one week using a bidding system. With the onset of severe drought in 2002 and until 2009 allocation trading has consistently been above 300,000 Ml (or 18 percent) annually. Since 2009, as the area emerged from drought, allocation levels increased and the carry-over provision was introduced, allocation water trading has increased further. As a result of the escalating use of the carry-over capacity, as irrigators attempt to "fill up" their allocation accounts in preparation for the following season, trading has more than doubled over the last two years. During 2010–11, 16.8 percent of all water available for the year was carried forward from 2009–10, but 49.8 percent of the water available for that year was carried forward to 2011–12 (GMW 2011). Trade during 2010–11 particularly reflects the influence spillable water accounts, introduced that year, had on irrigators' water management strategies.

Trading in the entitlement market was even slower to accelerate; it took 10 years before the volume traded consistently exceeded 1 percent of the entitlement base per year. When severe scarcity and escalating allocation prices emerged in the 2000s activities in the entitlement market began to accelerate, exceeding 3 percent for the first time in 2006–7. Initially there was a willingness to rely on allocation trading to offset declining seasonal allocations. Prices paid in the allocation market were reasonable and transfers were fast and reliable with the new water exchange. However, during 2002–3 prices soared and many farmers in desperate need to irrigate paid loss-making prices, and faced a weekly uncertainty as to whether they would be successful in their bid on the exchange. This caused much anxiety and resulted in an increased willingness to invest in water entitlements. Initially the entitlement market had limited impact on who owned water entitlements; only 6.3 percent of all entitlement changed hand over the first 10 years. However, by the end of 2007 about 20 percent of all entitlements had been traded. Since 2008–9 trade has accelerated strongly, driven primarily by the federal government starting to purchase entitlements in 2008.

Looking at the spatial redistribution of water, both within the GMID and to outside the GMID, which has caused the concerns expressed by the

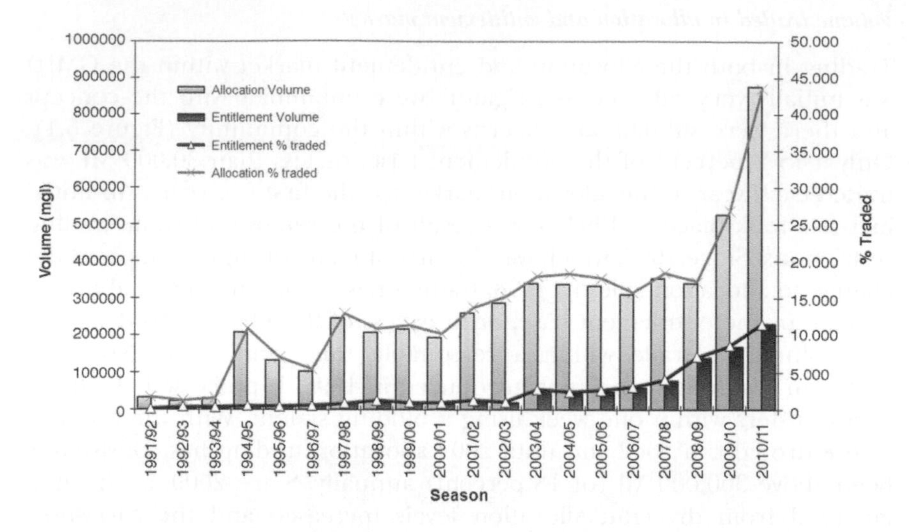

Figure 5.1 Development of trading within the Goulburn–Murray Irrigation District, 1991–2011

Note: The percentage traded is computed using the total volume of entitlement at the beginning of each year from July 1991 to June 2006, based on data provided by GMW. This time series was discontinued in 2006 and since July 2006 the entitlement volume is based on the annual report from GMW. In 2007 a policy change allowed district interrogators to separate their water right from the district entitlement to an entitlement held in their own name. Such separations are recorded in GMW annual reports as sales from a district entitlement into a new entitlement category, inflating the volume of real transactions. The volumes traded in this figure have been adjusted for this impact by reducing the total amount traded by the net increase in this entitlement category

community, the districts within the GMID had lost a total of 18.7 percent of all its water entitlements by June 2011. This water had been traded downstream to supply the growing horticultural industry (in particular wine grapes, olives, and almonds) in Sunraysia, Victoria and in SA. While all regions within the GMID have suffered a reduction in water entitlements, the magnitude of the reduction differs significantly across the GMID. Most parts have seen minimal reductions to date; however, the worst affected districts have lost up to a third of their water entitlements as of June 2011 (the percentages in this paragraph are computed based on data provided by GMW for trade up to 2005, and since then based on annual reports; see note to Figure 5.1).

Markets' ability to determine who gets access to use water each year

Another measure of the impact of the allocation market is the extent to which it determines who can use water each season. Initially trading only accounted for a relatively small proportion of total water use (Table 5.1).

There is a clear link between the allocation level and the importance of trading in determining who can use water that season. When allocation levels are at historical levels (e.g. 150–200%), trading only accounts for 3 to 5 percent, possibly representing that irrigators are making marginal adjustments to their water availability in response to fluctuating prices, supply and demand for the crops they are producing. However, when allocation levels dropped below 150 to 100 percent and even lower, and water prices increased correspondingly, the allocation market played an increasingly crucial role in determining who used the water. Allocation trading accounts for up to one fifth of all water use at 100 percent allocation, and in 2008–9 it approached half of all water use as allocation levels reached closer to 30 percent. Since then it has increased even further, as parts of the water purchased in a given season were not for use that season, but were to be carried over to the next season (Table 5.1).

From a slow start, it now seems that both entitlement and allocation markets are having a significant impact on who owns and who uses water both in the long term and short term. However, there is no doubt that the most significant reallocation of water takes place in the allocation market, where it is determined who can actually use water each season, rather than who owns the underlying right to receive water in the future.

Table 5.1 Relationship between seasonal allocations and extent of trade within the Goulburn–Murray Irrigation District

Season	*Goulburn system*			*Murray system*		
	Allocation (%)		*% of trade*	*Allocation (%)*		*% of trade*
	Opening	*Closing*		*Opening*	*Closing*	
1995/96	150	150	7	150	200	3
1996/97	200	200	4	200	200	3
1997/98	120	120	9	130	130	13
1998/99	40	100	13	95	200	5
1999/00	35	100	14	100	200	8
2000/01	48	100	16	200	200	2
2001/02	55	100	18	200	200	5
2002/03	34	57	24	129	129	16
2003/04	0	100	16	18	100	18
2004/05	0	100	18	42	100	22
2005/06	0	100	22	82	144	14
2006/07	0	29	37	76	95	20
2007/08	0	57	29	0	43	36
2008/09	0	33	53	0	35	42
2009/10	0	71	64	0	100	30
2010/11	0	100	105	0	100	102
2011/12	48	100	NA	21	100	NA

Source: based on Goulburn–Murray Water's records

Note: "% of trade" indicates total water trade for season as percentage of total water use

Market participation and the rate of markets adoption

When considering whether markets play an important role in determining access to water, thereby assisting farmers in managing their water supply, the level of market participation is an important measure (Figure 5.2).

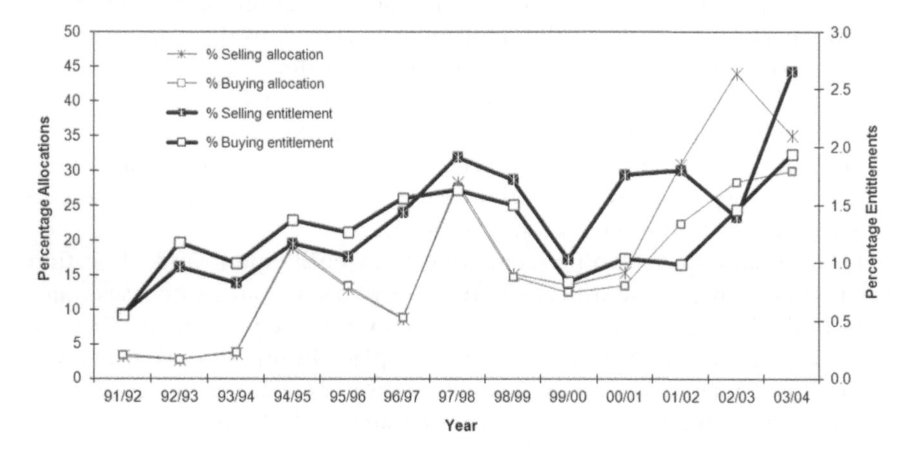

Figure 5.2 Farm businesses buying and selling allocations and entitlements annually

Market participation was initially very low, with less than 5 percent of farm businesses buying or selling allocations each year during the first three years. As Figure 5.2 illustrates, the participation rate has three peaks. A significant increase to 18 percent took place in 1994–5 following easing of trading restrictions. The next two peaks in 1997–8 and 2002–3 were caused by scarcity. The season of 1997–8 was the first time allocation levels dropped towards 100 percent, and 2002–3 was the first time it dropped below 100 percent (Table 5.1). Scarcity is clearly the dominant driver of activities in the allocation market.

During the initial years almost the same percentage of farm businesses were buying and selling water allocations. However, this changed so that during the early 2000s there were more businesses selling than buying. This development is driven by two forces. The first was the introduction of trading to outside the GMID in 1995 for allocation trading and in1997 for entitlement trading. However, such trade did not really flourish until increased wine grape prices caused the planting of wine grapes to boom in 1999. The second was the increase in water prices which caused more small entitlement holders to enter the market as sellers. At low prices it is not worthwhile for small entitlement holders to make efforts to sell their allocation water. However, as water prices increased it enticed sellers of smaller volumes into the market. On the other hand, with increased scarcity buyers are frantically buying water wherever they can find it, and they often have to buy many small parcels in order to cover their needs.

By 2004 about 30 percent of all farm businesses were buying allocations each year and 35 percent were selling. At this time it must be concluded that the allocation market is being used regularly and routinely by the majority of irrigators as a tool to manage their water use.

Participation in the entitlement market was much lower but increasing at about the same rate as in the allocation market. During the first year only around 0.5 percent of all farm businesses bought and sold water entitlements, and this only slowly increased to 1 percent in subsequent years. As it did in the allocation market, the participation rate first peaked in 1994–5 with the easing of trading restrictions. An important factor in this market was the enabling of trade between district irrigators and private irrigators, as there were a large number of unused entitlements among private irrigators.

During the first year of low allocation the participation rate increased to a peak of around 2 percent. Over the following four years the participation rate for sellers remained rather stable, with the exception of a drop in 1999–2000, while participation rate for buyers remained at a lower level of about 1 percent. The higher level of participation in the selling market, especially after 1999, is again associated with the boom in the wine grape industry. With the continued and worsening drought since 2001 participation rates, both as buyers and sellers, have increased to about 2 percent for buyers and 2.7 percent for sellers.

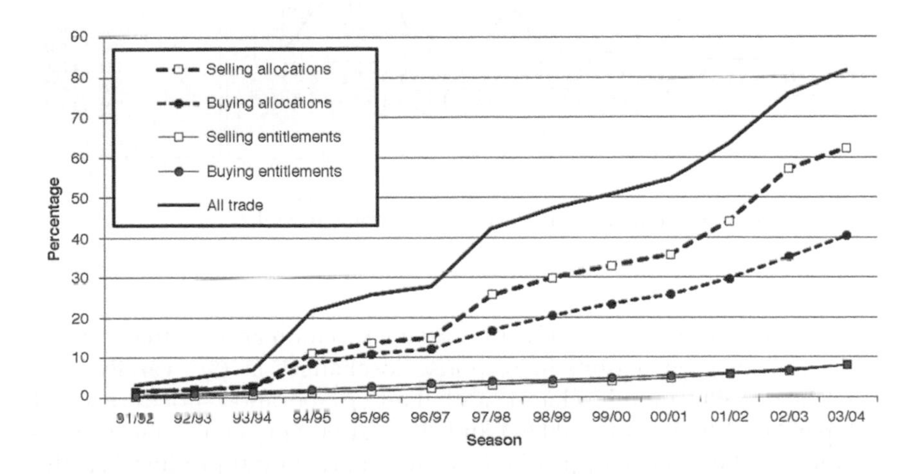

Figure 5.3 Adoption rate of water markets within the Goulburn–Murray Irrigation District, Victoria

When we look at the rate at which farm businesses adopted the use of water markets, a slow initial uptake is evident again. After the first three years less than 10 percent of all farm businesses had participated in some kind of trading (Figure 5.3). With the easing of trading restrictions in 1994 this jumped to more than 20 percent. By 1997–98, the first year of low allocation, over 40 percent had used the market at least once. Since then more and more farm businesses have taken to using the water market in response to the changing allocation policies and continued low allocation levels. At the end of the 2003–4 season some 81.4 percent of all farm businesses had participated in some kind of water trading: 9.6 percent had sold entitlements, 9.6 percent had bought entitlements, 49.8 percent had bought allocations, and 62.4 percent had sold allocations. This emphasizes that by 2004 the use of water markets had been widely adopted as a way of adjusting long-term and short-term water access.

Prices

Prices in both the allocation and entitlement markets have increased considerably over time with an annual growth of 20 percent and 12.5 percent respectively (Figures 5.4 and 5.5).

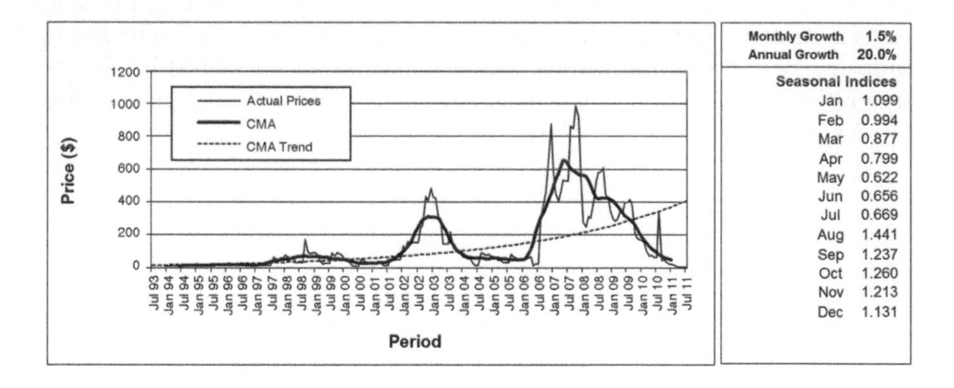

Figure 5.4 Water entitlements: ratio-to-moving average model

Allocation prices show significant seasonal variation and fluctuate widely within and between seasons in response to changes in the scarcity level driven by the allocation level and rainfall and evaporation. Prices reached \$500 per Ml (1,000 m^3) in 2002–3 and then \$1,000 in 2006–7, far exceeding what can be financially justified for most agricultural products. As the drought continued many irrigators and their banks realized that this was unviable, causing prices to drop as irrigators adopted mitigation strategies, such as reducing herds and areas with permanent plantings, and as the drought broke prices further declined (Loch *et al.* 2012).

Analyses of water prices and the price of commodities during the period from 1993 to 2003 show strong negative correlations, suggesting that irrigators pay higher prices for water while they receive lower prices for their commodities. This suggests that during times of severe scarcity farmers are not buying water to increase their profit in response to good commodity prices. Buyers are irrigators with significant investment in water dependent assets such as horticulture and dairy. They are buying to minimize their losses by protecting these assets in order to stay in business, with the expectation of better conditions in seasons to come. There are also many examples of banks lending farm businesses money to purchase water at these prices, with the knowledge that it will generate negative farm income, as they have an interest in helping these farm businesses to survive so that they can service their debt in years to come.

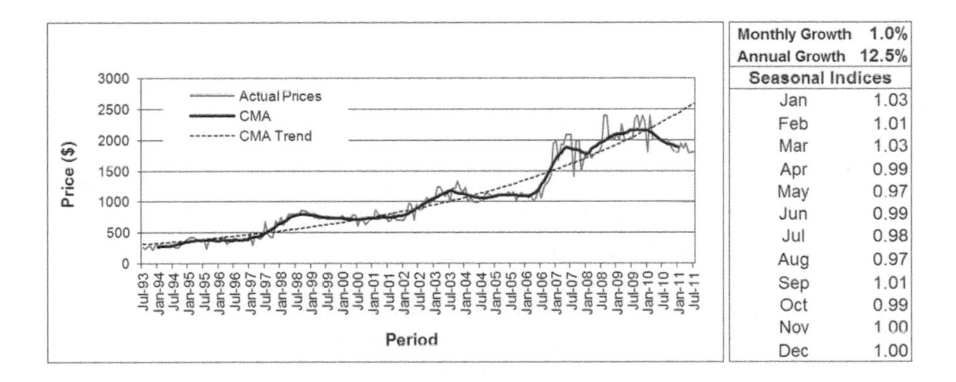

Monthly Growth	1.0%
Annual Growth	12.5%
Seasonal Indices	
Jan	1.03
Feb	1.01
Mar	1.03
Apr	0.99
May	0.97
Jun	0.99
Jul	0.98
Aug	0.97
Sep	1.01
Oct	0.99
Nov	1.00
Dec	1.00

Figure 5.5 Water allocations: ratio-to-moving average model

As Figures 5.4 and 5.5 indicate, entitlement prices have not increased nearly as much as allocation prices; they also do not show nearly the same fluctuations within and between seasons. The centered moving average seems to have three shifts in price level, with a fourth shift reflecting the first downward trend potentially commencing in 2010. The first price increase from below AU$500 to near AU$1,000 took place up until 1999, driven by high prices for wine grapes. After that, industry stagnated prices leveled out and even decreased slightly. The next shift to over AU$1,000 took place during 2002–3 and into 2003–4 as allocations dropped below 100 percent and allocation prices increased. Prices then leveled out again, and even declined slightly during the next two seasons of higher allocation levels. The third shift commenced in 2006, bringing prices in excess of

$2,000 by 2008 in response to even lower allocation levels, and driven by the entry of the federal government into the market. Because of this intervention, prices remained high into 2010, despite declining allocation prices and flooding. Since 2010 prices have started to decline due to full water allocations and increased precipitation.

Previous analyses suggest that the main drivers of entitlement prices are:

- scarcity, measured by the level of seasonal allocations and rainfall;
- the price of water allocations;
- monthly seasonal influence;
- commodity prices; and
- government water policy.

Analyses of cycle factors for water allocation and entitlement prices indicate that the two prices follow the same cyclical movement, but that allocation prices fluctuate more than twice as much as allocation prices. However there is some influence of commodity prices in the form of the price of wine grapes.

Return on investment

In a water management system where environmental and social water needs are overseen through statutory water planning processes, or the government participating in the water market as a buyer, it would be reasonable to pursue the most efficient use of the water available for consumptive use. It has been widely recognized that to maximize economic benefit from water more sophisticated and flexible market mechanisms and water derivative products will be needed. The changes introduced with the NWI, the Water Act 2007, and new state level legislation has created the foundation for the emergence of such mechanisms and products. Increased market participation and market activity suggests an increasing familiarity with the market institutions, and the link between entitlement and allocation prices, and the existence of market intermediaries suggest that the fundamentals for more sophisticated markets are emerging (NWC 2011; Wheeler *et al.*, 2013).

Analyzing the returns that could be obtained from investing in water entitlements will further indicate whether the fundamentals are there to introduce such markets and products. Figure 5.6 shows the return that could have been obtained from investing in water entitlements with the purpose of selling the allocation water during a five year holding period and selling it again with the expectation of some financial gain. This is similar to how investors operate in share markets and property markets.

The analysis indicates that return fluctuates depending on when the entitlement is purchased and the strategy followed when deciding when to sell the allocations. The returns depicted in Figure 5.6 are based on allocations being sold at the median price each year and should therefore be conservative. The bold line reflects the return that could have been achieved if the money had been invested on the Australian Share Index. Returns from investing in water entitlements over this 18-year period have consistently been well above what could have been achieved in the share market, with the exception of some short periods during 2007 and 2008. During the early years most of the return was generated by capital growth; however, since 2003 the balance between annual cash flows and capital gain has been more even. Even during the last two to three years, with falling allocation and entitlement prices, investments in water entitlements continued to outperform the share market, even increasing the gap. These findings further suggest that the fundamentals are emerging for more sophisticated markets and water products.

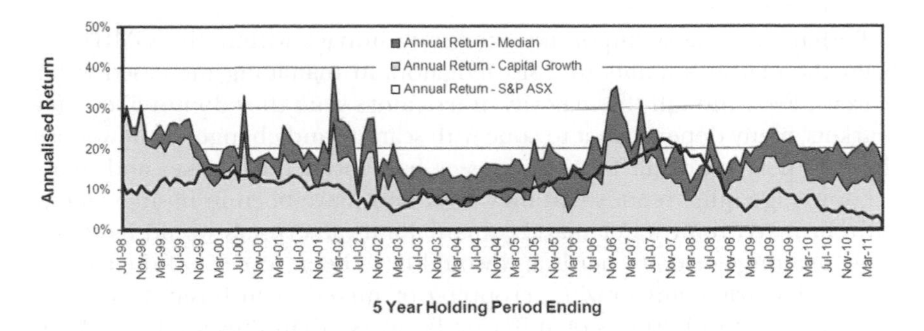

Figure 5.6 Median allocation and entitlement prices compared with capital growth and the Standard & Poor's Australian Securities Exchange Accumulation Index returns

Market outcomes and impacts

As previously discussed, trade was initially met with significant community concern and opposition driven by fear of the socio-economic impact they might cause. This section discusses the economic, social and environmental impacts of water markets identified by previous research, mainly carried out by the authors (Bjornlund 2002, 2004, 2006b, 2007, 2008; Bjornlund *et al.* 2011; Edwards *et al.* 2008; Wheeler *et al.* 2009).

Economic impact

The creation of water markets across the MDB has led to significant increases in Australia's gross domestic product (GDP), with the NWC (2010) suggesting gains of up to AU$220 million in 2008–9. Water in both markets has moved from lower value producing and less efficient users to higher value producing and more efficient users. This has increased both the volume of produce and the dollar value per unit of water used. Originally the dairy industry dominated the market as buyers. This dominance has declined over time due to falling commodity prices and industry deregulation. Since 2003 there is evidence that water is reallocated to farm businesses who concentrate on one product only (e.g. dairy, cattle, sheep or horticulture). It seems that high-value users have emerged within niches of production which were previously not considered high value.

The most significant reallocation has been associated with trading to outside the GMID. This water has traded downstream and been used for new or expanding horticultural industries with a very high gross margin per unit of water used; this could not have occurred without water trading. Without considering the regional impact of such reallocation, there is little doubt that overall the economic activity generated by the available water has increased.

Perhaps the most important economic impact within the GMID has been the market's ability to assist irrigators in managing increased water scarcity. Even though the majority of irrigators state they do not like water markets, many depend on it to cope with scarcity and change. By now more than 80 percent of all farm businesses have used the market and many acknowledge quite readily that they would not have been in business today without it. Without water markets many high-value users would have lost their plantings or dairy herds and would have been forced off their properties. Similarly many grazing, cropping or mixed farm businesses would have been out of business or suffering far worse than they are today. With only little water they would not have generated sufficient returns on their low-value products. However, with water markets they earned more income selling their allocation water than they ever could have using it. Many low-value producers have a gross margin of AU$50–100/Ml, but during this period they have received AU$300–1,000/Ml in the allocation market. So, even with an allocation of 50 percent they are at an advantage by selling some or all of their allocation water.

Recent research by Frontier Economics *et al.* (2007) and the NWC (2010) argues that the regional impact of water trading has been minimal. While they acknowledge that a lot of water entitlements have moved downstream, they argue that trading in the allocation market has moved water upstream, resulting in almost unchanged water consumption. However, this offset is likely to be temporal. Many downstream users bought entitlements adequate to fulfill their needs once their developments were in full production, and therefore they initially had excess water. This surplus will disappear as these enterprises are fully developed and plantings reach maturity. Then, at least in dry years, there is unlikely to be much water available to be traded upstream in the allocation market. The quantity of water available for such trade during years of "normal" supply remains to be seen and will depend on the allocation level on which these enterprises have based their water entitlement holdings. Analysis of Goulburn–Murray Water's annual reports since 2007 suggests that the GMID has mainly been a net seller of allocation water. Further, economic modeling indicates that the reallocation of water to the environment has had minimal economic impact (Grafton 2011; NWC 2010)

However, only a combination of annual cropping and permanent plantings can ensure the best possible economic outcome. If permanent plantings are developed to use all the available water during years of normal or high supply, then some of these will be lost during years of low supply, as nobody will be willing to sell allocation water. Only by a balance of annual and permanent productions can economic outcome be maximized so that during dry years annual producers stop irrigation and during seasons of high supply annual producers can use the excess water to generate extra income. Who owns the entitlements will depend on who wants to, or who can afford to, hold entitlements on their balance sheets. Will farm businesses with permanent plantings consider water entitlements as a good investment, with the prospect of future capital gain, and therefore buy enough water entitlements to secure supply in all years, and then sell back to the annual producers during years of ample supply? Or will annual producers predominantly in NSW, who historically hold most water entitlements, be reluctant to sell as they expect significant future capital gains?

Social impact

It has not been possible to quantify social impacts within exporting communities based on changes in census data. Census data indicate that rural communities have shown resilience and managed to maintain population and economic activity despite the export of water. However, qualitative evidence, gathered through in-depth interviews with key stakeholders, indicates perceptions of change for the worse. However, neither the research by Edwards *et al.* (2008) nor by Fenton (2007) could identify causation of these changes between water markets or the drought.

Fenton and Edwards *et al.* agree that trading in water entitlements has resulted in the abandonment of many farms, especially dairy farms and that much of this land now is left idle with the water sold off. Communities report negative impacts of this development. In some instances buildings and land are abandoned, resulting in infestations of weeds and pests; other properties are sold as dry land to city dwellers trying to find a cheap entry into a farming lifestyle. This causes a change in the composition of the population from traditional farming families with many shared values and aspirations, to a mixed population with an increased emphasis on city values and without shared aspirations. Community members resent this development and therefore are against entitlement trading. Irrigators that had initially sold all or part of their entitlement but remained in the community reported severe social repercussions stemming from their actions. However, this situation has changed as it is recognized that many had to sell due to financial hardship.

Community members report a changing population in traditional farming townships. Many retiring farmers move into the city after having sold their farm; that is seen as a good influence. Many traditional town dwellers are leaving as their jobs have disappeared and they are being replaced by people who have retired or are on various transfer incomes, in search of a cheaper place to live. As a result, behavioral patterns in local schools are changing and the willingness to participate in the community clubs and societies, which underpin the cultural life in the community, are declining (Barr 2009). Community members report reductions in sporting clubs making it difficult for children to participate in sporting activities. Finally, community members report increased difficulty in recruiting volunteers among farming wives, who traditionally have been the backbone of voluntary work in schools, sporting clubs and day care facilities. This trend is strongly supported by survey data showing a significant increase in off-farm work dependence (Wheeler *et al.* 2013). Today·most farming wives are working second jobs and therefore cannot participate in voluntary work. These communities are seeing their social fabric disintegrating, despite the fact that census data do not support economic decline.

The question is whether or to what degree water markets or the drought have caused these social impacts? The question is also whether these conditions would be worse still without water markets? If annual croppers could not have sold their water in the allocation market would they have been worse off? If farmers with permanent crops had not been able to buy water, would they have been at a disadvantage? Without water markets would there have been more bankruptcies, with people leaving the farm and the community in search of work? In the entitlement market, if dairy farmers had not been able to sell their land and water separately, would they have gone bankrupt and left in less favorable financial conditions? Now many of them have been able to move to the local towns, bringing their money with them and continuing to contribute to the community. Those who have only sold their water to clear their debt and now remain on the farm are still in the community and they continue to contribute. Without this flexibility would these people have been forced to leave the area all together? These are valid questions.

Our research clearly indicates that irrigators use the allocation market to manage scarcity and structural change. One group of irrigators has given up developing their farms to be viable. They use the allocation market to retain their lifestyle and stay within their community until the time of intergenerational change. Some sell all or most of their water to generate income, which combined with off-farm work and dry land farming, allow them to stay. Others buy a little extra water to keep their production going, which allows them to stay. A second group uses the allocation market to facilitate the development of their property to become financially viable. Some buy water allocations to expand their production and/or spend their available capital on improving their irrigation and drainage infrastructure. This group plans to buy more water entitlement once they have finished this process. Others already have the water to fully develop their farm; they sell their excess allocation while they develop their farm. Such sales help to finance the development. A third, much smaller group, consists of financially comfortable farmers; they buy or sell water in a more opportunistic way as they perceive they will be best off financially depending on current prices of commodities, farm input and water in the allocation market. From this perspective markets seem to assist farmers to stay on their farms and within their communities, contributing to community resilience.

Environmental impact

Environmental impacts of water trading are difficult to prove. Movements of water entitlements in a magnitude to cause real change have not really taken place. The farm plans implemented by entitlement buyers to ensure there is no negative impact have not been in place for long enough to prove their effectiveness or otherwise. Developments allowed against payment of bonds towards remediating negative impacts have not existed long enough for negative impacts to occur and therefore test whether these bonds are adequate and whether their commitments to remedy negative impacts can be legally enforced.

However, these arrangements have been criticized because of the inability to enforce obligations as well as due to the lack of monitoring of their implementation. On the positive side, both entitlement trading and allocation trading is moving water out of properties using less efficient irrigation and drainage infrastructure and management practices and into more efficient and better managed properties. Water has also moved out of poorer and degraded soils to better soils. Such reallocations may reduce the negative impact on river water quality. However, it has been argued that the move to more efficient irrigation reduces return-flow to the river and thereby reduces river flows with negative impact on the environment (Crase *et al.* 2009). In Australia the irrigators have the consumptive use to the water they extract. That is, if they improve efficiency they can increase their irrigated area or sell the saved water. The net effect of either action is less return flow to the river.

Modeling of trade-related changes in water use by the NWC (2010) revealed that those environmental effects are usually small compared with the impacts of drought and river regulation. Water trading has generally moved water downstream, leading to environmentally beneficial increases in flows at the ends of many tributaries. The NWC (2010) found that hydrological assessments indicated link between overall water trading patterns and key ecological assets in the Southern MDB.

Conclusions

Water markets in Australia have been aggressively pursued by policy makers to ensure more efficient and productive use of water, and to increase the provision of water for the environment. Irrigators have, after a slow start, widely adopted water markets. By now, markets have a significant influence on both who can use water and who owns the long-term entitlements to receive water allocations.

Regulatory developments in Australia have also been considerable. Among other things, they have provided:

- more efficient markets, more secure water entitlements, and an unbundling of the rights traditionally embedded in the water entitlements;
- adaptive statutory water planning processes that will ensure water for the environment and other public benefits, and define how much water is available for extractive and consumptive uses (though these are currently suspended); and
- wide scale government programs to use markets to reallocate water from consumptive to environmental use.

With these institutions in place the foundation is laid for the introduction of more sophisticated market instruments such as options, futures, entitlement leases and conditional contracts. Greater consideration also needs to be given to carry-over capabilities across regions, dam storage definition and the behavior of environmental water holders.

Once environmental and other public benefits are protected through a combination of regulatory processes and policy compensation, there is a strong argument for allowing markets to move the remaining water around between competing users, in order to achieve the best possible economic outcome. For this to happen a balance is needed between annual cropping, often considered as low-value uses, and permanent plantings, often considered high-value uses. The exact water use can be left to the market and determined by individual risk preferences and net benefits. The crucial point is that the adaptive water planning processes in place are sufficiently robust to ensure environmental and other public benefits under continued climate change.

References

Appelgren, B. and Klohn, W. (1999) "Management of Water Scarcity: A Focus on Social Capacities and Options." *Physical Chemical Earth (B)* 24(4): 361–73.

Barr, N. (2009) *The House on the Hill: The Transformation of Australia's Farming Communities.* Canberra, Australia: Land and Water Australia.

Bjornlund, H. (1999) "Water Trade Policies as a Component of Environmentally, Socially and Economically Sustainable Water Use in Rural Southeastern

Australia." Thesis, University of South Australia, Adelaide, Australia.
Bjornlund, H. (2002) "The Socio-economic Structure of Irrigation Communities: Water Markets and the Structural Adjustment Process." *Journal of Rural Society* 12(2): 123–45.
Bjornlund, H. (2003a) "Farmer Participation in Markets for Temporary and Permanent Water in Southeastern Australia." *Agricultural Water Management* 63(1): 57–76.
Bjornlund, H. (2003b) "Efficient Water Market Mechanisms to Cope with Water Scarcity." *The International Journal of Water Resources Development* 19(4): 553–67.
Bjornlund, H. (2004) "Formal and Informal Water Markets – Drivers of Sustainable Rural Communities?" *Water Resources Research* 40: W09S07.
Bjornlund, H. (2006a) "Increased Participation in Australian Water Markets." In G. Lorenzini and C. A. Brebbia (eds), *Sustainable Irrigation Management, Technologies and Policies*, 289–302. Southampton, UK: WIT Press.
Bjornlund, H. (2006b) "Can Water Markets Assist Irrigators Managing Increased Supply Risk? Some Australian Experiences." *Water International* 31(2): 221–32.
Bjornlund, H. (2007) "Do Markets Promote More Efficient and Higher Value Water Use? Tracing Evidence Over Time in an Australian Water Market." In C. A. Brebbia and A. G. Kungolos (eds), *Water Resources Management*, 477–88. Southampton, UK: WIT Press.
Bjornlund, H. (2008) "Markets for Water Allocations: Outcomes and Impacts." In *Proceedings from the International Conference, Water Down Under*. April. Adelaide, Australia: Engineers Australia.
Bjornlund, H. and Rossini, P. (2005) "Fundamentals Determining Prices and Activities in the Market for Temporary Water." *International Journal of Water Resources Development* 21(2): 355–69.
Bjornlund, H. and Rossini, P. (2007) "Fundamentals Determining Prices in the Market for Water Entitlements – An Australian Case Study." *International Journal of Water Resources Development* 23(3): 537–53.
Bjornlund, H. and Rossini, P. (2008) "Are the Fundamentals Emerging for More Sophisticated Water Market Instruments?" In *Proceedings from the 14th Annual Conference of the Pacific Rim Real Estate Society*, Kuala Lumpur, Malaysia, 20–23 January. Available at www.prres.net (accessed 11 July 2012).
Bjornlund, H. and Rossini, P. (2010) "Climate Change, Water Scarcity and Water Markets – Implications for Farmers' Wealth and Farm Succession." In *Proceedings from the 16th Annual Conference of the Pacific Rim Real Estate Society*, Wellington, New Zealand, 24–27 January. Available at www.prres.net (accessed 11 July 2012).
Bjornlund, H. and O'Callaghan, B. (2004) "Property Implications of the Separation of Land and Water Rights." *Pacific Rim Property Research Journal* 10(1): 54–8.
Bjornlund H., Wheeler S. and Cheesman J. (2011) "Irrigators, Water Trading, the Environment, and Debt: Perspectives and Realities of Buying Water Entitlements for the Environment." In Q. Grafton and D. Connell (eds), *Basin Futures: Water Reform in the Murray–Darling Basin*, 291–302. Canberra, Australia: Australia National University Press.
CoAG (1994) "Communiqué." Meeting of Council of Australian Governments, Hobart, Tasmania, 25 February. Available at www.pm.gov.au.
CoAG (2003) "Communiqué." Meeting of Council of Australian Governments, Canberra, 29 August. Available at www.pm.gov.au.

CoAG (2004) "Communiqué." Meeting of Council of Australian Governments, Canberra, 25 June. Available at www.pm.gov.au.

Crase, L., O'Keefe, S. and Dollery, B. (2009) "Water Buy-Back in Australia: Political, Technical and Allocative Challenges." 53rd Annual Australian Agricultural and Resource Economics Society Conference, 11–13 February, Cairns, Australia. Canberra, Australia: Australian Agricultural and Resource Economics Society.

Edwards, J., Cheers, B. and Bjornlund, H. (2008) "Social, Economic and Community Impacts of Water Markets in Australia's Murray Darling Basin Region." *International Journal of Interdisciplinary Social Sciences* 2(6): 1–10.

Fenton, M. (2007) *A Survey of Beliefs about Permanent Water Trading and Community Involvement in NRM in the Loddon Campaspe Irrigation Region of Northern Victoria.* Melbourne, Australia: North Central Catchment Management Authority.

Frontier Economics, Cummins, T., Watson, A., and Barclay, E. (2007) *The Economic and Social Impact of Water Trading.* RIRDC Publications No. 07/121. Canberra, Australia: RIRDC.

GMW (2011) *G-MW Annual Report 2010/11.* Tatura, Australia: Goulburn-Murray Water.

Grafton, Q. (2011) "Economic Costs and Benefits of the Proposed Basin Plan." In D. Connell and Q. Grafton (eds), *Basin Futures: Water Reform in the Murray–Darling Basin,* 245–62. Canberra, Australia: Australian National University E-press.

Gunatilake, H. M. and Gopalakrishnan, C. (2002) "Proposed Water Policy for Sri Lanka: the Policy versus the Policy Process." *Water Resources Development* 18(4): 545–62.

Hall, D., Lobina, E. and de la Motte, R. (2005) "Public Resistance to Privatisation in Water and Energy." *Development in Practice* 15(3–4): 286–301.

Howard, J. (2007) *The National Plan for Water Security.* Canberra, Australia: Prime Minister and Cabinet.

Loch, A., Bjornlund, H., Wheeler, S. and Connor, J. (2012) "Allocation Trade in Australia: A Qualitative Understanding of Irrigator Motives and Behaviour." *Australian Journal of Agricultural and Resource Economics* 56(1): 42–60.

MDBA (2010) *Guide to the Proposed Basin Plan: Overview.* Canberra, Australia: Murray–Darling Basin Authority.

MDBA (2011) *Proposed Basin Plan.* Report. Canberra, Australia: Murray–Darling Basin Authority.

MDBMC (1995) *An Audit of Water Use in the Murray–Darling Basin. Water Use and Healthy Rivers – Working Toward a Balance.* Canberra, Australia: MDBMC.

MDBMC (1996) *Setting the Cap.* Report of the Independent Audit Group. Canberra, Australia: MDBMC.

NWC (2010) *The Impacts of Water Trading in the Southern Murray–Darling Basin: An Economic, Social and Environmental Assessment.* June. Canberra, Australia: National Water Commission.

NWC (2011) *Australian Water Markets: Trends and Drivers, 2007–08 to 2010–11.* Canberra, Australia: National Water Commission.

Wheeler, S., Bjornlund, H., Shanahan, M. and Zuo, A. (2009) "Who Trades Water? Evidence of the Characteristics of Early Adopters in the Goulburn–Murray Irrigation District 1998–99." *Agricultural Economics* 40: 631–43.

Wheeler, S., Zuo, A. and Bjornlund, H. (2010) *Investigating Irrigator Water Use, Selling Water Entitlements, Exit Behaviour and Government Intervention in Water Markets in the*

Southern Murray–Darling Basin. Report prepared for the National Water Commission. Adelaide, Australia: CRMA University of South Australia.

Wheeler, S., Garrick D., Loch A. and Bjornlund, H. (2013) "Evaluating Water Market Products to Acquire Water for the Environment in Australia." *Land Use Policy* 30(1): 427–36 (doi:10.1016/j.landusepol.2012.04.004).

6 Trading into and out of trouble

Australian water allocation and trading experience

Mike Young

Introduction

In Australia, the right to manage and allocate water is defined as a public good, and all governments use a licensing system to allocate water. While there is a plethora of detail, as a general rule all licenses in an irrigation area are defined in the same way. Moreover, as Australia has an extremely variable climate, in some areas, two types of license exist – "high-security" and "general-security" licenses. In times of scarcity, allocations are made first to high-security licenses, then any water that is left over is allocated to general-security licenses.

Water within each of these security pools is allocated to entitlement holders on a pro-rata basis in proportion to the number of entitlement (shares) they have. Typically, shares are nominated in volumetric units and allocations announced as a percentage of water that volume that is now available for taking. Both water entitlements (shares) and water allocations are tradable. This approach makes it possible for some farmers to invest efficiently in permanent crops and others to take a more opportunistic approach, irrigating annual crops and pasture when water is available.

From a trading perspective, one of the most important reforms was a decision to convert most licenses from licenses to irrigate an area of land to volumetric entitlements. In the Murray–Darling Basin this happened many years ago – mostly in the 1960s and 1970s. Without metering and the allocation of volumes of water, many of the water trading arrangements in place today would not be possible.

Today, almost all water use is metered and use strictly enforced. Those who take more water than they are entitled to are penalized. Water theft is not a major problem.

Progressive unbundling and separation of function

Water trading in Australia began cautiously in the 1980s. Before the development of competition policies in 1993/4, water licenses were attached to land titles. While possible before that time, water trading was administratively difficult and time consuming. So much so, that many of the early inter-regional trades involved a person purchasing a farm with a water surplus and then arranging to transfer this surplus water to another farm, where this water could be used more productively. Once the transfer had been completed, which could take years, the farm with surplus water was sold with a lessor or no water entitlement.

In 1993/4, Australian governments decided collectively that it should be possible to hold a water license without having to own land and attach licenses to land titles. Arrangements were also put in place to make all water users pay the full cost of delivering water to them. The result, it was argued, would be a series of reforms that ultimately would allow water to trade to the place where it would contribute most to the economy – its highest and best use.

Since 1993/4, when Australia decided to allow water entitlements to be held independently of a land title, the volume of trade has increased significantly. Considerable investment and innovation has occurred. While there has been considerable angst, on the whole, irrigation communities are much richer than they otherwise would have been (Frontier Economics 2007; Young *et al.* 2006).

In regions like the Murray–Darling Basin two markets gradually emerged – a "temporary" water market and a "permanent" water market. Temporary markets involve the trade of a volume of water that has been allocated for use within a season. They are called "temporary" trades because, in the early stages of the development of trading, the only way a trade could be executed was temporarily transfer a license from one farm to another remove the water from the license and then transfer the license back to the original farms. A "permanent" trade involves the transfer of all or part of a license from one farm to another farm. Once the trade is completed, the buyer of the license is entitled to receive and use all future allocations made to that license.

Having separated water licenses from land titles, it soon became clear that further unbundling was needed and trading arrangements generally improved. Recognizing these needs, a host of other reforms to do with water accounting, planning and pricing a new intergovernmental agreement was negotiated in 2004. Known as the National Water Initiative, the ultimate result was the emergence of a regime that makes it possible to use separate instruments to pursue separate objectives (see Figure 6.1). Entitlement shares define a person's equity or permanent stake in the system. Trade in allocations facilitates efficient use of the available water resource at any point in time, while use licenses (approvals) are used to manage environmental externalities.

In the Murray–Darling Basin water entitlements are issued and water allocations made by state governments in proportion to the number of shares held. A pilot interstate water trading trial began in 1999, and following several reviews, was ultimately replaced by a permanent set of arrangements designed to facilitate trade among states. Figure 6.2 summarizes the rate of growth in water trading in Australia's Murray–Darling Basin from 1983 with data separated into intra- and inter-state trades.

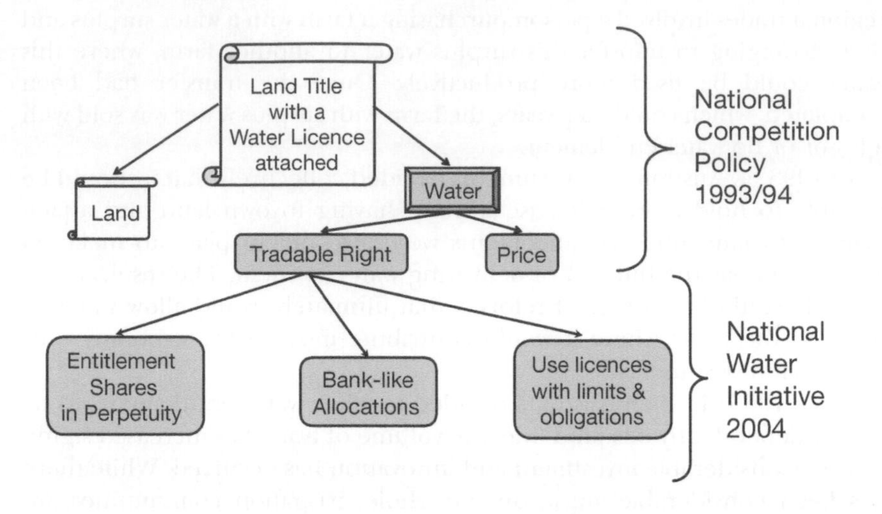

Figure 6.1 Evolution of water reform in Australia and the progressive separation of water licenses from land and their subsequent unbundling into separate components
Source: after Young and McColl (2005)

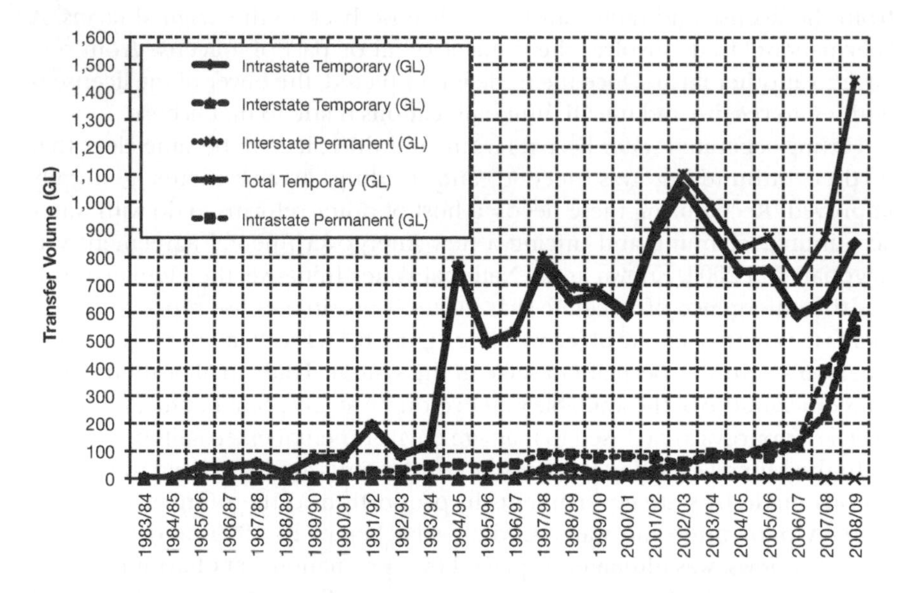

Figure 6.2 Growth in water trading in the southern connected portion of the Murray Darling Basin, 1983–2004
Source: Murray Darling Basin Commission, personal communication (2007)

Benefits and costs of water trading

There have been many assessments of the impacts of water trading on the economy and the environment. Many mistakes have been made and many lessons learned.

From an individual water entitlement holder perspective, the benefits of holding a water entitlement have been considerable. Recently, Bjornlund and Rossini (2007) have modeled the financial return from simply holding a water entitlement and selling all allocations made to the entitlement on the water market. The results, summarized in Chapter 5, show annualized returns in excess of 20 percent per annum. As noted above, this has stimulated significant innovation and the adoption of new technology (Young 2008).

Bjornlund and Rossini's data also show how, in the early stages of market development, the greatest returns come more from capital gains than from the use of water allocations. Returns from the annual sale of water entitlements follow once the appropriate suite of land use and other changes are put in place and market reforms completed. The most likely explanation of the reason why this system occurs is that capital markets are quick to see opportunity, but investment and development of the means to realize these new opportunities takes time.

While the early policy rhetoric focused on the benefits of letting water move to its highest and best use, in more recent times the emphasis has shifted to pointing the role of markets in facilitating structural adjustment and managing scarcity.

Arguably, the benefits of water trade during times of adversity are much greater than the returns from trade in times of abundance. Both Young *et al.* (2006) and Frontier Economics (2007) have found that water users state repeatedly that it was the ability to buy and sell water allocations that has enabled them to survive the long-drought that has plagued Australia.

As stated earlier, many of the necessary property right conditions and administrative arrangements for the development of water trading in Australia have been in place for many years. Unfortunately this has meant that trade was introduced without sufficient attention to the effects that this would have on return flows and groundwater use. As a result, a host of implementation problems have emerged. Indeed some of the problems Australia now faces are so severe that one could argue that the Australia's market approach should have been abandoned. However, this is not the case. Almost all analysts and all administrators are of the opinion is that the way forward is to trade out of trouble and retain access to the significant benefits from the development of allocation systems that are designed for trading.

Most analysts are also of the opinion that the costs of failing to get the fundamentals right have been high – especially at the system level. Markets operate according to rules and, when the rules for their operation are misspecified, they quickly reveal the extent of these flaws. The most commonly discussed problem is that of over-allocation.

In Australia over-allocation is said to occur "where with full development of water access entitlements in a particular system, the total volume of water able to be extracted by entitlement holders at a given time exceeds the environmentally sustainable level of extraction for that system" (CoAG 2004). Over-allocation occurs because either too many entitlements have been issued and/or allocations are being made to them at a rate that is not environmentally sustainable. Over-allocation also occurs when water-use efficiency is increased and, as a result, less water returns to the river. With the benefit of hindsight it has become clear that as efficiency increases allocations per entitlement need to be reduced. Similarly, as surface water becomes valuable those with access to groundwater started to use this water and as a result the flow of groundwater into rivers was reduced. Both these problems have emerged as serious problems in the southern connected Murray River System.

System-wide challenges

The water allocation and entitlement framework that Australia has developed emerged and evolved over time. No-one ever put together a grand design. In essence, existing entitlement regimes were bolted together (Young and McColl 2003a; 2003b). There is, however, some underlying theory which can be used to develop guidelines about the best way to specify water entitlement arrangements if one wishes to reap the benefit of trading without creating over-allocation and other water accounting problems.

One of the key building blocks for success is to recognize that markets have little, if any, role in making system-wide allocation decisions. Great care must be taken to separate system wide management issues from issues associated with the management of individual entitlements and allocations. If this is done then, individual entitlements and allocations can still be no matter how system wide plans are changed. This is possible if the rules about how much water is to be allocated are included in a plan and not in an individual entitlement. All the entitlement does is specify how large a share of all the water to be allocated will go to the holder.

In a trading environment it is critically important to understand that the role of the system manager changes to one of market facilitation. The role of government is, first, to set up the trading rules, then decide how much water can be allocated and make these allocations. Once this is done, the market takes over to determine where and how this allocated water will be used – constrained by a set of land use and land management rules designed to prevent unacceptable environmental problems from emerging.

Table 6.1 summarizes the structure that is now starting to emerge in Australia. Management of system-wide institutional arrangements is now being implemented using processes that are separate from those used to manage individual entitlements and allocations.

Table 6.1 An overview of the unbundled framework for the management of system-wide and individual attributes for the allocation of water in Australia

Scale	*Policy objective*		
	Distributive equity	*Economic efficiency*	*Environmental externalities*
System-wide strategic instruments	Plans and agreements that define entitlement priorities and rules for allocating water	Trading Protocols (Exchange rates, registers, etc.)	Plans that define protocols for the application of water to land; protocols for management of river flow, quality, etc.
Operational market instruments	Entitlements defined as unit shares	Allocations defined as volumes of water for use or trade within a specific period; in regulated systems, carryforward may be possible	Farm by farm water-use approvals that place limits on the way water may be applied to land with a view to managing local environmental and other impacts

Source: adapted from Young and McColl (2005, 2010)

Storage management

Before trading was introduced, centralized planning processes were used to determine how much water should be allocated to users and how much was kept in storage. Moreover, the regime that emerged was designed to take account of the fact that in most years a considerable amount of water would be left over at the end of a season and returned back to the allocation pool for use in the following year. The result was a regime that roughly optimized the amount of water kept in storage and that which was used in any year. But one could argue the reverse. The storage management regime and the nature of the supply infrastructure associated with it could have determined how much irrigation development occurred and what crops were grown.

When trading was introduced, however, it suddenly became possible for people to sell unused water and, as a result, storage management policies became sub-optimal. So much so that in States, like Victoria, Brennan (2007a, b) has shown that the costs of storage mismanagement have been greater than the benefits of trading. Her research has shown that, unless irrigators are allowed to carry forward unused water from one year to the next, trading will force virtually all allocated water to be used in each year and, as a result, increase supply variability and, through this, increase the severity of droughts. Working out how much water to save and how much to use is a process that requires continuous analysis and revision as conditions change. It also requires access to market and technical information that is poorly understood by government officials (Young and McColl 2007).

The solution to this problem is simple and requires the development of carryforward policies. Carry forward policies are policies that allow allocations with appropriate adjustments and delivery restrictions to be left in storage and used at a later date.

As a result of the growing recognition of the importance of allowing market processes to help optimize storage, all Murray–Darling Basin states have now introduced carry over arrangements (Young and McColl 2007). The most sophisticated carry forward arrangements can be found in Queensland where a range of continuous accounting processes have been put in place and markets established for delivery entitlements and storage capacity as well as for entitlements and allocations (Vanderbyl 2007).

Design challenges

Arguably one of the biggest mistakes that early Australian water licensing systems made was to assume that a similar mean volume of water would always be available. As a result, considerable effort was put into the definition of entitlement reliability and arrangements that would ensure that the requisite volumes would be available. In retrospect Australian planners should have paid much more attention to the development of management regimes that are designed to cope better with the emergence of long dry periods or for the adverse climate change. As a result, many systems have become over-allocated in the sense that too much water has been allocated to users and not enough for system maintenance and to the environment.

As the experience with the city of Perth's water supply shows, this is a serious error. In the 34 years since 1974, the water supply system used to supply much of the city of Perth has never once received its "mean" rainfall. As can be seen in Figure 6.3, the mean amount of water flowing into the system since 1974 has been less than half that received before then. Moreover, since 1974 inflows have never reached the previous average. As a result of experiences like these, Australian water managers are now coming to understand that systems do get drier and, when they do, inflows into rivers and storages drop dramatically.

The crude rule of thumb is that for every 1 percent decline in mean rainfall, storage inflow drops by, at least, 3 percent. In Perth's case, 14 percent less rain meant 48 percent less inflow and 20 percent less rain has meant 66 percent less inflow into their main storage (see Figure 6.3).

As a result, despite the fact that most systems state that the environment should be given first priority, Australia is now in the unenviable position of having to buy back water entitlements for system maintenance and for the environmental use. Recognizing the folly in doing this, the National Water

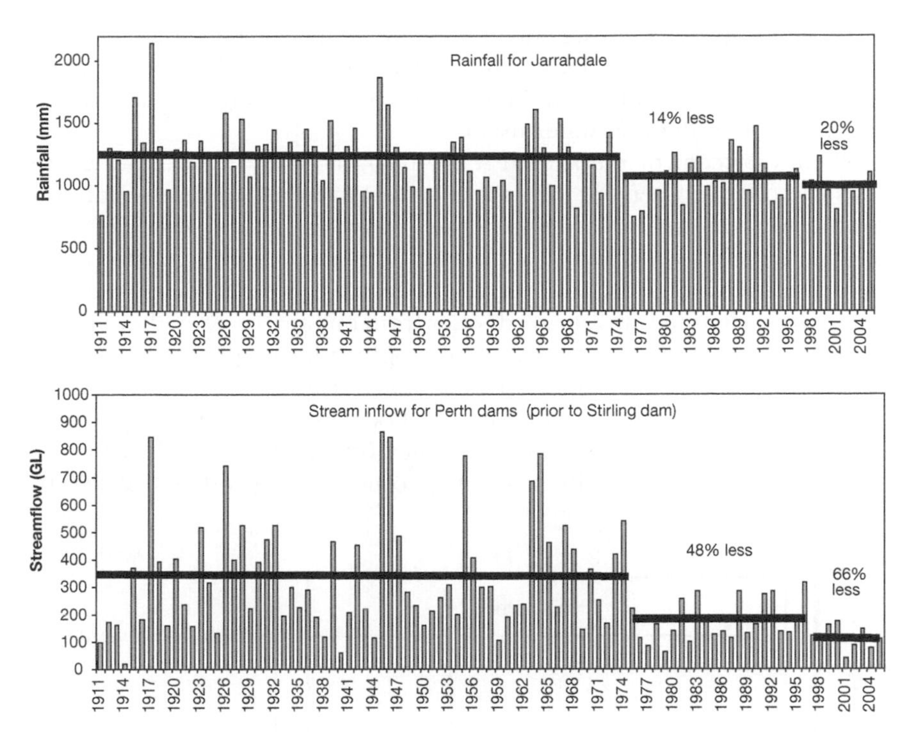

Figure 6.3 Decline in the volume of water flowing into Perth's main surface water supply system and its relationship with rainfall as measured at the Jarrahdale (the rainfall station most closely associated with it)

Initiative policy requirement is that environmental water entitlements be defined as shares so that in the future it will not be possible for allocations to be made in a manner that erodes the environment's interest.

Another challenge, particularly when there is a shift to a drier regime, as appears to have occurred in Australia, is to set aside enough water to enable water allocations to be conveyed from one end of a river system to another.

With the benefit of hindsight, it is now becoming clear that the first water allocation priority should be to put aside enough water to maintain the water level in a river at a minimum level so that entitlements at the end of the system can be honored. In regulated river systems and, as set out in Figure 6.4, the remaining non-flood water can then be shared between users and the environment. In the past Australia has used a combination of formal entitlements to irrigators and other water users and water sharing plan processes to determine how much water is to be allocated to the environment and water conveyance and how much allocated to users. As a result of the failure of this system, in the River Murray system, managers are now buying back entitlements from irrigators and moving to a regime where the environment will hold an entitlement of equivalent security as that was held by all other water users.

When x percent of the shared water resource is formally allocated to users and $(100 - x)$ percent formally allocated to the environment, it is not possible to favor one side over the other. Some water must always be allocated to the environment and some to all other users. To optimize water use under such a regime, however, it is necessary for both sides to have an unrestricted right to carry forward water from one year to the other and be empowered to manage inter-seasonal supply risk independently of one another. The result is a regime that is robust in the sense that it can be expected to with stand the test of time and be described as a regime that is likely to work well under all climatic regimes, including extremely dry ones. Neither side can erode the interests of the other (Young and McColl 2009).

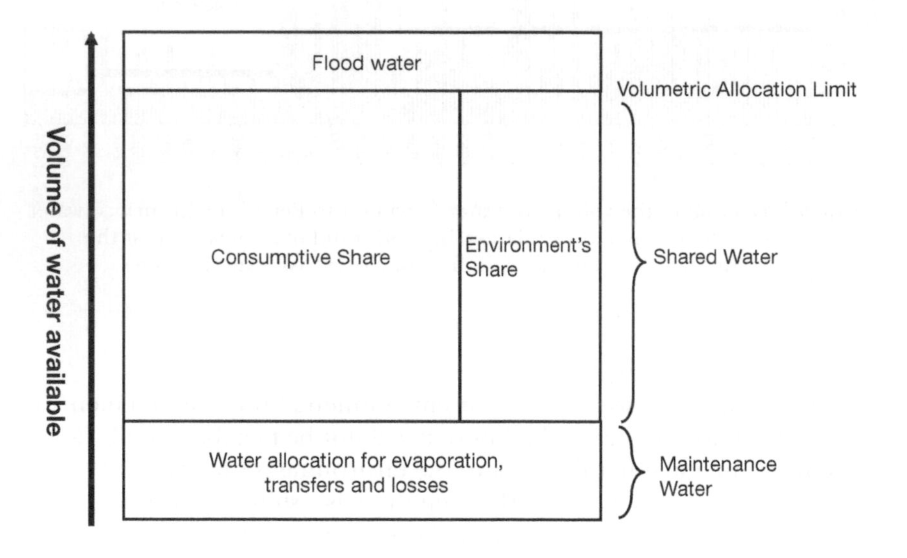

Figure 6.4 Indicative structure of a robust water entitlement regime
Source: Young and McColl (2009)

Preparing for trade

Over-entitlement

Another extremely painful experience that Australia has had to deal with is the difficult question of what to do with water entitlements which, before the introduction of trading, were either never or rarely used. In a non-trading environment, allocations to these unused entitlements have option value and were typically held as a hedge against future supply risk and to secure a future development opportunity. Hence many people sought to hold them. While these entitlements were not being used, however, the water assigned to them remained unordered and, unfortunately, was typically made available to other irrigators. As a result and with the passage of time, irrigators became accustomed to having access to more water than would be the case if all entitlements were fully used. When trading was introduced, however, it became possible for the holders of these entitlements to sell their unused allocations with two consequences. First, those who under the pre-trading regime were forced to buy water that was previously available to them at no charge, with the consequence that there could be a significant shift in wealth. Second, in systems where too many entitlements had been issued, governments needed to decide whether to reduce allocations per entitlement or find a way to buy back entitlements so that the system remains in balance.

With the benefit of hindsight, the advice now being given is to never issue entitlements to more than 100 percent of the available resource. When this is done, over-entitlement problems do not emerge.

Return flows

Unlike many allocation regimes used in the United States of America, Australia has typically allocated entitlements to take water and never specified an obligation to return a percentage of this water back to the system from which it was taken. When irrigation is technically inefficient, a significant proportion of this water tends to return back to the system and is then used by others or assigned to the environment. Once markets are introduced, however, there is a strong incentive for each irrigator to invest in water saving technology and, hence, reduce the amount of water that returns to the system (Young 2008). Once again, the result is the need to either reduce allocations to the environment and or all entitlement holders. In practice there are two ways of dealing with this return flow problem. Either:

- entitlements can be defined in net terms so that as technical efficiency of water use is increased the amount that can be pumped is reduced; or, alternatively,
- allocations per entitlement are decreased as the technical efficiency of water use in a system increases.

While Australia still has to formally determine which of these two approaches it prefers, in practice the approach taken in most systems is to more like the latter approach. As mean water use efficiency increases, allocations per entitlement and to the environment are reduced.

Interception

Still another water management issue that Australia has struggled with is the issue of how to account for all forms of water use including those that are unmetered and, in many cases, unmeterable. Two prominent examples of increases in unmetered water use are increases in small farm dam construction and increases in farm forestry. When processes like these are not included in the water allocation planning processes, the result can be a significant erosion of the amount of water that can be allocated to entitlement holders. If supply reliability is to be maintained, over-allocation problems emerge. Reluctantly as over-allocation and over-entitlement problems have worsened, Australian administrators are now in the process of developing arrangements that require those whose actions decrease runoff and, through this, river flow to offset these effects by purchasing and either not using or surrendering entitlements equivalent to the effects of these actions (Young and McColl 2009).

Interconnectivity

A related but even more challenging issue is the question of the best way to manage connected ground and surface water systems. Depending upon the direction of flow from one system to another, changes in the use of one part of these systems result in a change to the other. In the case of a river that tends to gain flow from a groundwater system that is connected to it, increases in the rate of groundwater extraction, for example, tend to reduce river flow. At the time of writing, Australian river and groundwater managers are still debating the best way to manage the effects of system inter-connectivity. Options under consideration and being tested include the definition of all groundwater extraction close to a river as extraction from the river and the assignment of groundwater shares to the river (Young and McColl 2008).

Register validation and allocation account development

As indicated earlier, Australian irrigation entitlement systems have evolved with time and in the early stages the licenses issued in each part of the system were subtly different. Unbundling these licenses into their various parts has enabled more standardization of the way that entitlements and allocations are defined and, in turn, made allocations and entitlements more fungible (in the sense that fewer types of entitlement and allocation are needed). As a result markets are much deeper and prices higher than they otherwise would have been.

In the process of doing this, in particular the separation of water licenses from land titles, it has been necessary to establish formal water license registers so that changes in the people entitled to use a water entitlement can be managed efficiently.

As a general rule, without going into the detail, register validation is a time consuming process. The final result, in Australia, has been a change to a regime where registers rather than pieces of paper are used to define ownership and an entitlement sale becomes, in effect, a contract to change the names recorded on a register. Under such a regime, ownership changes when and only when the name on the entitlement register is changed. Arrangements have also been put in place to allow entitlements to be mortgaged and other financial interests recorded on these registers.

As a general guideline, it is recommended that early attention be given to register validation as the process requires painstaking attention to detail. In Australia many of the names recorded on a license were not the same as those recorded on the land title with which it was associated. Wrestling with the difficult issue of who actually owned the license, the usual default position has been to assume that all people who have an interest in an area of land to which a license is attached have an interest in the water entitlement at the time the license is separated from the land title. All these legal entities have to consent to any dealing with the new entitlement.

Unbundling has also meant that formal water allocation accounts have had to be set up. The most advanced of these allocation accounting systems operate in a manner that is similar to the way bank accounts function. Whenever an allocation is announced, this amount is immediately credited to an account that can be accessed over the internet. In a similar manner, use is debited from the account as it occurs. Under such an arrangement, allocation trades involve the debit of water from one account and its credit to another. In the most sophisticated of these systems, account owners can use the internet to instantaneously transfer water from one account to another.

Market design and management

As markets develop and expand a host of management and transparency issues need to be managed. In particular, the administrative processes and procedures need to be fair and be seen to be fair. Great care needs to be taken to ensure that those involved are not accused, for example, of insider trading and all receive equally opportunity to access information.

As a result Australia has begun to limit policy announcements to one or two days per month and provide advance notice of when an announcement is likely to be made. Brokers have also begun to publish market price information. In an ideal world, at least daily trade and volume information would be available and all interested parties able to watch offers and bids being made.

In Australia brokers are not yet required to be registered, but codes of practice are being developed and recommendations for broker registration are becoming more common. As a general rule most brokers now hold any money received in a trust account and do not speculate in the market. One view is that the industry is capable of self-regulation (Allen Consulting Group 2007). At least one large broker, however, is of the view that formal registration is required (Waterfind 2007).

Market development

As opportunities for trade have expanded, different marketing formats have been tried and tested. The first of these tended to involve bulletin boards and notices in papers. People, typically a real estate agent, then began operating as brokers and specialists in handling the paperwork necessary to complete a sale. This was followed by the development of a formal market in Victoria known as Watermove. Watermove works by recording all offers to sell and buy as they are submitted during a week and then at the end of the week setting a single clearing price. Those who offered to sell water at less than that clearing price are paid the clearing price and those who offered to buy at more than that price only have to pay the clearing price.

More recently several internet based trading platforms have emerged. Two of these are run by independent water trading companies – Waterfind and the Water Exchange. Many irrigation water supply companies and at least one industry organization run similar services.

Supply and delivery charges

If a market is to produce an efficient outcome, then among many other things, water charging arrangements must be consistent both within and among regions. When charging arrangements are inconsistent, water entitlements and allocations will tend to move to the areas where supply arrangements are most subsidized.

In an effort to progress this issue and ensure that trading produces efficient outcomes, under the National Water Initiative all states are now required to move towards what is called "lower bound" pricing. Australia's National Water Initiative defines lower-bound pricing as:

> the level at which to be viable, a water business should recover, at least, the operational, maintenance and administrative costs, externalities, taxes or TERs (not including income tax), the interest cost on debt, dividends (if any) and make provision for future asset refurbishment/replacement. Dividends should be set at a level that reflects commercial realities and stimulates a competitive market outcome.
>
> (CoAG 2004)

In several states progress towards this goal has been assisted by an earlier Competition Policy Initiative that required administrative separation of water supply from water policy processes. As a result, governments were required to set up independent water supply companies, focus on policy and extricate themselves from the business of water supply and infrastructure management. While some of these companies are still government-owned entities several have been converted into businesses that are owned by irrigators. As a general rule this latter form of corporatization was achieved by issuing shares in proportion to the number of entitlements held. Figure 6.5 shows the result of corporatizing the Murrumbidgee Irrigation System. Administrative separation followed by the transfer of ownership and responsibility for water supply to irrigators has enabled irrigators to substantially reduce operating costs at a time when government bulk water charges have continued to rise.

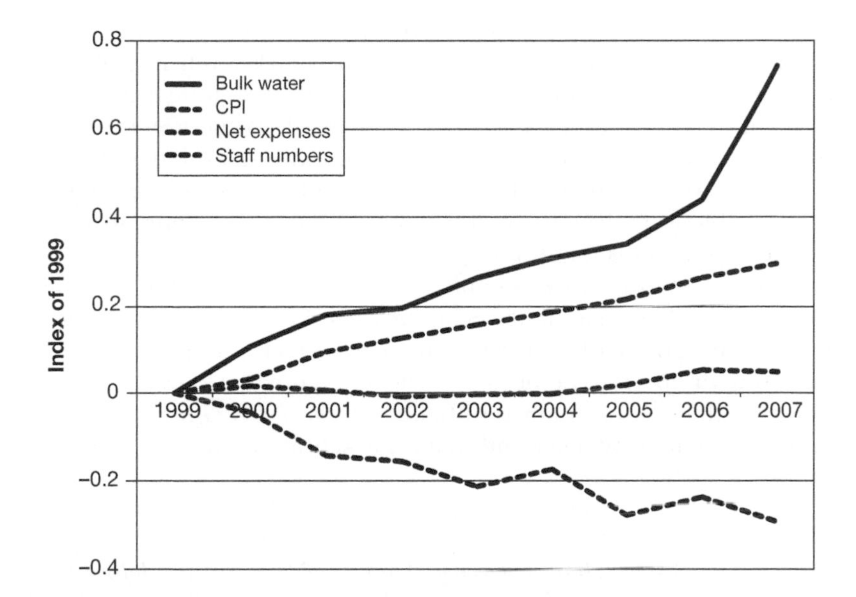

Figure 6.5 Index of water supply and delivery costs in real terms since the transfer of ownership and control of water supply assets to Murrumbidgee Irrigation compared with New South Wales government bulk water changes

While administrative separation from government has brought about significant savings to irrigators, it has also enabled these same water supply companies to protect their businesses by putting in place barriers to the trade of water out of their district. In response to this, Australia is now struggling with the question of whether or not to allow companies to set exit fees or termination fees and, if so, how to set them. In retrospect, it has become clear that every water user should operate under a formal supply contract and that exit arrangements should have been negotiated at the time ownership of the supply system was transferred to irrigators, and not left as a problem to be solved when it emerged (ACCC 2008; see also ACCC 2006).

Concluding comments

As can be seen from this chapter, the Australian experience in the development of water trading arrangements is rich and deep in insight. In retrospect it is clear that the development of water trading arrangements takes time and that sequencing is important. Moreover, it is critically important to begin with an entitlement and allocation regime that is designed for trade, avoiding the temptation to introduce trading without changing the allocation and entitlement regime.

A principled reform agenda is needed. Markets follow the signals they are given. Get the foundations wrong and the market will happily trade river and aquifer systems into trouble – so much so that, unless attention is simultaneously given to the putting in place an allocation system that has hydrological integrity, the costs of trading can outweigh the benefits. The good news, however, is that with good design and careful reform sequencing it is possible to realize many (if not most) of the economic, social and environmental benefits that trading can bring.

While many point to the economic benefits of moving water to its highest and best use, Australian water economists and policy maker now prefer to point to the power of markets in managing increasing water scarcity. Shifts to long dry periods do occur, and adverse climate change is possible. Getting by with a lot less water is difficult, but *much less difficult* when well-designed entitlement regimes and water markets are allowed to assist.

References

Allen Consulting Group (2007) *Improving Market Confidence in Water Intermediaries. National Water Commission, Canberra.* Waterlines Occasional Paper no. 3, July. Available at www.nwc.gov.au/__data/assets/pdf_file/0006/11022/improving-market-confidence-water-intermed-body-Waterlines-0707.pdf (accessed 11 July 2012).

ACCC (Australian Competition and Consumer Commission) (2006) *A Regime for the Calculation and Implementation of Exit, Access and Termination Fees Charged by Irrigation Water Delivery Businesses in the Southern Murray–Darling Basin.* Canberra, Australia: Australian Competition and Consumer Commission.

ACCC (2008) *Water Market Rules Issues Paper.* April. Canberra, Australia: Australian Competition and Consumer Commission.

Bjornlund, H. (2008) *Water Scarcity and its Implications for Land Management: Some Lessons from Australia.* London: Royal Institute of Chartered Surveyors.

Bjornlund, H. and Rossini, P. (2007) "An Analysis of the Returns from an Investment in Water Entitlements in Australia." *Pacific Rim Property Research Journal* 13(3): 344–60.

Brennan, Donna (2007a) *Managing Water Resource Reliability through Water Storage Markets.* Available at www.myoung.net.au/water/policies/Brennan_Storage_Markets.pdf (accessed 11 July 2012).

Brennan, Donna (2007b) *Missing Markets for Storage and their Implications for Spatial Water Markets.* Paper presented to AARES Conference, Queenstown, NZ, 14 February. Available at www.myoung.net.au/water/policies/Brennan_AARES07_storage_&_trade.pdf (accessed 11 July 2012).

CoAG (Council of Australian Governments) (2004) "Intergovernmental Agreement on a National Water Initiative." Agreement between the Commonwealth of Australia and the Governments of New South Wales, Victoria, Queensland, South Australia, the Australian Capital Territory and the Northern Territory. 25 June.

Frontier Economics (2007) *The Economic and Social Impacts of Water Trading.* Report prepared in association with Tim Cummins and Associates, Dr Alistair Watson, and Dr Elaine Barclay and Dr Ian Reeve of the Institute for Rural Futures, University of New England for the National Water Commission, Canberra. Available at www.frontier-economics.com/australia/au/publications/221 (accessed 11 July 2012).

National Water Initiative (2010) "National Water Initiative Pricing Principles: Regulation Impact Statement." Available at http://www.environment.gov.au/water/publications/action/pubs/ris-nwi-pricing-principles.rtf (accessed 25 July 2012).

Vanderbyl, Tom (2007) *Implementation of Continuous Sharing in Queensland.* Brisbane, Australia: SunWater. Available at www.myoung.net.au/water/policies/SunWater_Continuous_Sharing.pdf (accessed 11 July 2012).

Waterfind (2007) "Waterfind Supports National Regulation." Press release, 11 December. Adelaide, Australia: Waterfind.

Young, M. D. (2008) "The Effects of Water Markets, Water Institutions and Prices on the Adoption of Irrigation Technology." Paper presented to a conference on Irrigation Technology to Achieve Water Conservation, World Water Expo, Zaragoza, Spain, May.

Young, M. D. (2010) "Environmental Effectiveness and Economic Efficiency of Water Use in Agriculture, The Experience of and Lessons from the Australian Water Reform Programme. Background report prepared for OECD study". *Sustainable Management of Water Resources in Agriculture.* Available at www.oecd.org/water (accessed 11 July 2012).

Young, M. D. and McColl, J. C. (2003a) "Robust Separation: A Search for a Generic Framework to Simplify Registration and Trading of Interests in Natural Resources." *Agricultural Science* 15(1): 17–22.

Young, M. D. and McColl, J. C. (2003b) "Robust Reform: The Case for a New Water Entitlement System for Australia." *Australian Economic Review* 36(2): 225–34.

Young, M. D. and McColl, J. C. (2005) "Defining Tradable Water Entitlements and Allocations: A Robust System." *Canadian Water Resources Journal* 30(1): 65–72.

Young, M. D. and McColl, J. C. (2007) *Irrigation Water: Use It or Lose It Because You Can't Save It.* Droplet 6. Available at www.myoung.net.au/water/ (accessed 11 July 2012).

Young, M. D. and McColl, J. C. (2008) *Grounding Connectivity: Do Rivers have Aquifer Rights?* Droplet 13. Available at www.myoung.net.au/water/count.php?para=13 (accessed 11 July 2012).

Young, M. D. and McColl, J. C. (2009) "Double Trouble: The Importance of Accounting for and Defining Water Entitlements Consistent with Hydrological Realities." *Australian Journal of Agricultural and Resource Economics* 53(1): 19–35.

Young, M. D., Shi, T. and McIntyre, W. (2006) *Informing Reform: Scoping the Affects, Effects and Effectiveness of High Level Water Policy Reforms on Irrigation Investment and Practice in Four Irrigation Areas.* Technical Report no. 02/06. Mount Barker, Australia: CRC for Irrigation Futures.

7 The evolution of water markets in Chile

Guillermo Donoso

Introduction

In principle, in a competitive water use rights market, supply and demand are equated and the different resource valuations of different agents participating in the market converge to an equilibrium price. This would be the case if water use rights were homogeneous goods. However, several researchers have found that water use rights are heterogeneous, and the price varies according to their attributes (Crouter 1987; Colby *et al.* 1993; Bjornlund and McKay 2002; Ganderton 2002).

Water use rights markets have been implemented in Chile, the USA, and Australia. These cases have indicated that market mechanisms represent a good means to allocate water for two main reasons. First, it secures transfer of water from low-value to higher-value activities. Second, it puts the burden of information collection on water users and avoids problems of asymmetric information, common in centrally planned situations. However, as these experiences have shown, to operate properly, water markets require well-developed water conveyance facilities and the appropriate institutions to define water rights and water endowments contingent on water availability. It is also necessary to have a complete set of rules for trading in water endowments and in water use rights. Finally, institutions are needed to oversee trading activities and resolve conflicts when they arise.

Chile's Water Code of 1981 (WC) is the most important legal basis of Water Resources Management in Chile. This legislation protected water use rights (WUR), insuring ownership security, and allowed them to be freely traded so that reallocations would occur through the market of water use rights. Reforms of the water code focused on solving various problems, including the need to reconcile economic and competitive incentives with the protection of public interest, and give the state greater participation in the management of a complex resource that is key for development.

The existence of water use rights markets in Chile has been documented (Rosegrant and Gazmuri 1994; Ríos and Quiroz 1995; Hearne and Easter 1997; Donoso *et al.* 2002; Hadjigeorgalis 2004), particularly in basins characterized by water scarcity. Other studies have shown limited trading in the Bío Bío, Aconcagua, and Cachapoal Valleys (Bauer 1998, 2004; Hadjigeorgalis and Riquelme 2002). A lesson of these studies is that the operation of the WUR markets is variable across the country, and they depend significantly on the relative scarcity of water resources, the flexibility of the distribution infrastructure and water storage capacity, and the proper functioning of water user associations.

This chapter reviews Chile's water use rights markets and its water code regulations. The understanding of the evolution of water markets in Chile leads to the identification of challenges and water management lessons that must be considered in order to establish an effective water allocation mechanism based on a market.

Key features of the Chilean Water Code of 1981

The Water Code (WC) of 1981 maintained water as "national property for public use," but granted permanent, transferable water-use rights to individuals so as to reach an efficient allocation of the resource through market transactions of WUR. The WC allowed for freedom in the use of water to which an agent has WUR; thus, WUR are not sector specific. Similarly, the WC abolishes the water use preferential lists, present in the Water Codes of 1951 and 1967. Additionally, WUR do not expire and do not consider a "use it or lose it" clause.

The WC established that WUR are transferable in order to facilitate WUR markets as an allocation mechanism. Although private water use rights existed in Chile prior to 1981, the previous water codes restricted the creation and operation of efficient water markets.

The WC did not establish new institutions; however, it significantly modified their existing powers established in the WC 1967. Under the WC, the State reduced its intervention in water resources management to a minimum and increased the management powers of water use rights holders that are organized in water user associations.

The Directorate General of Water (Dirección General de Aguas, DGA), part of the Ministry of Public Works (MOP), is the main public institution and is responsible for monitoring and enforcing WUR. With its 15 regional offices, it collects and maintains hydrological data and the Public Registry of WUR (Registro Público de Agua, RPA). As the leading governmental agency in water resources management it develops and enforces national water policy. In general, the DGA has maintained a limited role in accordance with the paradigm of limited state interference by which the WC is inspired.

However, multiple central authorities (ministries, departments, public agencies) are involved in water policy making and regulation at the central government level. In Chile the number of actors involved in water policy-making is 15, the highest of OECD countries that were surveyed in a study on water governance (OECD 2011).

The WC establishes that WUR owners are responsible for water management. User management has existed in Chile since the colonial era, and currently there are more than 4000 water user associations (WUAs; Dourojeanni and Jouravlev 1999). Three types of WUAs exist in Chile and are recognized by the WC: water communities, canal user associations, and vigilance committees. Water communities are any formal group of users that share a common source of water. Canal user associations are formal associations with legal status that can enter into contracts. Vigilance committees are comprised of all the users and canal associations on any river, river section, or stream; they are responsible for administering water and allocating water to different canals.

Many of these WUAs have professional management (Hearne and Donoso 2005). The effectiveness of some of these institutions in managing irrigation systems and reducing transactions costs for water market transactions has been noted (Hearne and Easter 1995, 1997). However, according to the DGA and the Directorate of Water Works (DOH), a large percentage of these institutions have not updated their capacity to meet new challenges. Many managers of these user organizations do not have technical capacity and do not effectively communicate with their members. To address some of these concerns, the DOH and DGA have implemented programs to train WUA managers and directors (Peña 2000; Puig 1998).

The following sections describe the different types and characteristics of WUR under the WC, and the transfer of WUR regulation.

Recognized customary water use rights

Article 79 of DL 2603 recognizes customary WUR that were conceded previous to the WC. A water user shall be the owner of a customary water use right once their use over a certain amount of time is proven and ensuring that no third party effects or conflicts exist associated to this use. The specific details of this recognition are specified in in the transitory second article of the WC.

Thus, customary WUR exist and the ownership is guaranteed by the 1980 Constitution, even though they are not formally registered; the lack of formal registration does not imply lack of ownership security.

Regularization procedure for customary water use rights

The transitory second article of the WC establishes the procedure to inscribe these customary WUR in the Real Estate Registry (Conservador de Bienes Raíces, CBR). The regularization procedure has two stages: an administrative stage where the DGA publishes and informs other water users of the regularization request, and a judicial stage where the water user must legally demonstrate the existence of the customary water use right. The regularization of the customary water use right is finalized when the water use right is inscribed in the Public Registry of WUR (Registro Público de Agua, RPA) and in the Real Estate Registry (Conservador de Bienes Raíces, CBR), under the specifications established in the WC.

Specification of new water use rights

The WC specifies consumptive WUR for both surface and groundwater, and non-consumptive WUR for surface waters. Non-consumptive use rights allow the owner to divert water from a river with the obligation to return the same water unaltered to its original channel. Consumptive use rights do not require that the water be returned once it has been used. Consumptive and non-consumptive WUR are, by law, specified as a volume per unit of time. However, given that river flows are highly variable in most basins, these WUR are recognized in times of scarcity as shares of water flows.

In addition, consumptive and non-consumptive rights can be exercised in a permanent or contingent manner and in a continuous, discontinuous or alternating mode. Permanent use rights are rights specified as a volume per unit of time, unless there is water scarcity in which these WUR are recognized as shares of water flows. Contingent rights are specified as a volume per unit of time and only authorize the user to extract water once permanent rights have extracted their rights. Continuous rights are those use rights that allow users to extract water continually over time. On the other hand, discontinuous rights are those that only permit water to be used at given time periods. Finally, alternating rights are those in which the use of water is distributed among two or more persons who use the water successively.

Allocation of new water use rights

Initially, new water use rights are obtained free of charge, and the procedure for acquiring a new right started with an application that had to meet the following requirements: identification of the water source from which the water is to be extracted, specifying whether the source is surface water or ground water; definition of the quantity of water to be extracted, expressed in liters per second; yield and depth must be specified in the case of groundwater; specification of the water extraction points and the method of extraction; and definition of whether the right is consumptive or non-consumptive, permanent or contingent, continuous, discontinuous or alternating.

The administrative procedure requires that this application be published in the Diario Oficial, in a daily Santiago newspaper, and in a regional newspaper, where applicable. Previous to the WC reform of 2005, the DGA could not refuse to grant new water rights without infringing a constitutional guarantee, provided there was technical evidence of the availability of water resources and that the new use would not harm existing rights holders. If there is competition for solicited water rights, they are to be allocated through an auction with an award to the highest bidder. This allocation rule between competing WUR petitioners allows water to be allocated to its highest use value. The allocated water use right is registered in the DGA's RPA.

Peña *et al.* (2004) and Bitran and Saez (1994) point out that the absence of an obligation to use WUR led to a proliferation of WUR requests for speculation and hoarding purposes. Speculation and WUR hoarding led to non-real water shortages which created obstacles to the development of new investment projects due to the impossibility of acquiring new WUR. This was particularly evident in the case of non-consumptive WUR where entry barriers were created for new hydroelectric plants, discouraging competition in hydroelectric power generation. In fact, Riestra (2008) points out that of the 15,000 m^3/s granted in non-consumptive use rights, only 2,800 m^3/s were being effectively used. There is little concern about unused consumptive rights for water, given that, under a system of proportional use, all water is eventually distributed to users (Hearne and Donoso 2005). Dourojeanni and Jouravlev (1999) estimate the percentage of consumptive use rights that are unused to be less than one percent of the total allocated WUR.

The State, concerned about monopolistic behavior and supported by the antimonopoly commission, refused to grant new non-consumptive rights. In fact, the Constitutional Court established that the State could impose additional conditions on petitions for new water-use rights by reformulating the Water Code. This led to an amendment of the dispositions of the WC in 2005. Law no. 20,017 of 2005 amended the procedure to grant new WUR of the WC and introduced a non-use tariff (*patente de no-uso*). The amended water use right granting procedure for the DGA is as follows:

- the petitioner must justify the water flow that is petitioned and to clearly indicate the use that will be given to the water;
- WUR are only granted in accordance with the requirements of the use for which the WUR is solicited for;
- the DGA can limit the flow of an application for WUR if there is no equivalence between the amount of water requested and the use invoked by the petitioner;
- the DGA awards WUR to the petitioner if water is available and does not affect rights of other WUR holders, taking into account the relationships between surface water and groundwater;
- in the case of non-consumptive WUR, the withdrawal and restitution of the waters must not affect rights of other WUR holders over the same water source with respect to water quantity, quality, and temporality of use;

- the DGA has the obligation to establish minimum ecological flows, which can only affect new petitions of WUR, but not those granted previous to 2005; and
- a non-use tariff is introduced.

Non-use tariff

Due to the difficulties of monitoring the effective use of all WUR, the non-use tariff is applied to all consumptive WUR that do not count with water intake infrastructure and to all non-consumptive WUR that do not have water intake and return infrastructure (Law no. 20,017 of 2005, art. 129 bis 4–6).

Non-use tariff (τ) for consumptive and non-consumptive WUR is calculated as $\tau = \gamma Qf$ and $\tau = \gamma QHf$, respectively. Here γ is a constant that takes the value of 0.1 for all regions between Magallanes and Los Lagos, 0.2 for regions between O'Higgins and Araucanía, and 1.6 for all regions north of the Metropolitana (the coefficient γ increases for regions located further north to reflect increased water scarcity). Q represents the average water flow that is not used, measured in m^3/sec., f is a temporal factor that increases the non-use tariff if the water use right remains without use ($f = 1$ for years 1 to 5, $f = 2$ for years 6 to 10, and $f = 4$ for over 11 years without effective use), and H, only applied to non-consumptive WUR, is the difference between the water intake level and the level where the water is returned with a minimum value of 10 meters ($H \geq 10$).

Obligation to establish minimum ecological flows

At present there are two mechanisms for the establishment of ecological flows in the Chilean legislation: the environmental impact assessment system (Sistema de Estudios de Impacto Ambiental, SEIA) which proposes the establishment of an ecological flow as a mitigation measure, mainly for the construction of dams and setting minimum ecological flows in the act establishing new WUR by the DGA.

A major source of conflict between water users and the DGA is the diversity of criteria and methodologies under which minimum ecological flows are established (Vergara 2010). Law no. 20,417 of 2010, which created the Ministry of the Environment (Ministerio de Medio Ambiente, MMA), requires the Ministers of the Environment and Public Works to identify criteria by which to establish minimum ecological flows. This new regulation has not yet been approved.

Regulation of transfers of water use rights

The main regulatory measure established in the WC to control for potential negative effects on third parties and/or the environment due to the transfer of WUR between water users is when the transfer implies a change of water intake location, the transfer must be authorized by the DGA.

The analysis of potential third party or environmental effects associated with WUR transfers between water users is conducted by the DGA. Transfer requests, as well as new WUR petitions, are broadcast three times and published in a newspaper at the national and provincial levels. Additionally, the SEIA introduced in 1994 by Law no. 19,300 requires water users to mitigate or compensate environmental damages that may result from the transfer of WUR.

It is important to note that transfers of WUR that do not require a change in water intake location are not regulated. The Water Code of 1981 (WC) maintained water as "national property for public use," but granted permanent, transferable water use rights to individuals so as to reach an efficient allocation of the resource through market transactions of WUR. The WC allowed for freedom in the use of water to which an agent has WUR; thus, WUR are not sector-specific. Similarly, the WC abolishes the water use preferential lists, present in the Water Codes of 1951 and 1967. Additionally, WUR do not expire and do not consider a "use it or lose it" clause.

Chile's water use rights markets

The existence of water markets has been documented. Studies have shown active trading of WUR in the Limarí Valley, where water is scarce with a high economic value, especially for the agricultural sector (Hearne and Easter 1997; Donoso *et al.* 2002; Hadjigeorgalis 2004). Inter-sectoral trading has transferred water to growing urban areas in the Elqui Valley (Hearne and Easter 1997) and the upper Mapocho watershed, where water companies and real estate developers are continuously buying water and account for 76 percent of the rights traded during the 1993–9 period (Donoso et al 2002). Other studies have shown limited trading in the Bío Bío, Aconcagua, and Cachapoal Valleys (Bauer 1998, 2004; Hadjigeorgalis and Riquelme 2002).

A lesson of these studies is that the operation of the WUR markets is variable across the country, and they depend significantly on the relative scarcity of water resources, the distribution infrastructure and water storage capacity, and the proper functioning of WUAs. It should be noted that during the 2000s, the market was more active than in the previous two decades. This is largely due to a slow maturation in the public's knowledge concerning the new legislation. In a sense, the 1980s represented a preparatory stage in bringing the new code into full operation, in social, political and economic terms.

Table 7.1 presents consumptive WUR transaction data based on data of the RPA of the DGA, for the period 2005–8. The results for this four-year period show that there were 24,177 WUR transactions, of which 92.3

percent were independent of other property transactions, such as land. The value of WUR transactions independent of other property transactions is US$4.8 billion, which on average is US$1.2 billion per year.

Table 7.1 Consumptive water use rights transactions and prices for the period 2005–8

Region	*Total transactions of WUR*	*Transactions of WUR independent of other goods such as land*	*WUR transaction values (only WUR transactions independent of other goods) (US$ millions)*	*Average WUR transaction price (US$)*
I	568	564	20	36,121
II	153	131	216	1,652,519
III	16	15	8	530,933
IV	3,489	3,448	550	159,615
V	3,191	2,839	517	182,029
RM	4,804	4,226	2,312	547,095
VI	2,315	82,010	509	253,361
VII	6,518	6,159	622	101,059
VIII	2,330	2,162	29	13,432
IX	494	487	8	16,805
X	225	223	23	103,390
XI	68	68	0	2,588
XII	6	6	0	80,667
Total	24,177	22,338	4,817	215,623

Source: World Bank (2011)

Note: The RPA of the DGA has data only for the period 2005–9. The data for the year 2009 are incomplete

The most active WUR market is in the VIIth Region of Chile, Región del Maule, driven by a growing export oriented agricultural sector. The Metropolitan Region follows with 4,800 WUR transactions between 2005 and 2008. This region presents growing water demands from a growing urban population, industry, mining and agricultural sector. The third and fourth regions in terms of market activity are the Coquimbo Region (Region IV) and the Valparaiso Region (Region V), characterized by greater water scarcity than the previous two regions and a high demand of water for the export oriented agricultural sector.

The main participant in the WUR market is agriculture, buying and selling 57 percent and 68 percent of total transactions, respectively. Thus, despite price trends, most water moves not from agriculture, but within agriculture. This pattern reflects the high value of water to the agricultural fresh fruit and wine exporting sector, which has increased its surface area between the Region of Atacama (Region III) and Maule (Region IV). Despite a legal separation between land and water rights, many Chilean farmers maintain that water and land should not be separated. This traditional integration of land and water has kept many farmers from offering water for sale without also selling the corresponding land. Also, the agricultural sector in Chile has continued to grow, often at a rate greater than the rest of the Chilean economy (World Bank 2003; ODEPA 2004). Because of this growth, the value of water in irrigation has remained high and farmers have little incentive to sell water.

The average WUR price is US$215,623 per WUR. WUR prices in the north of the country are greater than in the south, which indicates that the market at least in part reflects the relative scarcity of water. The highest WUR price is in the Antofagasta Region, characterized by desert conditions, where mining is the main economic activity dependent on water.

On the other hand, a water right is essentially a claim on a proportional share of available water in any year. Donoso *et al.* (2011) show that as water flow increases water use rights prices decrease. Thus, in a high-flow (low-flow) year, the relative absence (presence) of scarcity decreases (increases) the price of WUR. A suggested area for further research is how agents form expectations on water flow and how these impact WUR trading and prices. Additionally, this result emphasizes the importance of the DGA's role in providing adequate and timely information for potential WUR buyers and sellers.

WUR prices present a standard deviation of US$100,460,800 per WUR; price dispersion is lower in the more active WUR markets. Thus, Chilean WUR markets are characterized by a large price dispersion for homogeneous WUR (Cristi and Poblete 2010). This large price dispersion is due, in great part, to the lack of reliable public information on WUR prices and transactions. Given the lack of reliable information, each WUR transaction is the result of a bilateral negotiation between an interested buyer and seller of WUR where each agent's information, market experience and negotiating capacity is important in determining the final result. Donoso *et al.* (2011) estimated a reduced form WUR price model which captures the importance of non-productivity factors such as an agent's water rights market experience and expected price. These results are consistent with those of Bjornlund (2002), who found that factors that influence the negotiating process and the agent's negotiating power significantly influence water rights prices in Australia. Donoso *et al.* (2011) show that buyers (sellers) with previous WUR market experience obtain lower (higher) WUR prices due to their greater negotiating ability. Also, consistent with Bjornlund's (2002) result, their model indicates that the agent's awareness of prevailing WUR market prices is a significant factor that explains water WUR price. This has important policy implications, given that in Chile WUR markets do not have a price revealing mechanism. This result suggests that it is necessary for the DGA to design and implement a WUR price revealing mechanism. Some work has been done with the creation of a national database of water rights transactions which registers volume traded and transaction prices. However, this database is incomplete and not kept up to date, which makes it difficult to obtain data on WUR prices.

According to Peña (2010), as a result of the WC 2005 reform, combined with the performance of the Antitrust Commission, the monopolistic distortion due to speculation and non-consumptive WUR hoarding has been reduced. In turn, Jouravlev (2010) notes that as a result of the reform of 2005 (together with other measures), WUR that still are not used are generally no longer a major obstacle to the development of the water basin, and it is likely that non-use of WUR will continue to reduce in the future due to the projected increase in the non-use tariff. Along the same lines, Valenzuela (2009) notes that the non-use tariff has operated as an incentive for the return of WUR; an equivalent of 65 m^3/s has been returned, which represents 1 percent of the total WUR affected by the non-use tariff. However, Cristi (2010), states that the effect of the non-use tariff has been limited, based on evidence that during the year 2009 only 2.08 percent of the WUR subject to the non-use tariff were returned or began to be used. Thus, further research is required to determine the effectiveness of the non-use tariff.

It is important to point out the major flaws with respect to the design of the non-use tariff. The non-use tariff is only applied to WUR for which the water intake infrastructure has not been constructed. However, the mere existence of water intake infrastructure does not necessarily ensure that waters are used in practice. It can be applied only to the registered WUR and regularized customary WUR that are contained in the records of the RPA. As was already mentioned, the majority of WUR are not registered. The non-use tariff, as defined at present, is a disincentive to hold WUR for environmental purposes that do not require water extraction. A holder of WUR for this type of environmental use will be subject to payment of the non-use tariff.

A major challenge of the WUR markets in Chile is how to ensure optimal water use without compromising the sustainability of rivers and aquifers. The sustainability of northern rivers and aquifers is comprised due to the over-provision of WUR related to the practice of allocating WUR based on foreseeable use. The foreseeable use considers the probable effective water extraction of different sectors. For example, an agricultural WUR does not extract water in winter months, whereas a mining WUR extracts water all year round. In this case, the authority would consider a lower pressure on water resources of an agricultural WUR with respect to the pressure of a mining WUR. This administrative practice commits the mistake of not considering the transferable nature of WURs. Thus, when water scarcity increases and inter-sectoral WUR transactions increase, water resources will be overexploited and unsustainable. The allocation of new WUR based on foreseeable use for nearly 15 years has led to an underestimation of the actual use of water of the conceded WUR, causing a considerable impediment to reach the sustainability of water resources and associated ecosystems. This problem is more serious in northern basins characterized by arid areas. Additionally, the over-provision of WUR gave rise to increased water insecurity as WUR are transferred to users with a more intensive water use.

On the other hand, increased consumptive WUR market activity has generated increased conflicts with downstream users due the existence of WUR defined on return flows. The consumptive WUR entitles the holder to totally consume the water taken in any activity. However, in practice, almost all consumptive WUR holders generate significant return flows (leakage and seepage water) that are used by downstream customary WUR holders. At present it is not known how many regularized or non-regularized customary WUR are dependent on return flows. Thus, it is extremely difficult for the DGA to foresee potential third party effects associated with WUR transfers that alter return flows. For this reason, in those rivers with large return flows, such as the Elqui and Aconcagua rivers, located north of Santiago, there have been cases where the Junta de Vigilancia have tried to prohibit the transfer of WUR from agricultural users to non-agricultural users in earlier sections of these rivers, in order to protect users who depend on return flows (Rosegrant and Gazmuri 1994).

The WC did not pay much attention to the sustainable management of groundwater because groundwater extraction was marginal during the early 80s. Recognizing the need to improve groundwater management regulation due to increased groundwater pumping, the 2005 amendment of the WC introduced procedures to reach a sustainable management of underground water resources. The main provisions are: (i) extraction restrictions when third parties are affected, (ii) the authorization for the DGA to impose the installation of extraction measurement equipment in order to monitor effective extractions, (iii) the establishment of areas subject to extraction prohibitions and restrictions, and (iv) the need to consider the interaction between surface water and groundwater when analyzing the petition of new surface or groundwater WUR.

World Bank (2011) concludes that these groundwater regulations have not been fully implemented over time and thus, there exist various problems associated with groundwater management. A major concern is the general lack of information about groundwater and insufficient knowledge about its dynamics, in particular its interaction with surface waters. There are significant gaps in the registry of wells, extraction and quality measurements, recharge balances, and identification of pollution sources. In general, information systems are not linked to the measurement and monitoring of aquifers to estimate groundwater withdrawals. An effective information system is a prerequisite to be able to control and sustainably manage an aquifer.

An additional challenge for a sustainable groundwater management is the fact that at present ground and surface waters are managed independently despite their recognized interrelations. This implies that there is no conjunctive management of surface and groundwater, which has proven to be an effective adaptation mechanism for climate change.

There are, in general, no WUAs that manage groundwater user rights; the only exception is in some sections of the over-exploited Copiapo aquifer. According to the WC, there should exist a groundwater WUA at least for all aquifers that have a restriction or prohibition declaration by the DGA. The fact that users have not yet organized themselves in groundwater WUAs to take over the management of groundwater reflects the lack of understanding of a large proportion of users of the long term effects that uncontrolled exploitation of aquifers may cause. In the absence of groundwater WUAs, the WC establishes that the DGA is responsible for controlling and monitoring groundwater withdrawals. However, evidence has shown that the DGA does not have the necessary resources (human, technical, and financial) to monitor all groundwater extractions

Jouravlev (2005), based on a survey of the literature on WUR markets in Chile, concludes that these markets have helped to (i) facilitate the reallocation of water use from lower to higher value users (e.g. from traditional agriculture to export-oriented agriculture and other sectors such as water supply and mining), (ii) mitigate the impact of droughts by allowing for temporal transfers from lower value annual crops to higher valued perennial fruit and other tree crops, and (iii) provide lower cost access to water resources than alternative sources such as desalination.

The analysis of the problems that have been resolved through the market of water use rights, indicates that the use of this allocation mechanism has allowed users to consider water as an economic good, internalizing its scarcity value; constitutes an efficient reallocation mechanism which has facilitated the reallocation of granted WUR; has permitted the development of new high value economic activities such as mining in areas in the semi-arid northern region of Chile where this resource is scarce, by buying WUR from agriculture; the solution of problems associated to water deficits derived from a significant increase of water demand, caused by the significant population growth in the central region of Chile; and the solution or water scarcity problems when a quick response has been required (Donoso 2006).

The problems that water use rights market have not been able to resolve are: water use inefficiency in all sectors, not only in the agricultural sector, environmental problems, and the maintenance of ecological water flows.

The elements that have hindered WUR market effectiveness are the lack of WUR and WUR market information, lack of regularization of customary WUR, existence of transaction costs, and lack of a rapid, efficient controversy resolution system.

Lack of water use rights and water use rights market information

The wide dispersion of prices documented by Donoso *et al.* (2002), Donoso (2006) and Cristi and Poblete (2010), among others, is an indication of the limited information that buyers and sellers have access to on WUR transactions and prices. WUR markets have lacked transparency.

A centerpiece of the information system on WUR is the RPA. This provides the DGA with the necessary information on WUR to enable it to effectively fulfill their functions, and water users with the required data for an efficient water management and planning, as well as WUR market information. However, as discussed previously, the RPA is incomplete; only 20 percent of all WUR and 50 percent of market transaction cases are registered (World Bank 2011).

The main reason why the RPA is incomplete is that only regularized and formally inscribed WUR can be registered. Moreover, the record is not updated because the CBR and users rarely transmit to the DGA market transaction data, even though the 2005 amendment of the WC requires all CBR to inform WUR market transaction data.

Lack of regularization of customary water use rights

Only registered rights can be bought, sold, and mortgaged, and thus, the fact that most WUR remain unregistered impedes thc transfer of water. However, most WUAs maintain their own registries in order to effectively distribute water to rights owners. However, these registries do not imply legal title. The DGA is also responsible for maintaining the RPA, which contains information on all water use rights that are granted by the DGA. This RPA also contains hydrological and water quality data, information on WUAs and water withdrawals, and all transactions.

Since the promulgation of the WC, efforts have been made to regularize, register WUR, and grant title for WUR in order to resolve overlapping claims to water. This is especially important for WUR that were redistributed under the Agrarian Reform and might be contested by previous owners. Estimates of WUR that are not registered range from 60 to 90 percent (Dourojeanni and Jouravlev 1999). This can be in part explained by the fact that courts have protected unregistered rights, and thus undermined the registration requirement (Baucr 1998, 2004).

Thus, the regularization procedure has not been effective. This lack of regularization and registration can be explained by several reasons (World Bank 2011). One is that there is a lack of incentives and penalties for holders of WUR to regularize and register their customary WUR. In particular, the second transitory article of the WC does oblige users to regulate their WUR; however, there are no impediments to exercise their rights even though the WUR are unregistered. Regularization procedures are complex and lengthy, due to the complexity and rigor of the verification process.

However, it is also due to an excessive judicialization of proceedings. Of the customary WUR that have been certified by the DGA since 1981, between 40 and 65 percent are still awaiting a court ruling (World Bank 2011).

The regularization procedures have generated an important proportion of the current water use conflicts that must be settled by the DGA and the judicial courts. Given that the regularization procedures were established 30 years ago, the difficulty of verifying the validity of the customary water use has significantly increased (World Bank 2011).

Existence of transaction costs

Transaction costs associated with the WUR market have increased due to the difficult process of finding potentially suitable buyers or sellers. Given the lack of public information on water transactions, WUR prices and WUR market activity, in general, interested water rights buyers usually contact WUAs or lawyers specialized in water law for information on potential water rights sellers. It is also common for individuals or companies with WUR to hire consulting services to assess and value their rights.

Additionally, there is no WUR price-revealing mechanism. An initiative to reduce transaction costs is the research project entitled "Development of an Electronic Market for Water in Chile" that is developed in collaboration with the DGA, academic institutions and CORFO.

There are no quantitative and empirical studies to gauge the real magnitude of these costs at a national level. One exception is a study conducted by Hearne and Easter (1995), who established the total net earnings associated with the transfer of water rights, and thus established the estimated transaction costs. However, this analysis was conducted only for the Elqui and Limarí River basins. The study determined that, in these basins, significant net earnings were achieved; hence, the benefits obtained from the transactions were considerably greater than the cost incurred.

On the other hand, unavoidable transaction costs provide an explanation for why water markets emerge in some cases and why they have not in a large proportion of water basins. The unavoidable transaction costs associated to the need to modify water distribution infrastructure will be involved in any type of water reallocation and are independent of water management institutions.

In the Maipo River basin, for example, there is a rigid canal infrastructure (the main water distribution system is made up of a structure that separates the flow into multiple off-takes). The cost of modifying this structure has been assessed at approximately 10 percent of the value of the right in section one of the Maipo (1SMaipo; Donoso *et al.* 2002). This percentage diminishes as the total volume of water transferred increases.

Transactions carried out in river basins with flexible-pipe distribution systems occur with much greater frequency. Unavoidable transaction costs in the Paloma System, for example, are considerably lower than those of the Maipo, due to the existence of a more flexible water distribution infrastructure (consisting primarily of variable gates) and the consequent lower cost of modifying it.

Lack of a rapid, efficient controversy resolution system

The WC establishes that any conflicts that may arise among water users, must be solved by the board of directors of the WUAs that arbitrate and such decisions may be enforced with the assistance of the public force. However, WUAs have problems solving controversies since they have weak, inappropriate capacities to solve problems (Bauer 2004). Thus, the majority of conflicts have ended in the judicial system.

Additionally, Chilean lawyers and judges are not always formally educated on the water law. The legislation on WUR is not, in general, taught in the law schools . When proceedings are carried out to solve these conflicts, the judges must resort to the DGA to obtain further information (Bauer 2004). Therefore, a major challenge to improve water allocation through WUR markets is to increase the conflict resolution capacity.

Conclusions

The WC is the most important legal basis of the Water Resources Management in Chile. This legislation protected WUR, insuring ownership security, and allowed them to be freely traded. Although the WC was successful in promoting investments related to water and improving water use efficiency in many economic sectors, it also led to new difficulties, which were partially addressed in the reforms of 2005.

Reforms of the WC focused on solving various problems, including the need to reconcile economic and competitive incentives with the protection of public interest, and give the state greater participation in the management of a complex resource, that is key for development. Additionally, the reform centered on avoiding the concentration of WUR through speculation and hoarding. There is still no comprehensive analysis of the impact of the 2005 reform; however, there are indications that it has been able to solve some problems while others still require attention.

The clear enabling factor that allowed for the implementation of the water use rights market in Chile was Chile the tradition and culture of managing water resources with water use rights that dates back to colonial times. Additionally, water management under the Water Code of 1951 allowed for limited transfers of WUR, thus, the idea of employing economic instruments for the efficient management of water resources was not a foreign one for Chile.

The strong legal security of WUR is a second important enabling factor. This has encouraged private investment and improvement of water use efficiency across sectors. Additionally, an enabling factor is the separation of WUR from land ownership; this reduces transaction costs associated with water inter-sectoral water use rights transfers.

The elements that have hindered WUR market effectiveness are the lack of WUR and WUR market information, the lack of regularization of customary WUR, existence of transaction costs, and the lack of a rapid, efficient conflict resolution system.

This review of Chile's WUR markets and WC regulations leads to the identification of lessons that must be considered in order to establish an effective water allocation mechanism based on a WUR market. The main lessons are the following:

- A cultural context of the society consistent with the economic paradigm of solving inefficiencies of free access goods based on the establishment of property rights; society's acceptance is a prerequisite for a successful WUR market.
- The existence of water scarcity; when water is not scarce, there is no need to reallocate WUR.
- It is essential that WUR be clearly specified, ownership secure, and formally registered.
- The period during which the water can be used; the lack of this definition has caused conflicts between consumptive and non-consumptive users.
- It is important that regulations and conditions for WUR trade and transfers be explicit, and designed to accommodate transfers quickly and at low cost; this also requires that there be an effective conflict resolution mechanism.
- The issues of total resource use, unused entitlements, and environmental and in-stream needs should be addressed prior to the introduction of trade; otherwise, the authority will probably not be able to satisfy minimum ecological flows.
- The need to consider the issue of unused water. Failure to address this issue with adequate regulatory mechanisms can lead to inefficient and unsustainable water uses and opens up for speculation and WUR hoarding behavior and negative economic impacts.
- Adequate regulations that address externalities and potential damage to third parties due to WUR transactions. As Bjornlund and McKay (2002) point out, a balance between private market forces and government regulation to protect third party interests, including environmental concerns must be considered.
- A complete registry of WUR holders is necessary; lack of information on water use right holders increases transaction costs.
- An efficient information system is needed that considers an efficient flow of market information such as data on transactions and a price-revealing mechanism; this is especially important where the ability of

well-informed and well-financed buyers to benefit from ill-informed and poor sellers is present.

- Detailed information and models of both surface and groundwater resource availabilities are required.
- Efforts should be made to remove obstacles to the spatially free movement of water. For example, a flexible infrastructure that allows for the transfer of WUR at low costs;
- Strengthening and capacity-building of WUAs is necessary; successful WUAs have proven to be instrumental in facilitating WUR markets and reducing transaction costs.

References

Bauer, C. (1998) *Against the Current: Privatization, Water Markets, and the State in Chile.* Boston, MA: Kluwer Academic Press.

Bauer, C. (2004) *Siren Song: Chilean Water Law as a Model for International Reform.* Washington, DC: Resources for the Future Press.

Bitran, E. and Sáez, R. (1994). "Privatization and Regulation in Chile." In B. P. Bosworth, R. Dornbusch, and R. Labán (eds), *The Chilean Economy: Policy Lessons and Challenges.* Washington, DC: The Brookings Institution.

Bjornlund, H. (2002) "Signs of Maturity in Australian Markets." *New Zealand Property Journal* 2002(July): 31–46.

Bjornlund, H. and McKay, J. (2002) "Aspects of Water Markets for Developing Countries: Experiences form Australia, Chile and the US." *Environment and Development Economics* 7(4): 769–95.

Colby, B., Crandall, K. and Bush, D. (1993) "Water Right Transactions: Market Values and Price Dispersion." *Water Resources Research* 29(6): 1565–72.

Cristi, O. (2010) *Diagnóstico Económico.* Report. Santiago, Chile: DGA.

Cristi, O. and Poblete C. (2010) *Qué nos Dicen los Conservadores de Bienes Raíces del Mercado de Aguas en Chile.* Working paper. Santiago, Chile: Escuela de Gobierno, Universidad del Desarrollo.

Crouter, J. P. (1987) "Hedonic Estimation Applied to a Water Rights Market." *Land Economics* 63: 259–71.

Donoso, G. (2006) "Water Markets: Case Study of Chile's 1981 Water Code." *Ciencia e Investigación Agraria* 33(2): 157–71.

Donoso, G. Montero, J. P. and S. Vicuña (2002) "Análisis de los Mercados de Derechos de Aprovechamiento de Agua en las Cuencas del Maipú y el Sistema Paloma en Chile: Efectos de la Variabilidad de la Oferta Hídrica y de los Costos de Transacción." In A. Embid (ed), *El Derecho de Aguas en Ibero América y España: Cambio y Modernización en el Inicio del Tercer Milenio.* Spain: Civitas.

Donoso, G. Melo, O. and Jordan, C. (2011) "Estimating Water Rights Demand and Supply: Are Buyer and Seller Characteristics Important?" Departamento de Economía Agraria, Facultad de Agronomía, Pontificia Universidad Católica de Chile, Santiago, Chile.

Dourojeanni, A. and Jouravlev. A. (1999) *El Código De Aguas De Chile: Entre La Ideología Y La Realidad.* Recursos Naturales e Infraestructura 3. Santiago, Chile: CEPAL.

Ganderton, P. (2002) "Western Water Market Prices and the Economic Value of

Water." 2nd Congress of Environmental and Resource Economists, Monterrey, Mexico.

Hadjigeorgalis, E. (2004) "Comerciando con Incertidumbre: Los Mercados de Agua en la Agricultura Chilena." *Cuadernos de Economía* 40(122): 3–34.

Hadjigeorgalis, E. and Riquelme, C. (2002) "Análisis de los Precios de los Derechos de Aprovechamiento de Aguas en el Río Cachapoal." *Ciencia e Investigación Agraria* 29(2): 91–9.

Hearne, R. and Donoso, G. (2005) "Water Institutional Reforms in Chile." *Water Policy* 7(1): 53–69.

Hearne, R. and Easter, K. W. (1995) *Water Allocation and Water Markets: An Analysis of Gains-From-Trade in Chile.* Technical Paper no 315. Washington, DC: World Bank.

Hearne, R. and Easter, K. W. (1997) "The Economic and Financial Gains from Water Markets in Chile." *Agricultural Economics* 15(3): 187–99.

Jouravlev, A. S. (2005) "Integrating Equity, Efficiency and Environment in the Water Allocation Reform." Presentation at the International Seminar Water Rights Development in China, Beijing.

Jouravlev, A. S. (2010) "Código de Aguas de Chile: Desafíos Pendientes." Presentation to the Directorate General of Water, Santiago, Chile, 15 July.

ODEPA (Oficina de Estudios y Políticas Agrarias) (2004) *Agenda Agroalimentaria y Forestal 2004–2006.* Available at www.odepa.gob.cl (accessed 28 March 2012).

OECD (2011) *Water Governance in OECD Countries: A Multi-Level Approach.* OECD Studies on Water. Paris, France: OECD Publishing.

Peña, H. (2000) "Desafíos a las Organizaciones de Usuarios en el Siglo XXI." In *Proceedings III Jornadas de Derechos de Agua.* Santiago, Chile: Universidad Católica de Chile.

Peña, H. (2010) "Copiapó: Realidades, Desafíos y Lecciones." Presentation for Seminario Nacional AHLSUD Capitulo Chileno A.G.

Peña, H., Luraschi, M. and Valenzuela, S. (2004) *Water, Development, and Public Policies – Strategies for the Inclusion of Water in Sustainable Development.* Santiago, Chile: South American Technical Advisory Committee, Global Water Partnerhsip.

Puig, A. (1998) "El Fortalecimiento de las Organizaciones de Usuarios para una Gestión Integrada de los Recursos Hídricos." Paper presented at the International Conference on Water and Sustainable Development, Paris, 19–21 March.

Riestra, F. (2008) "Pago de Patente por no Uso de Derechos de Aprovechamiento." Santiago, Chile: Dirección General de Aguas (DGA).

Ríos, M. and Quiroz, J. (1995) "The Market of Water Rights in Chile: Major Issues." *Cuadernos de Economía* 32: 317–45.

Rosegrant, W. M. and Gazmuri. R. (1994) *Reforming Water Allocation Policy through Markets in Tradable Water Rights: Lessons from Chile, Mexico and California.* EPTD Discussion Paper no 6. Washington, DC: International Food Policy Research Institute.

Valenzuela, C. (2009). *La Patente por la No-utilización de las Aguas en Chile: Origen, Diseño y Primeras Experiencias en su Implementación.* Santiago, Chile: Universidad de Chile.

Vergara, A. (2010) *Diagnóstico de Problemas en la Gestión de Recursos Hídricos: Aspectos Institucionales para una Futura Propuesta de Modificaciones Legales, Reglamentarias y/o de Prácticas Administrativas.* Report. Santiago, Chile: DGA.

World Bank (2003) "Country Data Tables." Available at www.worldbank.org/data/countrydata/countrydata.html (accessed 20 March 2012).

World Bank (2011) *Chile: Diagnóstico de la Gestión de los Recursos Hídricos*. Washington, DC: World Bank. Available at http://water.worldbank.org/node/83999 (accessed 12 July 2012).

8 The experience of water markets and the market model in Chile

Carl J. Bauer

Introduction

What can the Chilean experience teach us about water markets and sustainable development? Chile's experience is uniquely valuable to help answer the following question in international water debates: Is a free-market approach to recognizing water as an economic good compatible with the broader and long-term goals of integrated water resources management? I will argue that the Chilean experience shows that the answer is no. Other economic approaches, however, including market instruments, are nonetheless essential.

This paper draws on my previous work in Chile (Bauer 2004, 2005) to respond to several policy questions:

1. What criteria should we use to evaluate the efficiency of water markets? How should these criteria be incorporated in market instruments?
2. Can the economic efficiency of water allocation be compatible with social and environmental sustainability?
3. How do water's physical characteristics affect its commoditization?
4. How useful are mathematical models for decision-making in this field?

Chile's free-market Water Code was passed in 1981. Since then Chile has become the world's leading example of the free-market approach to water law and economics – the textbook case of treating water rights not merely as private property but also as a fully marketable commodity. Other countries have recognized variations of private property rights to water, but none have done so in as unconditional and deregulated a manner as Chile. Because the 1981 Water Code is so paradigmatic an example of free-market reform, people around the world have disagreed about whether the "Chilean model" has been a glorious success or a disastrous failure or something in between.

The predominant view outside of Chile is that Chilean water markets and the Chilean model of water resources management have been a success. This has been the view of many economists in the World Bank, the

Inter-American Development Bank, and related institutions, who have encouraged other countries to follow Chile's lead in water law reform. Since the early 1990s, these proponents have promoted a simplified description of the Chilean model and its results in other Latin American countries and in the wider international water policy arena. Although they sometimes recognize flaws in the model, their general tendency has been to play down the importance of those flaws and instead to emphasize the model's advantages.

For example, a senior water adviser at the World Bank has publicized Chilean water markets as a prominent international example of "good practice" in managing water "as an economic resource" (Briscoe 1996; Briscoe *et al.* 1998). After praising "the genius of the water market approach" (Briscoe 1996: 21), Briscoe concludes that in the case of Chile, the "system of tradable water rights and associated water markets is a great achievement and is universally agreed to be the bedrock on which to refine Chilean water management practices" (Briscoe *et al.* 1998: 9). Similarly, a paper for the Global Water Partnership (Rogers and Hall 2003) describes Chile as "a world leader in water governance." The authors recognize that "many mistakes with openness, transparency, participation, and ecosystem concerns were made in the hurry to get effective water markets established," but they conclude that "the system is adaptive and now these concerns are being addressed 20 years after the initial laws were passed" (Rogers and Hall 2003: 30–31). This optimism about the prospects for correcting the Water Code's flaws rests on ignorance of Chile's political and constitutional system, as I have argued in detail elsewhere (Bauer 2004).

The Chilean model of water markets and water rights trading is different from other countries in at least one essential way. In other countries that have allowed or encouraged water markets, in varying degrees and circumstances, these markets have been a policy instrument within the larger context of water law and regulation. In Chile this order is reversed: water resources management takes place in an institutional context that has been shaped by and for water markets. The Chilean Water Code is so *laissez-faire* that the overall legal and institutional framework has been built in the image of the free market, with strong private property rights, broad private economic freedoms, and weak government regulation.

When we look at Chilean water markets, therefore, we are also looking at the Chilean model of water management in general, to a greater extent than in other countries. My analysis here emphasizes an institutional economic perspective by showing how legal rules and political forces have affected property rights, economic incentives, and market performance.

Water reforms and international debate

Since the early 1990s there has been increasingly urgent international debate about a global water crisis. Growing demands and competition for water resources worldwide have caused increasing scarcity and conflict. As these problems have become worse, they have led to a broad international consensus about the need for major reforms in how water is used and managed. Such reforms point to more "integrated" water resources management, which is supposed to be holistic, comprehensive, interdisciplinary, and sustainable over the long term (Global Water Partnership 2000). Reaching this goal will require good "water governance," another open-ended term that refers to the social, political, and institutional aspects of water management.

One of the core principles of integrated water resources management (IWRM) is that water should be recognized as an "economic good." But people disagree about what that means and what the policy implications might be. There is a range of economic perspectives, which vary in how they address the institutional context and foundations of markets – in simple terms, the rules of the game (Bauer 2004, 2005). With respect to the idea of water as an economic good, two general economic perspectives are most important. One is that water should be treated as a scarce resource; the other is that water should be a private and marketable commodity. Both positions agree that allocating water involves difficult choices and trade-offs, for which economic incentives can be powerful tools. But they have quite different implications for the role of markets and for legal, institutional, and regulatory arrangements, as the example of Chile shows clearly.

The institutional arrangements that are critical to IWRM are those required for defining and enforcing property rights, coordinating different water uses within shared river basins, internalizing externalities (both economic and environmental), resolving conflicts, and monitoring compliance. These processes are all inter-related by their nature. Moreover, they have unavoidable social and political aspects, since they allocate costs and benefits among different people and determine who gains and who loses. Because different economic perspectives shape the design of these institutional arrangements, the stakes are high. For all these reasons, Chile's 30 years of experience with the water market model offer valuable lessons for both national and international water policy reforms.

Empirical results in Chile

If we look back over the 30 years since the passage of the 1981 Water Code, and particularly the period since Chile's return to democratic government in 1990, two major trends stand out. First, there has been a great deal of political and policy debate about water rights and water markets in Chile, but there has been much less academic or empirical research – by which I mean research attempting to be objective rather than tied to particular political positions. Much of the academic research has been done by North Americans, and much of the policy research has been done by people in international organizations. The relative lack of empirical research by

Chileans is both a cause and an effect of the highly politicized nature of the domestic debate about water policy.

Second, two related issues, or sets of issues, have dominated discussions in both policy and academic arenas:

- What should be the basic legal rules defining property rights to water? How have the current rules (which are extremely *laissez-faire*) affected the economic incentives for water use? Should the law be changed, whether for economic or social reasons? (A limited reform was approved in 2005, as I discuss below.)
- How have water markets worked in practice? What have been their strengths and weaknesses, and how might they be improved by legislative or regulatory reforms?

Both sets of issues have been hotly contested in Chile. Outside the country, in contrast, nearly all attention has been directed to the second set: Chilean water markets. The distinction is important: it points to the narrow focus that has dominated international evaluation of the Chilean model.

We can assess the empirical results of the 1981 Water Code in two ways: (i) by comparing them with the law's original objectives; and (ii) by comparing them with the issues now considered critical for integrated water resources management. The law's original objectives, of course, were determined by the strong political and ideological views of the Chilean military government and its civilian advisers. Many people both inside and outside Chile do not share those views. In comparing the law's results with those objectives, therefore, I do not mean to imply that the objectives themselves were necessarily correct.

Comparison with the Water Code's original objectives

It is important to recall a few general points about the Water Code's political background and legislative history (Bauer 1998, 2004). When the Water Code was being drafted in the late 1970s and early 1980s, the government's primary concerns were irrigation and agriculture. Relatively little attention was given to non-agricultural water issues, other than to assume that free markets would reallocate some agricultural water supplies to non-agricultural uses. As far as other issues were concerned, free-market economic theory was simply applied to water management without much effort to adapt it to the specific characteristics of water resources. The best example of this is the *laissez-faire* approach to coordinating water uses within river basins. Another example is the inadequate definition of the rules governing the new category of non-consumptive water rights. Finally, issues of water quality and environmental protection were left out of the Water Code almost entirely. IWRM, in short, was not yet on policy-makers' radar screens.

It is also important to keep in mind that some of the fundamental elements of the Chilean water law model have been determined by the 1980 Constitution more than by the 1981 Water Code. One such element is the constitutional provision that declares water rights to be private property, which enjoys very strong protection from government intervention. A second such element is the basic structure of the overall institutional and regulatory framework, particularly the limited power of the government administrative agency, the Dirección General de Aguas (DGA), and the strong and broad authority of the judiciary. This framework is what determines the resolution of water conflicts and related institutional functions.

The Water Code, together with the Constitution, has been fairly effective in achieving several of its original priorities, mainly those having to do with strengthening private property rights.

- The legal security of private property rights has encouraged private investment in water use and infrastructure. The level of this investment has varied in different parts of the country; it has allowed new mining development, especially in northern Chile, and the planting of high-value fruits and vegetables for export.
- The counter-reform in agrarian land tenure has been consolidated.
- Government regulation of water use and water management has been tightly restricted.
- The freedom to trade water rights has allowed reallocation of water resources in certain circumstances and geographic areas.
- The autonomy of private canal users' associations from government has been affirmed, which in some cases has encouraged these organizations to improve their administrative and technical capacity; however, these organizations operate only within the agricultural sector and rarely include non-agricultural water users.
- The creation of non-consumptive water rights has encouraged hydroelectric power development, first by government enterprises and later by private companies - though not without serious and uncompensated impacts on other water users, particularly irrigators.

The Water Code has been much less effective in achieving other original objectives, particularly those having to do with the smooth operation of water markets and market incentives. This observation is now widely accepted by Chilean water experts, although not by the law's strongest political supporters. Note that some of these results represent a lack of success rather than failure; others are more clearly negative.

- Market incentives to promote more efficient water use, particularly within the agricultural sector, have not worked as expected. Irrigation efficiency remains low nationwide, and in the few areas where it has increased, the change reflects factors other than the water market –

namely, investment to improve crop yields or reduce costs of labor and canal maintenance. Investment in these areas has been encouraged by the legal security of property rights, but not by market incentives to sell unused or surplus water rights. Such rights are rarely sold.

- Continued government subsidies have been required for the construction and maintenance of irrigation works at small, medium, and large scales.
- Examples of significant market activity, as indicated by the amount of water resources reallocated or by the frequency of water rights transactions, remain limited to a few areas of the desert north and the metropolitan area of Santiago.
- The definitions of water rights remain vague or incomplete; the legal and technical details are inadequate and confusing in most of the country.
- The idea that water market forces would benefit peasants and poor farmers by improving their access to or ownership of water supplies has generally failed. If anything, the water market seems to have harmed many of these farmers more than it has helped them, although there is not enough evidence to make strong generalizations.
- Reliance on private bargaining to coordinate different water uses and resolve river basin conflicts, particularly between consumptive and non-consumptive water rights, has failed. Neither DGA nor the courts have adequately or reliably redressed the problem.
- In the hydroelectric sector, non-consumptive water rights have been subject to problems of speculation, concentrated ownership, and private monopoly power.

The mixed performance of Chilean water markets is due to the variety of factors that shape these markets' wider social, institutional, and geographic contexts. Institutional arrangements – the rules of the game – have been among the most important of these factors. The Water Code's *laissez-faire* definition of property rights has obviously had a strong impact on the specific economic incentives and disincentives that are faced by water users and water rights owners. A core lesson here is that private property rights and market forces do not always go together. Many water rights owners (or would-be owners) consider legal security to be much more important that market pricing and exchange (Bauer 2004).

Comparison with integrated water resources management today

The most negative results of the Water Code have involved issues that were of little concern in Chile in 1981 but have emerged as ever more critical since the early 1990s. These are the economic, environmental, and social problems that are at the heart of contemporary international debates about integrated water resources management and water governance:

- coordination of multiple water uses in managing river basins; integrated management of surface water and groundwater;
- resolution of water conflicts through either judicial or non-judicial processes;
- internalization of both economic and environmental externalities;
- clarification, enforcement, and monitoring of the relationships among different property rights and duties, e.g. the relationship between consumptive and non-consumptive water rights;
- environmental flows and ecosystem protection; and
- public assistance to poor farmers to improve social equity in matters of water rights and water markets.

These problems are inter-related and must be addressed within a common institutional framework. Under the current framework in Chile, these issues have generally been addressed in an ad hoc or ineffective manner and in some cases they have not been addressed at all. Many of the flaws in the existing framework have been widely recognized by Chilean water experts, regardless of their political viewpoints. In some areas the empirical research is still missing or insufficient to draw definitive conclusions, although the available evidence indicates at least that there have been problems.

Research on the Chilean Water Code has had strengths and weaknesses. The relatively single-minded focus on water markets and water rights trading means that we understand how they have worked much better than before, and this is an important step forward. This focus has also had significant blind spots, however. Our improved understanding of markets has come at the cost of other issues of water management and institutional performance that are at least as important. There is much more to water resources management than simply the allocation of resources – whatever the allocation mechanism may be. In part this is because the notion of "efficiency" of allocation is often a misleading objective, to which I will return in my conclusions.

Because the Water Code did not address the broader economic, environmental, and social problems that are important today, it may be unfair to criticize the code for its failure to solve them. But that is not the point here. Rather, the current legal and institutional framework – which is determined by the Constitution as well as by the Water Code – has shown itself incapable of handling or adapting to these unforeseen problems. The current framework, as we have seen, is characterized by a combination of elements that reinforce each other to maintain the status quo: strong and broadly defined private economic rights; tightly restricted government regulatory authority; and a powerful but erratic judiciary that is untrained in public policy analysis, reluctant to intervene in issues with political implications, and committed to a narrow and formalistic conception of law. The problems of water management will only get worse as the demands and competition for water increase, putting ever more pressure on the existing institutional framework.

Water Code reform in Chile

In view of the mixed results and the problems identified, what are the prospects for improving the current framework in Chile? Chile's legislature finally approved some changes to the Water Code in 2005, after nearly 15 years of political debate and stalemate. The debates were ideologically charged, revolving around fundamental issues such as the nature of private property, the institutional framework for markets, and the limits of government regulation. Since Chile returned to democracy in 1990, three successive governments of the Concertación coalition tried to moderate the neoliberal approach of the Water Code. Over that period, the scope of the government's proposed reforms narrowed steadily in response to strong political opposition from conservative political parties and private sector business interests. At the same time, the government's own position on water markets gradually became more favorable (Bauer 2004, 2005).

The 2005 reform consists mainly of incremental improvements in water law and administration, including provisions to improve water rights title information and record-keeping, to strengthen management of groundwater, to strengthen the DGA's regulatory authority over future grants of water rights (but not over existing rights), and to begin to address the problem of minimum ecological flows. These provisions are worthwhile but have limited scope (Bauer 2009).

The most important and controversial change was the establishment of "fees for non-use," which must be paid to the government annually by any water rights owner who has not yet put his or her new rights to concrete use. The goal of these fees is to prevent private speculation, hoarding, and monopoly power over water rights – all problems for which the 1981 Water Code was criticized – and to encourage water's productive use. The fees were designed to apply primarily to non-consumptive water rights – reflecting the high priority placed on hydropower development – and to effectively exempt most consumptive water rights, which are used for irrigation or urban supplies. In their current form, the fees for non-use make it costly for owners of water rights to allocate them to environmental flows (Prieto and Bauer 2012).

Implementing the fees for non-use was expected to activate water markets by inducing people to give up their unused water rights, which the DGA would then auction to new owners. We do not know how well this new system has worked, since it is still early and empirical research is lacking (Valenzuela 2011).

In the bigger picture and from an international perspective, however, the 2005 reform was decidedly modest. It tinkers with the existing legal rules and institutional framework but barely touches the core principles of private property rights, market forces, and a weak state. River basin governance and coordination of multiple water uses remain untouched. Hence, the reform does very little to improve the capacity for integrated water management. In fact, when the reform was finally passed, it was partly because of Chile's on-going electricity crisis, not because of broad political consensus about water policy. The urgent need to stimulate hydropower development helped to overcome the political opposition to modifying the water law. Whether the reform will have much concrete impact on the water rights system or on water governance remains to be seen (Bauer 2009).

Any additional reforms to water law are politically unlikely in Chile for years to come. Water issues are still hotly contested in Chile and some politicians continue to call for more complete reform, sometimes called the "nationalization" of water, but in the current context this is a rhetorical position rather than a practical alternative.

Lessons for international water reforms

Chile's experience shows the problems that can flow from implementing a free-market water law. Its narrow economic approach has led to policies and institutional arrangements that cannot meet the challenges of integrated and sustainable water resources management. To avoid such outcomes, international efforts to reform water policies must foster a broader and more interdisciplinary approach to water economics, with more legal, institutional, and political analyses of markets and economic instruments. My hope is that this analysis of the Chilean experience will help raise the level of international debate about IWRM, particularly about its economic and institutional aspects.

The Chilean model has had two main economic benefits: first, the legal security of private property rights has encouraged private investment in water use, for both agricultural and non-agricultural uses; and second, the freedom to buy and sell water rights has led to the reallocation of water resources to higher-value uses in certain areas and under certain circumstances. These are important benefits, even though market incentives themselves have been only partly functional in practice, and they are the kind of results that pro-market policies hope to deliver. Many other countries would improve their water use and management if they could achieve similar results.

However, those economic benefits are directly linked to a legal, regulatory, and constitutional framework that has proven not only rigid and resistant to change but also incapable of handling the complex governance problems of river basin management, coordination of multiple water uses, water conflicts, and environmental protection. These more complex problems, of course, are precisely the fundamental challenges of integrated water management. In addition, peasants and poor farmers in Chile have for the most part not received the economic benefits, which indicates that social equity is another weak point of the current framework.

The strengths of the Chilean model, in other words, are also its weaknesses: the same legal and institutional features that have led to the model's success in some areas have effectively guaranteed its failure in others. The ability of the institutional framework to resist significant reform is the strength of rigidity. Its regulatory limitations, incomplete price signals, and lack of balance have been built to last. Even if we put the most positive spin on the model's economic benefits, for the nation as a whole these flaws are a high price to pay, particularly over the long term.

This analysis points to the inaccurate manner in which the Chilean model has often been described in international water policy circles, particularly by the economists who have been the model's strongest supporters (see the introduction to this chapter). The Chilean experience is cited as an example of successful free-market reforms, and although supporters may recognize some problems in the model, they present them as secondary issues that do not affect the overall positive assessment. The absence of effective institutions for managing river basins or resolving water conflicts is mentioned as an afterthought, or played down as a separate issue that can be addressed later, or seen as acceptable in light of the model's presumed advantages. According to this viewpoint, the problems have been identified and the Chilean government is in the process of solving them. In this way the spotlight remains on water markets and water rights trading, which are supposedly the model's strong points. The message to other countries is that they can adopt the Chilean model of water markets without also adopting its deeper institutional weaknesses in other areas of water resources management.

Such assessments are mistaken or misleading and are based on insufficient knowledge about Chilean politics and institutions. The flaws in the Chilean model are structural: they are integral parts of the same legal and institutional arrangements that underlie the water market. These flaws are not separable from the rest of the model. On the contrary, they are the necessary institutional consequences of the free-market reforms of property rights and government regulation. The aspects of the model that privatize water rights so unconditionally and define them as freely tradable commodities are inextricably connected to the aspects that weaken and restrict the regulatory framework. This is not a theoretical matter: in Chile these structural connections have been demonstrated in practice over the past 30 years, both by the mixed empirical results of the Water Code and by the long and difficult process of attempted Water Code reform.

Another way of putting this argument is to return to the familiar image of IWRM and sustainable development as a tripod, whose three legs are economic efficiency and growth, social equity, and environmental sustainability. The Chilean model of water management has one strong economic leg and two weak social and environmental legs, making it unbalanced overall. The social and environmental legs cannot be strengthened without weakening the economic leg in ways that are politically and constitutionally very difficult to do. Moreover, even the economic leg is weaker than it appears, because the ineffective mechanisms for resolving conflicts and internalizing externalities also reduce economic efficiency and growth, especially over the long term. Because the Chilean approach to managing water as an economic good puts all the emphasis on water as a private good and tradable commodity, it is very difficult to recognize or enforce the other aspects of water as a public good.

These are sobering lessons for people in other countries and in international organizations who are interested in water law and policy reforms. The most obvious warning is that other countries should not copy or closely follow the Chilean water law model, at least not without a thorough understanding of the model's weaknesses as well as its strengths. Both the flaws and the rigidity of the Chilean institutional framework should be made clear to other countries interested in the Chilean approach. The problems for IWRM should not be presented as secondary, as separate from the water market, or as readily solved. The failure of the World Bank and other boosters to make this clear in their advocacy of the Chilean model has been highly irresponsible.[1]

Many of the details of the Chilean model are unique to Chile and have been shaped by the local political and economic context, particularly by the Chilean Constitution. It is also probable that this charter has an unusual weight and importance in Chile that is matched by few constitutions in other countries, since the Chilean Constitution dictates the basic rules for economic institutions, as well as for the political system. It might seem possible to create an "improved" version of the Chilean Water Code by keeping the model's better features and avoiding its more serious flaws. According to this argument, if Chile went too far in the free-market direction and is now locked in place by its history, politics, and constitution, that is Chile's problem – it does not prevent other countries from learning from Chile's mistakes.

From an institutional perspective, however, the prospects for such an improved version in other countries are slim. Regardless of the specifics of the Chilean case, any country that tries to follow the *laissez-faire* economic approach of Chilean water law will necessarily confront similar institutional and political problems. How is it possible to create a legal and institutional framework that provides such strong guarantees for private property and economic freedom, and such wide scope for free trading of water rights and private decision-making about water use, without also severely restricting government regulation and legislative reform? If a country does not want to grant the judiciary such broad powers to review the actions of government agencies, how else can those agencies be prevented from interfering in water markets and property rights? If private economic rights are so strong and public regulation is so weak, through what institutional mechanisms other than the courts can conflicts be resolved effectively? How much room can there be for environmental protection in such a framework, and how can the level of that protection increase over time, given the strength of vested property rights?

If other countries want to follow the Chilean approach to water economics and markets, they will have to adopt a legal and institutional framework that is functionally equivalent to Chile's. If instead a country chooses a stronger regulatory framework or places more conditions on private rights, that country is, by definition, no longer following the Chilean economic approach. Hence, one of the deeper lessons of the Chilean water model is to show how different economic perspectives have different consequences for institutional design. The Chilean experience shows the lasting problems that result when a narrow economic perspective is combined with the political power to design legal institutions in the image of the free market.

In international debates about integrated water resources management, the principle that water should be recognized as an economic good should not be thought of as a separate or independent component of legal and policy reforms. In particular, countries and governments should not make the mistake of thinking that they can implement reforms in two steps, by first adopting a free-market approach to water economics as a straightforward initial step, and then turning their attention to the remaining problems of IWRM and water governance. At that later point, their hands will already be tied by a definition of property rights that has major political and institutional implications. On the contrary, reformers should put greater and earlier efforts into mechanisms for resolving conflicts and reflect carefully on how to define and enforce property rights to something as complicated as water.

We do not want to throw the baby out with the bathwater. I am arguing against unregulated markets and narrow economics, not against all use of market-based instruments in water management. The goal of a broader and more interdisciplinary approach to water law and economics is to build on conventional neoclassical principles, not to reject them. We can get the most benefits from markets by recognizing their limits and not asking them to handle problems beyond their proper scope. That is why I think we need to revive older approaches to institutional economics, approaches that are rooted in qualitative and historical analyses and that draw on politics, law, and other social sciences in order to understand so-called "economic" issues. Such an approach is especially important in developing countries, where social and institutional contexts are significantly different from the developed countries, where most economic theory has originated and continues to be shaped.

The current global water crisis is driven by growing scarcity and growing conflict, two water problems that are ever more tightly bound together. Although economic principles can be powerful tools for dealing with water scarcity, legal and political institutions are the key to resolving water conflicts. Moreover, scarcity is not simply a physical problem – rather, it depends on social context and is often driven by social factors more than physical factors (Aguilera *et al.* 2000). This further underlines the importance of legal and political institutions in shaping the use of economic principles. In short, the Chilean experience confirms the need for a more critical and interdisciplinary approach to water law, economics, and policy.

Conclusion

I want to close by returning to the four policy questions that I posed at the beginning of the chapter. Although the Water Tribune's mission is to promote innovative solutions and "new models" for dealing with water problems, part of this mission is to "recover forgotten valuable knowledge" (Expo Zaragoza 2008). In that spirit I recommend that people take another look at the older schools of historical and institutional law and economics, especially as they grappled with notions of nature and property.

Most of this chapter addressed questions related to efficiency of water markets and water allocation. My analysis drew heavily on John Commons (1924), Daniel Bromley (1989), and Karl Polanyi (1944), to name a few, who showed that law, politics, and society determine economic value, as they decide the "rules of the game" that shape how markets work and what incentives people face. Bromley's argument about economic efficiency still seems devastating to me, decades after I first read it in graduate school: economic efficiency cannot serve as an objective guideline for public policy because it assumes an initial set of institutional arrangements and distribution of wealth, yet dealing with those institutions and that distribution are the essence of public policy.

Water's mobility, changeability of form, and environmental functions make it a hard resource to commoditize, except by causing significant social and environmental costs. My question is whether the recent global interest in valuing and protecting "ecosystem services" is a trend that runs counter to the commoditization of water, or whether the two trends are instead two sides of the same coin. Here too I think Polanyi (with his mention of Aristotle) and the environmental historian William Cronon (1991) are good points of reference for the gap between physical reality and abstract forms of value.

Finally, I am no expert on mathematical models or their efforts to quantify aspects of the world and the human condition. But it is clear that in such models it is the basic assumptions, definitions, and categories that influence the results as much as the mathematical content itself. I prefer telling stories, and I think story-telling is more important for how people make decisions, about water management or most other things.

Note

1. In 2011 the Bank issued a diagnosis of water management in Chile that agrees in most respects with my description here (World Bank 2011). The report is in Spanish and its impact on other Bank operations is not clear.

References

Aguilera, F., Perez, E. and Sanchez, J. (2000) "The Social Construction of Scarcity: The Case of Water in Tenerife (Canary Islands)." *Ecological Economics* 34: 233–45.

Bauer, C. (1998) *Against the Current: Privatization, Water Markets, and the State in Chile.* Boston, MA: Kluwer Academic Publishers.

Bauer, C. (2004) *Siren Song: Chilean Water Law as a Model for International Reform.* Washington, DC: RFF Press (also published in Spanish as *Canto de Sirenas: El Derecho de Aguas Chileno como Modelo para Reformas Internacionales,* Colección Nueva Cultura del Agua no. 13, Bilbao, Spain: Bakeaz).

Bauer, C. (2005) "In the Image of the Market: The Chilean Model of Water Resources Management." *International Journal of Water* 3: 146–65.

Bauer, C. (2009) "Dams and Markets: Rivers and Electric Power in Chile." *Natural Resources Journal* 49(3/4): 583–651.

Briscoe, J. (1996) "Water as an Economic Good: The Idea and What it Means in Practice." Paper presented at World Congress of International Commission on Irrigation and Drainage, Cairo, Egypt.

Briscoe, J., Anguita, P. and Peña, H. (1998) "Managing Water as an Economic Resource: Reflections on the Chilean Experience." Environment Department Paper 62. Washington, DC: World Bank.

Bromley, D. (1989) *Economic Interests and Institutions: The Conceptual Foundations of Public Policy.* New York: Basil Blackwell.

Commons, J. (1924) *Legal Foundations of Capitalism.* New York: Macmillan.

Cronon, W. (1991) *Nature's Metropolis: Chicago and the Great West.* New York: W.W. Norton.

Expo Zaragoza (2008) "An Expo Without Sell-By Date." *Water Tribune* p. 5.

Global Water Partnership (2000) *Integrated Water Resources Management.* Technical Advisory Committee Background Paper 4. Stockholm, Sweden: Global Water Partnership.

Polanyi, K. (1944) *The Great Transformation: The Political and Economic Origins of our Time.* Boston, MA: Beacon Press.

Prieto, M. and Bauer, C. (2012) "Hydroelectric Power Generation in Chile: An Institutional Critique of the Neutrality of Market Mechanisms." *Water International* 37(2): 131–46.

Rogers, P. and Hall, A. (2003) *Effective Water Governance.* Technical Background Paper 7. Stockholm, Sweden: Global Water Partnership.

Valenzuela, C. (2011) "Efectos de la Aplicación de la Patente por No-utilización de los Derechos de Aprovechamiento de Aguas." B.S. thesis, Universidad de Chile, Santiago, Chile.

World Bank (2011) *Chile: Diagnóstico de la Gestión de los Recursos Hídricos.* Washington, DC: World Bank.

9 Breaking the gridlock in water reforms through water markets

Experience and issues for India

Nirmal Mohanty and Shreekant Gupta

Introduction

In recent decades India has witnessed rapid growth in demand for water, particularly in domestic and industrial sectors, due to population growth, urbanization, industrialization, and rising incomes. This growth in demand has not been matched by an increase in supply. The problem is compounded by pollution of water, which has reduced its suitability for various uses. Under these circumstances, it is more important than ever before to use water efficiently. It is also necessary to anticipate and address inter-sectoral conflicts over allocation and use of water.

The standard approach so far has been to advocate reform of water pricing across sectors to reflect the scarcity value of water and the cost of service provision. Nevertheless, major users of water particularly of irrigation water have resisted these reforms so far, resulting in inefficiency in water use, persistently low quality of water services and in some cases, loss of sustainability.

In this context, economic theory tells us that markets increase economic efficiency by allocating resources to their most valuable uses. In other words, if certain conditions are met, markets provide the correct incentives and lead to efficient resource use. Therefore, one way to change the incentives so that water users support the reallocation of water, and to achieving a more efficient allocation of water is through *water markets*. These allow water users to buy and sell water, thus changing the whole incentive structure and breaking the logjam of water pricing reforms – when water users can gain from reallocation they would be willing to sell water or pay a higher price for new supplies.

The aim of this chapter is to explore the reforms necessary for water markets to evolve in a manner capable of addressing India's emerging challenges. The following section elaborates on the concept of water markets and their rationale. We discuss the deficiencies of the current systems of water allocation and how water markets could be an improvement over them. In particular, we draw out the advantages of water markets over administered efficiency pricing (i.e. pricing marginal units of water at their marginal cost). We then review experience of informal water markets in India, before discussing the emerging challenges in the context of the current management framework. After describing briefly how formal water markets could be an improvement over informal markets, we identify the legal and institutional measures required to establish formal markets and recent reform measures in this direction.

Water markets and their rationale

Usufructuary rights to water have evolved either explicitly through laws and regulations or implicitly through conventions. These water rights are generally based on one of three systems: first-come, first-served allocation (also known as *prior appropriation* rights), allocation based on proximity to flows (or *riparian* rights) and *public* allocation (Sampath 1992; Holden and Thobani 1996; Haddad 2000). Whereas queuing for water is the basic approach of the prior appropriation doctrine, the location of one's land determines water rights under the riparian doctrine. Under this approach whoever owns land along (above) the water has the right to ownership/reasonable use of the water. Finally, public allocation involves publicly administered distribution of water: "Under this system, public authorities decide how to allocate water using guidelines or laws establishing priorities and often specify the uses to which the water can be put" (Holden and Thobani 1996: 2).

Most developing countries follow variants of the last approach where essentially the rights are allocated free – though there may be a charge for water use (typically based on the amount of irrigated area), the water rights themselves are obtained without charge.[1] The track record, however, of administered systems of water allocation has not been impressive – water is typically underpriced and wastefully used and the delivery is high cost and unreliable (see Holden and Thobani 1996 for details).

While this is well known, the important point to note here is that none of these systems fulfill the conditions for well-defined property rights to water, which in turn are essential for water markets to exist. In this context, the question could well be asked "why not use administered efficiency-based pricing of water as an intermediate policy between managed quantity allocation and water markets?"

There are three reasons why water markets could be preferred to administered efficiency pricing (i.e. pricing marginal units of water at their marginal cost):[2]

1. Reduction in information costs – whereas it is theoretically possible to devise and implement a system of administered prices which would lead to efficient allocation of water, the information requirements are demanding and may require experimentation by trial and error.
2. Perhaps more important, if the value of prevailing usufructuary water

rights (formal or informal) has already been capitalized into the value of irrigated land, then imposition of administered pricing is (correctly) perceived by right holders as expropriation of those rights. In effect, this would result in a capital loss for irrigated farms. This could explain the strong resistance by these groups to establishing administered efficiency prices. Under the water market system, establishment of transferable water rights would formalize the existing situation (where irrigated land is more expensive), rather than being viewed as a usurpation of these rights. Thus, their implementation should be more feasible politically as compared to administratively imposed water pricing reforms.

3. The administrative solution presumes "far-seeing, incorruptible, influence-free" (Holden and Thobani 1996: 5) administrative bodies that are able to design and implement the "correct" prices. In practice, this may often not be the case: these bodies could be captured by interest groups or they may be short-sighted and unable to estimate future demand, or they may be unable to set and collect appropriate water charges. Advocates of administrative approaches, however, often ignore these problems.

For water markets to work, property rights to water must be private, exclusive and transferable (Bauer 1997). In this context, secure ownership provides an incentive to invest in greater productivity of the resource, while freedom to exchange provides the flexibility to reallocate the rights according to changing demand and other conditions. The role of the state should be minimal in this setting and should be restricted to protecting property rights, enforcing contracts, and reducing transaction costs and barriers to exchange. In fact, it can be argued that much of the current inefficiency in the water sector in India is due to excessive state regulation and subsidies which have distorted patterns of water use. As a corollary, then, freer markets would help in "getting the prices right," and in strengthening the incentives to conserve water as demand increases since any water saved could be sold.

Another important rationale for water markets is the *relationship between markets and liberty*. In contrast to non-market allocation which gives the state leverage in non-economic spheres as well, private property creates a space for individuals where the state cannot trespass. "Private property [the necessary precursor to markets] has thus been viewed . . . as a bulwark against the dictatorial authority of governments" (Cooter and Ulen 1996: 109).

In this context, by creating entitlements where none existed earlier, these markets can be a *tool for empowerment*. Holders of water rights would be sought after (irrespective of their socio-economic status), by those who would like to buy these rights.

Similarly, *environmentalists can also purchase water rights* in order to preserve a valued wetland or to increase a waterway's flow. Without a market mechanism, environmental groups would have to depend on the state to achieve the same end (Haddad 2000). In fact, this has already happened in several states in western United States – the Oregon Water Trust, the Washington Water Trust, and Nevada's Great Basin Land and Water are three recent groups that have used water markets to acquire water rights and convert them into instream flows. The Oregon Water Trust for instance has been able to increase flows on more than 25 different streams and rivers flowing into the Columbia River (Landry 1998). In sum, the private space created by market mechanisms can be used to achieve socially valued ends.

Before we review the Indian experience with water markets, it is important to distinguish between formal and informal markets. Under the former, water rights are clearly and universally assigned, with legal validity for freely negotiated sale of these rights. In case of informal markets, there is neither clear assignment of rights nor legal sanction to trade. Thus, in formal water markets enforcement of trades occurs by recourse to legal and institutional measures, whereas in the case of informal markets (which simply arise from spontaneous response of water users to changes in demand-supply situations), such recourse is not possible (Easter *et al.* 1998). Also, formal markets are often defined with respect to water rights, while informal markets operate for volume of water.

Extent of water markets in India

Water markets that exist in India are informal and are generally limited to localized water trading between adjacent farmers and the practice is quite common for groundwater. Although found in many parts of India, the occurrence of groundwater markets is not uniform. While water markets are widespread in Gujarat, Punjab, Uttar Pradesh, Tamil Nadu, Andhra Pradesh and West Bengal, they are most developed in Gujarat. The extent of area irrigated through water markets, which is often considered to be a surrogate for the magnitude of water trading, varies across regions as well as over time depending on a number of factors such as rainfall, groundwater supply, cropping patterns, and the cost and availability of electricity (Saleth 1994). In water-scarce pockets of Gujarat, Tamil Nadu, and Andhra Pradesh, a substantial area is irrigated through groundwater markets.

Several micro studies illustrate the degree of variation in use of water trading in India. In terms of area irrigated through groundwater markets, estimates vary from 80 percent for Northern Gujarat (Shah 1993) to 60 percent in Allahabad district in Uttar Pradesh (in a 16-village sample study; Shankar 1992) to 30 percent in the Vaigai basin, Tamil Nadu (Janakarajan 1994). Some studies report no water trading in their study area (Shah 1993; see also Shah 2009).

There is no systematic estimate at the national level of the magnitude of water trading. The area irrigated through water markets has been projected to be about 50 percent of the total gross irrigated area with private lift irrigation systems (Shah 1993). Other estimates, using a methodology based on pumpset rental data, put the figure at 6 million hectares or 15 percent of the total area under groundwater irrigation (Saleth 1999).[3] Assuming a net addition to output of US$230/ha/year (based on the difference between the average irrigated and rainfed yields as reported by Government of India 1999), Saleth estimated the total value of output due to water sales at US$1.38 billion per year (Saleth 1999).

Nature and characteristics of informal water markets in India

A review of the functioning of informal water markets in India can improve our understanding of the market and provide useful insights, which could form the basis for designing formal markets.

Localized and fragmented

As stated earlier, water markets in India are mainly limited to the irrigation sector – that is, one irrigator selling water to another irrigator. Water trading in India is localized, fragmented and is over short distances and periods. Unlike in Chile, western USA, and Australia, the institutions, legislation and regulatory framework do not exist in India for more formal transactions. In some rare cases, however, water purchases for non-irrigation uses have been reported. For example, brick manufacturers purchasing water have been reported by Shankar (1992).

Mainly driven by surplus supply

Most water sales do not involve any reduction in irrigation by sellers (Saleth 1999). Most of the sellers are large farmers owning deep wells and large capacity pumpsets and the buyers are usually small farmers without wells or pumpsets, though there are non-poor farmers who rely on groundwater markets due to farm fragmentation or inadequacy of water in own wells. By providing access to use of groundwater and irrigation assets to resource poor farmers, groundwater markets have promoted equity.

Monopoly power

The existing informal markets are small and unbalanced and are typically characterized by a weak bargaining position for buyers. Buyers often do not have a choice because of low density of wells, compounded by uneven topography and potential for seepage losses (Shah 1993), which gives sellers a degree of monopoly power. Further, there is evidence of buyers being tied down to sellers from contiguous plots, as sellers can and do refuse conveyance of water through their plots to other possible suppliers

(Janakarajan 1993, 1994). Monopoly power helps sellers not only in raising prices but also in compromising the quality of service they offer.

Influenced by social factors

Social factors and agrarian relations sometimes determine the development of water markets. For example, in Bihar it has been found that it was the water buyers' position in the social network, particularly their social proximity to sellers – rather than their ability to pay – that determines their access to water (Wood 1995). Moreover, there were several cases of price discrimination with prices being lowered for favored clients. In Paldi village in Gujarat, there is evidence of many water transactions being "bundled into existing landlord–tenant relations" (Dubash 2000). Thus, out of 20 wells sampled, 11 sold water – five separately and six to tenants.

Widely varying terms of payment

Terms of water payment vary widely and differ by crop and by season. Payments can be made through cash transaction or non-cash contracts. Cash payments are made on the basis of time, volume or area irrigated. Hourly price ranges between 3 rupees in West Godavari district of Andhra Pradesh to 45 rupees in Mehasana district of Gujarat (Shah 1993). Non-cash contracts, which typically take the form of sharecropping (i.e. seller collects a water rent in the form of a share of the buyer's output), are not uncommon.[4] They have been found to be incentive-compatible (Aggarwal 1999). These contracts work as "double-sided" incentive, providing the seller an incentive to ensure that water supply is timely and reliable and the buyer an incentive not to shirk in the application of labor. Sometimes the market displays a feudal character. In Tamil Nadu, there are cases where water buyers have to offer labor services such as operating the pump and irrigating the well-owners' fields for a paltry sum or none at all (Janakarajan 1993, 1994).

Groundwater overexploitation

There is some evidence of decline in groundwater table caused by competitive water withdrawal due to intense water marketing activities (Moench 1992). Under the current legal system, there is "open access" to groundwater (see below) and every landowner has an incentive to pump as much as possible, since what it is not pumped remains to be appropriated by someone else. This results in "the tragedy of commons where each user tries to maximize his/her own share winds up lowering everyone's share. When groundwater gets lowered it increases costs for all as they need to deepen their wells and require more powerful motors" (Planning Commission 2007). Further, since farmers are faced with zero marginal cost for pumping (see below), over-exploitation cannot be avoided. Clearly,

Box 9.1 Large-scale purchase of groundwater by Chennai Metro Water

Chennai Metro Water is a state-owned water utility in Chennai, a city in south India. Between 2001 and 2004, when drought conditions prevailed, Metro purchased water from farmers in the peri-urban areas using PVC pipes (through annual contracts) as well as water tankers (without contracts; on ad hoc basis). These were voluntary transactions and varied from 13 percent of Metro's supply during 2002 to 58 percent in 2004. Rates were fixed through negotiations at about 26 rupees per hour of supply (equivalent to about 1 rupee per kl as compared to reportedly 48 rupees per kl from desalination plant currently under construction in Chennai). Under the contract system, farmers pumped water into PVC pipes laid by Metro, which was collected and distributed by the Metro. In addition to conveyance; the cost of power used in pumping was also borne by the Metro. Under the tanker system, Metro paid about 5–6 rupees per kl and also bore the cost of transportation. The practice has been kept in abeyance since 2005, when rainfall and availability from other sources improved.

This was perhaps the only instance, where a formal segment of economy accessed groundwater from farmers on a large scale. The terms of these transactions were attractive not only for farmers, who managed to sell large quantities by reducing cropped area in dry seasons, but also for Metro, which could tide over a period of extreme scarcity.

The practice had a severe impact on the economy of the peri-urban areas that supplied water to Metro, however. The prospect of high income through water sales prompted several farmers to resort to over-pumping leading to water tables falling to extremely low levels, affecting all farmers including the sellers. Further, the change in land-use pattern led to falling agricultural employment and serious livelihood problems (Janakarajan *et al.* 2007). Many landless laborers and small farmers migrated to cities in search of employment, putting additional pressure on cities' strained infrastructure.

An important lesson from the Chennai experience is that although groundwater from peri-urban areas is an attractive proposition (for both urban local bodies and prospective sellers) and large scale transfers are feasible, the consequences for such transfers on the peri-urban areas can be severe, unless there is a sound framework to manage extraction in a sustainable manner.

Source: partially based on discussion with Chennai Metro Water officials under the current system, it is natural to expect overexploitation of groundwater in water-scarce areas; the presence of groundwater markets only exacerbates it. In addition to reducing ecological sustainability, one important side effect of this phenomenon is that poor farmers who do not have the resources to deepen their wells are driven out of farming.

Ineffective and iniquitous regulation

Only a few states that are severely affected by groundwater extraction have opted for groundwater acts. These legislations rely typically on state imposed control mechanisms (permits for digging new wells over limited area and limited period of time, spacing and depth norms), with little emphasis on cooperative management. While there is no legal basis for trading, restrictions on withdrawal can potentially affect trading indirectly. These regulations, however, have been ineffective because of poor enforceability resulting from inadequate supervisory resources with states to deal with the large number of wells that are in operation.[5] Individual farmers can also render some of these regulations ineffective. For example, restrictions on the number of tube wells can be overcome by increasing the power of the pumpsets. The legislations are also iniquitous: while recognizing the rights of those who already own wells, they exclude others (Planning Commission 2007).

Irrigation–electricity nexus

Groundwater pricing is indirect and is determined mainly by the cost of electricity which accounts for the dominant part of marginal cost of pumping water. For electric pump sets, charges are levied on a flat basis per month in proportion to the horsepower of the pump set.[6] As marginal cost is zero, most farmers tend to use water inefficiently. Further, state power utilities find it difficult to raise flat tariff for years on end under pressure from Chief Ministers of states, who see pricing of power for agriculture as a "powerful instrument in populist vote bank politics" (Shah *et al.* 2007). Thus, "below cost, often free, and unmetered electricity supply for agriculture has contributed to an erosion of electricity distribution systems and also encouraged wasteful groundwater use" (Dubash 2007). To the extent water markets have encouraged excessive pumping, they have contributed to the deterioration in the financial health of the power sector.

Limited interface with formal sector

There are limited instances of the formal sectors of the economy (such as cities and industrial establishments) accessing groundwater market. This is typically done through water tankers, which are usually operated by small units encompassing around ten people; the impetus for operators to enter into this activity is the prior possession of at least one factor of production: land, tankers or wells (Llorente and Zerah 2003). The water so distributed is not subject to any quality control. Municipalities hire water tankers from private firms to supply water to residents in emergency situations and also in newly constructed residential areas where waterpipe network is unsuitable and inadequate. Large industrial establishments also use this service. According to a survey conducted in Delhi in 1998, about 25 percent of industries and institutions surveyed were found to be dependent on private tankers on a fairly regular basis (Llorente and Zerah 2003). Even though private water tankers play a key role in supply to some users, they account for a small fraction of cities' total water supply. In a rare instance, the water utility for the metropolitan area of Chennai (Chennai Metropolitan Water Supply and Sewerage Board or Chennai Metro Water) had scaled up its reliance on groundwater purchases from farmers; the experience of Chennai Metro Water is discussed in Box 9.1.

Emerging challenges

To examine the context in which the role and nature of future water market will get determined, it is important to take note of the emerging challenges. Water scarcity is already a significant issue and is likely to get aggravated in future: "Indeed, by 2025, by many accounts, much of India is expected to be part of that one-third of the world which is expected to face absolute water scarcity" (Shah 2007). Signs of growing water scarcity are already visible. For example, the share of blocks (i.e. assessment units) of the country that are classified as over-exploited has increased from 5 percent in 1995 to 15 percent in 2004, making overexploitation of groundwater a matter of serious concern (Planning Commission 2007).[7]

Scarcity has also manifested itself in the form of growing conflicts, which have reached several levels: user groups, sectors, rural and urban areas, political parties, states, groups and individual farmers. These conflicts are likely to worsen and pose "a significant threat to economic growth, social stability, security and health of the ecosystem and the victims are likely to be the poorest of the poor as well as the very sources of water: rivers, wetlands and aquifers" (Gujja *et al.* 2006).

The second important challenge, moving forward, is the changing nature of water demand: "Industries and cities (which both require water and produce wastes) are growing rapidly. Rural life is changing, with more than half of the people in rural Punjab and Haryana no longer engaged in agriculture" (Briscoe and Malik 2007). According to official estimates, the demand for water from domestic and industrial sectors is projected to rise much faster in the coming decades than from the agricultural sector (see Table 9.1), implying that at the margin water will have to be increasingly reallocated toward domestic and industrial sectors and away from agriculture.

Table 9.1 Projection of demand for water (km[3])

	1997–8		*2010*		*2050*	
	Amount	*Share (%)*	*Amount*	*Share (%)*	*Amount*	*Share (%)*
Irrigation	524	83	550	78	718	67
Domestic	30	5	43	6	101	9
Industries and power	39	6	56	8	148	14
Others	36	6	54	8	111	10
Total	629	100	702	100	1,077	100

Notes: Amounts for 2010 and 2050 are averages of "high" and "low" scenarios for the respective years. "Others" includes inland navigation, environment and evaporation loss
Source: Government of India (1999)

Both of these issues call for significant changes in water resource allocation and improvement in water use efficiency, particularly in the irrigation sector which accounts for a dominant share in total water use. Will these be possible within the management framework currently being pursued? In surface irrigation, which accounts for about 60 percent of total irrigation use, water user contributions represent less than half of actual operational and maintenance expenses and in some states, only 5 percent (World Bank 1999a), reflecting low tariff (due to political reasons) and unwillingness among farmers to pay in view of poor irrigation services.[8] Not surprisingly, a large part of the canal network is in disarray. In many parts of India, farmers at the "head end" (typically rich) get excessive water and tend to use water inefficiently, while those at the "tail end" (typically poor) hardly get any water. Reform efforts in recent decades have focused on cost recovery. Without addressing the accountability question, exhortations to increase cost recovery, however, have largely failed. The surface irrigation sector is thus caught in a logjam, in which services are poor, farmers do not pay and service quality declines. Groundwater, in contrast is in private domain and capital costs and operational and maintenance costs of extraction – albeit highly subsidized – are borne by the well-owners. As stated earlier, zero marginal cost together with the open access common property character of groundwater and ineffective regulations have led to unsustainable extraction. While water is being inefficiently used in irrigation sector, rising demand from cities and industries are being met from new supplies mainly to avoid conflict situations.

Clearly, the current sector framework is not in a position to cope with the problems at hand. Further, focus on tariff reforms, while necessary for financial sustainability of infrastructure, will not be able to fully address the emerging issues, because the tariff revision required to reflect scarcity value and attain allocative efficiency would be much higher than what would be politically feasible.[9] and can be effectively implemented. Given limited supplies and growing and changing demands, "the need is obviously for a management framework which stimulates efficiency and which facilitates voluntary transfer of water as societal needs change" (Briscoe and Malik 2007).

Introducing formal water markets in India

In India, there has been no explicit policy statement in favor of water markets. At the same time, though there is no legal basis for informal markets to exist and function, the state has followed a policy of non-interference *vis-à-vis* such markets. Under such a dispensation, when informal markets have grown and served a useful purpose, why do we then need formal markets? Some of the major benefits that a formal market is expected to yield are:

- This would allow water transfers to take place on a large scale and also between sectors, thus allowing a reallocation of water to higher productive use. For instance farmers instead of producing low-value, water intensive crops might sell water to a neighboring city if it fetches them a higher price. At the same time, the possibility of large-scale inter-sectoral transfers could postpone or make unnecessary construction of costly hydraulic infrastructure. La Serena city, for example, was able to meet its water needs by purchasing water rights from farmers and this was attained at much less cost as compared to building a dam.
- Second, the nature of informal markets is such that trading cannot be regulated. In contrast, since property rights are well defined in formal markets, trading can be regulated. Regulation can lead to better resolution of the negative side effects of trading such as aquifer depletion or monopoly creation or equity issues – which have acquired significant proportion in India – more effectively.
- Third, formal markets, based on an explicit water right system, can help potential investors and water companies gain secure long-term access to water. An important factor inhibiting private investment in water sector in India, particularly in urban water supply, is that potential investors do not feel confident of meeting their service obligations in the absence of secure long-term access to water.
- Fourth, legally well-defined and registered property rights reduce transaction costs involved in water trading. These costs include monitoring and enforcement costs, conveyance costs, and costs of designing contracts. Low transaction costs would encourage trade and thereby expand the scope of the market.

Clearly, formal markets that retain and extend the potential gains of informal markets and counteract many of their negative features, are preferable. Before introducing formal water markets with tradable property rights, however, some legal and institutional issues would have to be resolved. These are discussed below.

Legal and institutional measures

Manage surface water on a river basin basis

Indian law treats all surface water as state property. The fragmentation of basins by state boundaries and lack of cooperation between states is a critical issue for interstate water development and allocation. In the absence of legal clarity on what individual states' shares are, conflicts between states have grown. It is therefore important to introduce necessary legal arrangements to facilitate the management of surface water on a river basin basis.

Clarify legal position on individual usufructuary rights for surface water

There is also a lack of clarity on individual usufructuary rights for surface water, as the legislation has failed to devise a system for providing secure, defensible and enforceable surface water rights. Although courts have upheld the riparian rights – individuals abutting upon a (natural) stream can use water without disturbing a similar benefit to other riparians – as natural rights, individualized rights of abstraction and use of such water can only be established through time-consuming litigation (World Bank 1999b). Furthermore, states' sovereign rights over surface water have in the past been challenged in courts by riparian landowners, who claimed that their rights had been infringed upon by the government in pursuit of its irrigation projects (World Bank 1999b).[10] Unless surface water rights are better clarified and in favor of individuals, conflict and litigation will grow in the future and formal water markets will not be possible.

Separate rights to groundwater from rights to land

Under the law of riparianism applicable in India, ownership of groundwater accrues to the owner of the land above. By virtue of these laws, groundwater is "attached like chattel" to land property and cannot be transferred separately from the land to which it is attached (Singh 1992). This has constrained the potential for inter-sectoral allocation. To establish an active water market, rights to water use must be authorized separate from land.

Establish limits for withdrawal of groundwater

Under the current laws, there are no quantitative limits on groundwater withdrawal by individual users. This provision together with the provision of tying land rights with water rights has serious equity implications, because it allows larger farmers with higher pumping capacity and deeper tube wells to have a disproportionate claim over water than others. Further, sellers can get a payment from the very group whose water rights get infringed by the seller's activities (Saleth 1994). Besides, withdrawal limits will promote efficient water use. Furthermore, in a theoretical sense, an efficient operation of a market is critically dependent on the prior existence of an effective legal institution of property rights establishing the initial resource endowments of individuals. There is therefore a need to specify water withdrawal limits by individuals in volumetric terms.[11] Although establishing individual withdrawal limits can promote equity and efficiency, ecological sustainability requires collective withdrawal limits keeping in view annual recharge.

Broaden the market

It is not enough to give users the option to buy and sell water. Such options should be as numerous as possible to make the market competitive. This requires not only institutional and organizational changes but also improved canal infrastructure to make sure that trading can take place over a larger area – for example, by joining different systems. Similarly, management may have to be improved so that buy and sell orders are easily executed. Improved control structures are also necessary which would allow managers to easily increase the flow in one canal and decrease it in another.

Create conflict-resolving institutional arrangements

Further, institutional arrangements are needed for resolving conflicts over water rights. Committees of water users comprising elected representatives of the community created for cooperative management can play this role for disputes among their members. The institution to resolve conflicts among user groups and between user groups on one hand and states' irrigation departments on the other must have necessary independence and capacity.

Attempt to establish formal market: the case of Maharashtra

While the measures outlined above would be required in an ideal situation, not all of them are possible to implement at the current stage, especially those relating to groundwater. Yet it is possible to introduce formal water markets in a limited way, as has been demonstrated by the state of Maharashtra. Maharashtra has embarked on a pioneering reform initiative, which involves the creation of a formal water market (only in its irrigation command areas). The program is at an early stage and the rules of the game are still unfolding; yet it is useful to examine how it works and its implementation issues.[12]

The program rests on four pillars. First, the State Water Policy, 2003 sets priority among uses as well as principles of tariff setting. Second, the Management of Irrigation Systems by Farmers Act 2005 mandates the creation of Water User Association (WUA), a legal body with elected representatives. Third, the Maharashtra Water Resources Regulatory Act (MWRRA), 2005 ("the Act") has established a water resource regulatory authority ("the Authority") outside the Government. The primary objectives of the Authority are to determine, regulate and enforce distribution of entitlements (surface water) between uses and within each use, act as dispute resolution authority, set criteria for trading entitlements and fix bulk tariff. Finally, a separate bill for groundwater is being enacted with focus on regulation of groundwater extraction and creation of institutions for cooperative management in areas suffering groundwater depletion.

How does the system work? Allocation among uses is subject to priorities set in the State Water Policy. Under the Policy, as amended in May 2011, drinking water is given the top most priority followed by agriculture and industry respectively. (Earlier, industry enjoyed higher priority over agriculture; this meant that a part of agriculture's quota given at planning stage could be reallocated to industry at the implementation stage. After the policy amendment, this is no longer possible. A cut in agriculture's quota can only be undertaken to boost allocation for drinking water.) By government order, 15 percent of reservoir capacity is allocated for drinking purpose among urban and rural local bodies and 10 percent for industrial purposes at the planning stage. Allocation to these two categories is not affected by reduced availability in reservoirs except in drought years. Allocation to irrigation, however, is subject to availability in reservoirs even in ordinary years. (Thus while the right to irrigation water is permanent, its quantum varies from year to year.) The basis for allocation within a category of use also differs across categories. While allocation within local bodies (for drinking water) and industries is based on "first come, first served basis" and considerations of "reasonable use," allocation for agriculture is according to land ownership. In irrigation, entitlements are distributed to WUAs by specifying (i) volumetric allocation per hectare and (ii) volumetric quota for each WUA (based on land under WUA's jurisdiction). Mirroring allocation among WUAs, allocation to each farmer within a given WUA is made proportional to his or her land ownership.

To ensure transparency, gauge registers (i.e. measuring devices installed to deliver volumetric supplies to WUAs) are made available with both the Irrigation Department and WUAs for independent verification and independent regulators are appointed by the Authority to test check gauge readings. Allocation at the individual farmers level is operationalized through the instrument of "irrigation passes," which are sanctioned by the concerned WUAs. In matters relating to distribution of entitlements among WUAs, the Authority is the appointed body for dispute resolution, while WUAs are authorized to resolve disputes between individual farmers within their respective jurisdiction.

As regards trading of entitlements, the Act provides an enabling framework, although the contours of trading are still being worked out.[13] According to the Act, River Basin Agencies (RBAs) are the appointed bodies with which trades are to be registered. They have the authority to deny any proposed transfers on grounds of damage to third party rights or incompatibility with operation of projects.

The framework was initiated through 6 pilot projects in 2006, which were later scaled up. Currently, there are about 240 projects in the implementation program with around 800 WUAs. The experience gained from these projects is being used in framing rules to operationalize the new legal framework. Two important elements of recent experience are worth noting. First, even though the Act provides for determination of rights to subsurface water as well, it is not being pursued, as setting and monitoring of such rights have been found to be extremely difficult in practice. Difficulties arise from technical reasons, as aquifer flows are not very predictable. Further, such flows are continuous and beneath the land belonging to several people; a new well, for example, can affect "safe yields" of those who already own wells in the same vicinity. Given that wells are very large in number and widely scattered, enforcement would also be difficult. This has forced the government to formulate a separate bill for groundwater. Second, a large part of the canal network is in a state of disrepair and rehabilitating the network is costly and time consuming. Since well-functioning canals are a prime requirement for WUAs to be effective, WUAs has taken over distribution operations yet. While WUAs have begun to receive their entitlements, water distribution is still being predominantly done by the Water Resources Department.

During this limited period of implementation, a number of issues have cropped up, significantly:

- Large-scale investment is necessary upfront to have meaningful peoples' participation. The same holds for almost all states of the country. Do states have the necessary resources?
- There is no clarity on whether allocation to individual farmers constitutes legal rights. The current interpretation is that the Act provides legal rights to WUAs, but not to individual farmers. Without legal rights at the individual level, the benefits of trading would be limited.
- Although the Act allows trading across sectors, the present thinking is to limit trading to "among irrigation users" only. There would then be no scope currently for industries and local bodies to purchase water rights from WUAs. If on the other hand this position is only an intermediate stance before inter-sectoral trading is allowed, the bigger question arises: would sale of water by farmers to industries not entail a misuse of agricultural subsidies, which aim at increasing food security?
- According to the Approach Paper of the Authority, only up to 50 percent of Entitlement can be traded. This may in some cases prevent

farmers from reaping the full benefits of water saving techniques. For example, a farmer wishing to produce the same output as before, may do so with less than 50 percent of her Entitlement by using water saving techniques. The trading restriction would mean that she would then be left with some surplus water, which she would not be allowed to trade.
- In case of dispute resolution, it is not clear as to what should be the nature of compensation. Should it be monetary or in terms of larger allocation to the aggrieved party in the next irrigation rotation within the same season or even in the next season.

Despite these implementation issues, the reform program is a step in the right direction. The issues that have arisen can be resolved in the light of the lessons learnt from other emerging countries that have adopted similar reforms, but within the context of India's own political economy surrounding water. The reform program is appropriately moving in a gradual manner by building consensus among stakeholders and taking into account the administrative and political realities. A broad consensus has already emerged in the following three significant reform areas. The first relates to transparent distribution of entitlements. Several farmers who are at the tail-end of the irrigation system will benefit from this. Second, unlike in the past, allocation and tariff would no longer depend on crops grown. Entitlements are now given to WUAs depending on the land under their jurisdiction. This together with the new practice of volumetric supplies would create appropriate incentives. Third, the Water Resources Department would be subject to discipline because of active involvement of WUAs. While these three will yield significant benefits, the scope of trading in its current, restrictive form will not. Of course, as experience is gained and the market is better understood by water users, it would be easier to build consensus to widen the scope of trading.

Summary and conclusion

In India, water markets have been informal, confined mainly to trading of groundwater between irrigators. These markets are localized, fragmented and primitive, displaying feudal characteristics in some instances. Although these markets have led to some efficiency gains and have expanded the access to irrigation for many resource poor farmers, gains have been limited. Large scale transfers of groundwater from peri-urban areas to cities have created serious problems. Further, with difficulties in implementing effective regulation, these markets have in many instances compounded the problem of overexploitation. Meanwhile, the problem of scarcity is growing and the sector profile of demand for water is changing. These challenges have emphasized the need for significant changes in water allocation and efficiency improvement.

In this context, it is imperative that India explores the option of formal water markets that assign property rights to individual users, vastly expand the scope of trading and make large scale inter-sectoral water transfers possible. Since formal water markets have legal basis, they can be subject to adequate regulation. The markets will be of significant relevance to the urban and industrial sectors, which have been suffering from acute shortages of water, but have not been able to access informal markets. While tariff rationalization can improve the financial sustainability of infrastructure, they would need to be complemented by formal water markets to address the fast growing and changing water demand. A beginning has already been made in water markets by piloting Entitlement projects in the irrigation command areas of Maharashtra, albeit with limited scope for trading. With gain in experience and stakeholder consultations, the scope of trading can be expanded to a point that cities and industries can benefit. Given the practical difficulties in determining and monitoring property rights to groundwater for millions of wells in India, formal markets in groundwater is (appropriately) not being attempted.

Acknowledgement

This chapter is a significantly revised and updated version of an earlier publication, "Water Reforms Through Water Markets: International Experience and Issues for India," by Nirmal Mohanty and Shreekant Gupta, which appeared in the *India Infrastructure Report 2002* (pp. 217–25), published by Oxford University Press, New Delhi (www.idfc.com/pdf/report/IIR-2002.pdf). Reproduced with permission of the Infrastructure Development Finance Company Ltd (IDFC).

Notes

1. The water rights themselves under any of these systems are defined volumetrically as a share of the stream or canal flow or of the water available in a reservoir/lake, or in terms of shifts or hours of availability at a certain intake.
2. This discussion is based on Holden and Thobani 1996, and Rosegrant and Binswanger 1994.
3. It is assumed that pumpset rentals inherently involve water sales for all fixed pumpsets permanently fitted to wells or connected to electric power lines.
4. For example, in some parts of Gujarat water is provided to tenants by the land and well owners, where the buyer receives one-quarter of the crop, while the seller receives three quarters. Of the three quarters that water sellers receive, half is on account of land and one quarter is on account of water.
5. According to the report on the third Census of the Minor Irrigation Schemes (2005), India had about 18.5 million wells in 2001; the number is expected to have gone up since then.
6. As the number of tube wells increased in the 1970s and 1980s, states found metered billing costly and difficult (due to rampant meter tampering and corruption at the meter reader level) and switched to flat tariff system (Shah *et al.* 2007).

7. An over-exploited block is one where groundwater draft is higher than groundwater availability and the water table shows significant long-term decline in pre- or post-monsoon or both.
8. In 1997, Punjab made water and power free for irrigation!
9. Besides, the huge agricultural subsidies given by industrialized countries are compelling the farmers of developing countries including India to demand for subsidies on water, energy and other agricultural inputs.
10. Both the Madras High Court in 1936 and the Bombay High Court in 1979 have established that the Government's sovereign rights do not amount to absolute rights.
11. In addition to legislative efforts, quantification of ground water rights in an operational context requires technological changes.
12. Based on discussions with Mr. A. Sekhar of the Maharashtra Water Resources Regulatory Authority, Mumbai, Maharashtra, India
13. The Authority has, in June 2011, floated an Approach Paper indicating the contours of trading for public consultations.

References

Aggarwal, Rimjhim M. (1999) "Risk Sharing and Transaction Costs in Groundwater Contracts: Evidence from Rural India." University of Maryland, College Park, Maryland. Available at www.arec.umd.edu/libcomp/ARECLib/Publications/Working-Papers-PDF-files/96-17.pdf (accessed 25 July 2012).

Bauer, Carl J. (1997) "Bringing Water Markets Down to Earth: The Political Economy of Water Rights in Chile, 1976–95." *World Development* 25(5): 639–56.

Briscoe, John and R.P.S. Malik. (2007) "India's Water Economy: An Overview." In J. Briscoe and R. P. S. Malik (eds), *Handbook of Water Resources in India*. Oxford, UK: Oxford University Press.

Cooter, R. and T. Ulen (1996) *Law and Economics*. St. Louis, MO: Harper Collins.

Dubash, Navroz K. (2000) "Ecologically and Socially Embedded Exchange: Gujarat Model of Water Markets." *Economic and Political Weekly* 35(16): 1376–85.

Dubash, Navroz K. (2007) "The Electricity-Groundwater Conundrum: case for a Political Solution to a Political Problem." *Economic and Political Weekly* 42(52): 45–55.

Easter, K. William, Ariel Dinar and Mark W. Rosegrant (1998) "Water Markets: Transaction Costs and Institutional Options." In K. William Easter, M. W. Rosegrant, and Ariel Dinar (eds), *Markets for Water: Potential and Performance*. Dordrecht, The Netherlands: Kluwer.

Government of India (1999) *Report of The National Commission for Integrated Water Resources Development*. New Delhi, India: Ministry of Water Resources.

Gujja, Biksham, K J Roy, Suhas Paranjape, Vinod Goud and Shruti Vispute (2006) "Million Revolts in the Making." *Economic and Political Weekly* 41(7): 570–74.

Haddad, Brent M. (2000) *Rivers of Gold: Designing Markets to Allocate Water in California*. Washington, DC: Island Press.

Holden, Paul and Thobani, Mateen (1996) *Tradable Water Rights: A Property Rights Approach to Resolving Water Shortages and Promoting Investment*. Policy Research Working Paper no. 1627. Washington, DC: World Bank.

Janakarajan, S. (1993) "Triadic Exchange Relations: An Illustration from South India." *Institute of Development Studies* 24(3): 75–82.

Janakarajan, S. (1994) "Trading in Groundwater: A Source of Power and Accumulation." In M. Moench (ed.), *Selling Water: Conceptual and Policy Debate*

Over Groundwater in India. Ahmedabad, India: VIKSAT/Pacific Institute/Natural Heritage Institute.

Janakarajan, S., John Butterworth, Patrick Moriarty, and Charles Batchelor (2007) "Strengthened City, Marginalized Peri-urban Villages: Stakeholder Dialogues for Inclusive Urbanization in Chennai, India." In John Butterworth, Raphaele Ducrot, Nicolas Faysse, and S. Janakarajan (eds), *Peri-urban Water Conflicts*. Technical Paper Series 50. Delft, The Netherlands: IRC International Water and Sanitation Centre.

Landry, Clay J. (1998) *Saving Our Streams Through Water Markets: A Practical Guide*. Bozeman, Montana: Political Economy Research Center.

Llorente, Marie and Marie Helene Zerah (2003) "The Urban Water Sector: Formal Versus Informal Suppliers in India." *Urban India* XXII(1).

Moench, Marcus (1992) "Chasing the Water Table: Equity and Sustainability in Groundwater Management." *Economic and Political Weekly* 27(51–2): A171–7.

Planning Commission (2007) "Report of the Expert Group on Groundwater Management and Ownership." New Delhi, India: Planning Commission.

Rosegrant, Mark W. and Hans P. Binswanger (1994) "Markets in Tradable Water Rights: Potential for Efficiency Gains in Developing Country Water Resource Allocation." *World Development* 22(11): 1613–25.

Saleth, R. Maria (1994) "Groundwater Markets in India: A Legal and Institutional Perspective." *Indian Economic Review* 29(2): 157–76.

Saleth, R. Maria (1999) "Water Markets in India: Economic and Institutional Aspects." In K. William Easter, M. W. Rosegrant, and Ariel Dinar (eds), *Markets for Water: Potential and Performance*. Dordrecht, The Netherlands: Kluwer.

Sampath, R. K. (1992) "Issues in Irrigation Pricing in Developing Countries." *World Development* 20(7): 967–77.

Shah, Tushaar (1993) *Groundwater Market and Irrigation Development*. Bombay, India: Oxford University Press.

Shah, Tushaar (2007) "Institutional and Policy Reforms." In J. Briscoe and R. P. S. Malik (eds), *Handbook of Water Resources in India*. Oxford, UK: Oxford University Press.

Shah, Tushaar (2009) *Taming the Anarchy: Groundwater Governance in South Asia*. Washington, DC: RFF Press.

Shah, Tushaar, Christopher Scott, Avinash Kishore, and Abhishek Sharma (2007) *Energy-Irrigation Nexus in South Asia: Improving Groundwater Conservation and Power Sector Viability*. IWMI Research Report 70. Colombo, Sri Lanka: IWMI.

Shankar, Kripa (1992) *Dynamics of Groundwater Irrigation*. New Delhi, India: Segment Books.

Singh, Chhatrapati (ed.) (1992) *Water Law in India*. New Delhi, India: Indian Law Institute.

Thobani, Mateen (1998) "Meeting Water Needs in Developing Countries: Resolving Issues in Establishing Tradable Water Rights." In K. William Easter, M. W. Rosegrant, and Ariel Dinar (eds), *Markets for Water: Potential and Performance*, 35–50. Dordrecht, The Netherlands: Kluwer.

Wood, Geoffrey D. (1995) *Private Provision after Public Neglect: Opting Out with Pumpsets in North Bihar*. Bath, UK: Centre for Development Studies, University of Bath.

World Bank (1999a) *The Irrigation Sector*. New Delhi, India: Allied Publishers.

World Bank (1999b) *Initiating and Sustaining Water Sector Reforms: A Synthesis*. New Delhi, India: Allied Publishers.

10 Areas of conflict and the role of water trading

The case of Spain

Antonio Serrano Rodríguez

Introduction

Public authorities need to ensure the availability of water for life (the right to be supplied with quality water and to available water resources to be able to ensure the environmental sustainability of our natural heritage for future generations). Public authorities need to do so without overlooking that water is also a production factor, and market factors are extremely important. With this approach, this chapter addresses the management experience of the Spanish Ministry of the Environment in the context of a changing framework characterized by: the new dynamics caused by climate change, with periods of severe droughts; the impact of the new statutes of Spain's Autonomous Communities, which are modifying the traditional organization of river basin management competencies; and some indecisive public policies both in Spain and elsewhere in the European Union *vis-à-vis* the international economic situation. This context may require reconsideration in Spain of the role of agriculture and irrigated land whose effectiveness and efficiency needs to be a precondition. This is the backdrop for the overview in this chapter of the management and intervention tools needed for Spain to be able to achieve the aims of European directives ensuring environmental sustainability and a balanced spatial development in the different Autonomous Communities of Spain. In this chapter we argue how during the 2004–8 drought period in Spain, market tools, duly combined with the internalization of the external effects of the different water uses and activities, constituted efficient mechanisms for dealing with the drought situation, so that the essential resources for drinking water supply and for sustaining the natural heritage at risk, can be guaranteed.

An integrated approach

An integrated approach in water policy in Spain may need to take two things into account: first, that there is a public right to water in Spain, meaning that the public authorities are obliged to ensure the availability of water for life (i.e. the right to a quality water supply and the availability of water resources that ensure the environmental sustainability of our natural heritage for future generations); and second, that water is considered also a valuable a production resource in a framework in which there is no capacity to insure that all the costs of making water available are accounted for; a framework that is inefficient in many respects due to how water use rights are conceptualized (the concession concept) and how the applicable tariffs are calculated, according to current legislation.

On the basis of this, we reflect upon the use of economic instruments in water management in order to help understand the importance and consequences of incorporating all the costs of making water resources available to users, in a context of change characterized by: the new dynamics arising from climate change (already with lower rainfall in certain regions of Spain); the impact of the new Statutes of Spain's Autonomous Communities which are modifying the traditional organization of river basin management competencies in Spain; and certain indecisive public policies both in Spain and elsewhere in the European Union *vis-à-vis* an international, economic situation which seems to call for the role of agriculture to be reconsidered.

This may explain, in part, why several specific regions in Spain have highly significant water use problems and conflicts. The issues at stake in some of the regions with significant conflicts may require consideration of water pricing, considering all water costs (including the amortization of infrastructures, and the environmental and opportunity costs). It may also require consideration of other water allocation instruments and, in particular, how the public institutions that manage trading with exclusive water rights, currently operate – these being mechanisms that enable water resources to be allocated more efficiently. The creation of public water banks, where the respective authorities analyze the demand and available supply of water at different levels may be another instrument to foster greater transparency and improving the allocation of the water resources, providing an additional solution for conflicts.

Water planning (the Water Framework Directive River Basin Management Plans) is an essential part of improving water policy and a necessary condition of using economic instruments for water management. The plans would need to be coordinated with regional plans and (as expected) specify the maximum water resources available (and their quality) in each Autonomous Community in times of drought. This maximum is the basis for allocating/redeeming concessions to insure the improvement of welfare and the overall achievement of public goals. Within this planning framework, future scenarios would focus on the problems arising from the need to achieve an adequate allocation of resources, in uncertain, risk scenarios, rather than on solving structural shortages by building new infrastructures.

Areas of conflict and public policies

In Spain a major concern is the structural mismatch between the available supply and the water demands (at existing water charges). A mismatch that is unsustainable during the periodic droughts that characterize certain Spanish regions. This has as a consequence that in some areas there is severe environmental impacts (over-exploitation and pollution of aquifers and surface and coastal waters, the deterioration of ecosystems, loss of biodiversity, etc.) and significant socio-economic losses in agriculture. In part due to the fact that there has been certain increases in irrigated areas that were given water use rights (concessions) that exceed the sustainable volume available today. This situation causes frequent socio-political conflicts between regions and water uses (mainly farming and energy uses) in a situation in which the opportunity costs for society as a whole, of making water available, are not internalized. Furthermore, the respective governments and interested parties staunchly defend the interests of the uses of their Autonomous Communities, which may clash with long-term, general interests.

From the latter viewpoint, the second-generation statutes of the Autonomous Communities may have significantly increased the obstacles to finding an efficient socioeconomic and environmental solution within the desired framework of integrated river basin management and planning. In the next drought, (that is predicted for the second decade of the twenty-first century), the solutions will unfortunately be undoubtedly more complex and the conflicts, more intense.

The specific regions where problems are greatest, and conflicts may be more significant because of the supply and demand trends, are as follows (see also Figure 10.1):

- Ebro Delta in Catalonia;
- Júcar and Vinalopó irrigated land;
- Segura basin;
- Doñana;
- Upper Guadiana;
- Upper Tagus basin.

This is despite the fact that great efforts have been made in each of these regions as regards both planning and public investment and management. Among this there are two paradigmatic cases that may be considered because they help understand the types of challenges facing water policy and the type of foreseeable conflicts arising from a mismatch between potential supply and demand. These are the Ebro Delta and the Tajus Water Transfer.

Today's challenges of the Ebro Delta are those typical of the deltas formed at the mouths of major rivers around the globe. The Delta is made

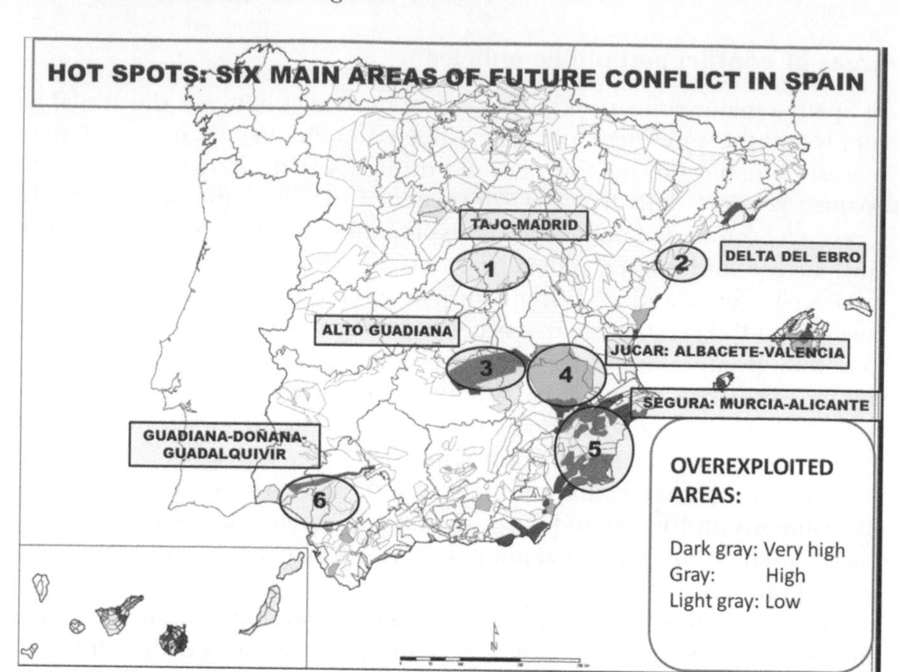

Figure 10.1 Hot-spots: main areas of future conflict

of fluvial silt giving rise to ecosystems of great value and highly productive lands. However, the stability and sustainability of the Ebro Delta ecosystem, in which water plays a key role, is seriously affected by a variety of factors: an increasing regulation of the tributaries of the Ebro with growing numbers of reservoirs and expanding irrigated areas upstream, leading to a reduction in water flows and sediments; more agriculture in the delta itself, with an increased use of fertilizers and pesticides, land subsidence, and salinization of aquifers; urban water pollution, and increasing urbanization; and higher sea levels.

The PIPDE (Ebro Delta Integral Protection Scheme) implemented in the Ebro Delta in 2006, applied a territorial model that considered using natural resources sustainably making compatible the conservation of the Delta's ecological values with the region's socioeconomic development. The PIPDE envisaged actions to be incorporated in the Programme of Measures of the Ebro River Basin Plan in implementation of the Water Framework Directive (Directive 2000/60/CE) aiming at ensuring good status water (in quantitative, physical, chemical and ecological terms) in the Lower Ebro river, the Delta and the coastal water bodies affected.

The PIPDE made a diagnosis of the situation in the Delta and took stock of the programs that were under way or envisaged in the short/medium term, with an investment of 426 million euros from different departments. In compliance with the environmental aims of the Water Framework Directive, the PIPDE elaborated the methodology for the calculation of the environmental flows in the Ebro Delta (this served as a general methodology to be applied to all river basins in Spain) and it established measures: to determine the ecological flow; avoid subsidence, Delta degradation and aquifer salinization; to improve the physical habitat and ecosystems in the area; to define and implement a sustainable farming model; to guarantee the functions of the riversides as biological corridors of the river; and, particularly, to restore the environment and clean up pollution in the Flix reservoir (polluted for more than a hundred years by the dumping of industrial waste).

The PIPDE was prepared in collaboration between the different authorities and departments and involved a policy of transparent information and public participation. This allowed reduction of the extremely tense situations that were in the area and facilitated building solutions for the future on the basis of consensus. The Spanish Ministry of the Environment and the "Generalitat de Catalunya" signed a General Collaboration Protocol and incorporated the MAPA (the Spanish Ministry of Farming, Fishing and Food) into the PIPDE together with the Department of Farming, Livestock and Fishing, and the Departments of Territorial Policy and Public Works of the "Generalitat de Catalunya." Likewise, before drafting the PIPDE and implementing and coordinating its actions, the representatives of the local bodies in the Ebro Delta area were consulted and asked to participate, as were also the social organizations and users (the Generalitat had created the Ebro Area Sustainability Committee). In spite of the participatory approach and the search for policy consensus in a particularly fragile area, where water is of prime importance for the physical and environmental stability, the plan encountered considerable difficulties because of regional interests. Although the process involved the preparation of a spatial plan this was not by itself sufficient The repeated and on-going demands from upstream regional governments for new irrigated areas – together with the considerable water reserves of 6,550 hm^3 demanded by the Community of Aragón and the demands for the Ebro transfer made repeatedly by the Autonomous Communities of Valencia and Murcia – has developed into a political conflict whose main risk is the possibility of permanent damage to the Ebro Delta: an important element of our common world heritage.

The Tagus transfer: higher water use productivity, inter-regional water competition and environmental impacts in Murcia, Madrid and the Upper Guadiana

The Tagus–Segura water transfer infrastructures connect four water basins (the upper basins of the Tagus and Guadiana, and the Júcar and Segura basins). In some of these areas can be found irregular/illegal water extractions, over-exploited aquifers, and expanding irrigated areas, in addition to considerable urbanization pressure and growing numbers of golf courses, with direct or indirect support from the respective Autonomous Communities' governments and institutions. All of these factors accentuate the potential conflicts in areas characterized by major pressures.

In Ciudad Real, in the area of aquifers 23 and 24, and in Murcia, in the areas receiving water from the Tagus–Segura Transfer, one of the reasons for the present problem concerns Spain's considerable economic growth in the latter half of the 1970s. This led to a great increase in groundwater extractions which have continued, with varying volumes and degrees of control of their legality, until the present day, to be able to satisfy the demands for water, particularly for farming. These extractions were carried out mainly by the private users at a time of a legal vacuum regarding groundwater, and having as a consequence the over-exploitation of aquifer that has continued to the present day.

Figure 10.2 shows the main areas of conflict about the transfer related to the over-exploitation of aquifers and the excessive demands made on available resources.

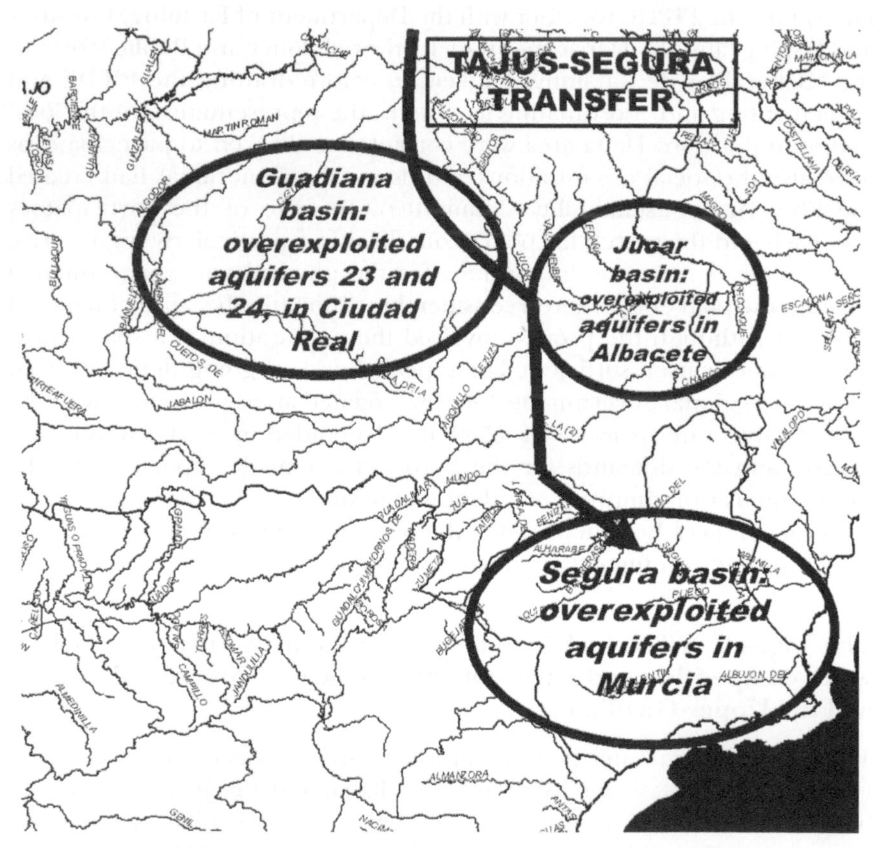

Figure 10.2 The Tagus–Segura transfer and the over-exploited aquifers 23 and 24, in Ciudad Real, Albacete, the Júcar basin and Murcia

The Tagus–Segura transfer plays a key role in understanding the evolution and present state of different regions in Spain. It was designed and commissioned in 1979 and it explains many of the changes in land use and the spatial policies implemented. The Transfer is an infrastructure thought up and designed in a time past to deal with long-standing demands of the eastern coastal areas of Spain and also, when the project materialized in the 1960s, to be able to mitigate the growing over-exploitation of aquifers. The main working assumptions for the transfer were:

1. The potential maximum of irrigated land in the Upper Tagus basin (the ceding basin) was to be 15,000 ha, requiring 15,000 m^3/ha/year (i.e. a total of 225 hm^3/year).
2. The average reserves in the Upper Tagus Basin were 1350 hm^3/year (with a variation between 400 and 2,000 hm^3/year). A transfer of 1,000 hm^3/year was considered possible and was planned in two phases: 600 hm^3/year in phase 1, with the addition of another 400 hm^3/year in phase 2.
3. The aim of the 600 hm^3/year would enable the irrigated area in the Levante to be doubled.
4. It would be possible to safeguard the ecological conditions of the Tagus's River by insuring that the flow in Aranjuez would be above 6 m^3/sec.

Based on these, the Council of Ministers in 1970 and in 1973 approved the transfer of 600 hm^3/year to be allocated to irrigation (400 hm^3), and drinking water supply (110 hm^3) considering that there would be some transfer losses (90 hm^3). The region of Murcia was to receive 260 hm^3/year for irrigation and some 70 hm^3/year for drinking water supplies, the rest being for Alicante and Almería.

According to the information at the time of the design of the infrastructures the available surface water resources in the Region of Murcia in the late 1970s was some 390 hm^3/year, of which 350 hm^3/year were for irrigation and about 40 hm^3/year for drinking water supply. Groundwater abstraction totaled about 505 hm^3/year, which were well above the renewable resources of 223 hm^3/year. Irrigation returns and wastewater re-used made available an additional 110 hm^3/year. The total available resources were then approximately 720 hm^3, but some 1000 hm^3 were used (i.e. there was a total over-exploitation of groundwater of about 280 hm^3/year). Part of this over-exploitation was caused by the expectations created by the transfer itself, since in some cases land began to be irrigated before the water was actually transferred and water rights effectively allocated. This was compounded by the fact that the transfer management tests began on 25 May 1979, during a drought period that lasted until 1984, as a result of which far less water was transferred than envisaged (less than 200 hm^3/year).

The inventory carried out by the Agriculture Ministry in 1978–9 calculated that there were 145,347 ha of irrigated land in the province of Murcia: the same figure used in 1980 for the analysis in the Spanish Water Plan primer. The figure given in the 1981 Official Farming Statistics by Province was that by then there were 156,205 ha of irrigated land. This was 7.5 percent higher than the previous figure. In 1982, when the transfer began, these irrigated land area continued to grow. The estimation is that in the early 1980s, at the end of the first phase of the transfer, the area in the Region of Murcia irrigated with any type of water (i.e. surface water, groundwater or wastewater), was 185,118 ha. This meant a 27 percent increase since the late 1970s.

By 1983–4 water resources in the region had increased with 330 hm^3/year from the Tagus water transfer. This did not, however, reduce aquifer over-exploitation and water consumption climbed to some 1,325 hm^3/year. Twenty years later, in 2003, available surface water had increased by almost 10 percent, renewable groundwater was about 200 hm^3/year, transferred water had increased to 550 hm^3/year and irrigated water re-use and wastewater re-use available was 200 hm^3/year, in addition to another 200 hm^3 produced by desalinization. This meant total available resources of 1540 hm^3/year (i.e. 16 percent more than in 1983, and enough to irrigate some 80,000 ha and supply a town with a population of 700,000 and medium-sized industry). At this point, however, there were requirements of additional water because of a shortfall in irrigation water of more than 500 hm^3/year.

It is important to mention that the 600 hm^3/year, was transferred in only one year. The average transfer on record has been 360 hm^3/year, with some 355 hm^3/year transferred to the Segura. This reality is very different from what was expected and provides evidence of the socio-politic tension and conflicts related to the Tagus water transfer. A source of water which, despite having higher prices than surface water in the Segura Basin (the Tagus transfer was charged at approximately 0.18 euros/m^3 at current values), is provided at prices far lower than the actual cost of the system as a whole – and far below the price of more than 1 euro per cubic meter paid for irregular, internal transfers between individuals.

The Official Geographers' Association has produced a study entitled *Spatial Processes in the Segura Basin.* This study has looked into the evolution of land uses in the Segura Basin and their impact on water use. The report's findings explained the main processes in the region and showed how the dynamics and trends in land use reflected the paradox of the Segura basin: an increasing shortage of resources alongside a significant increase in irrigated land and a very important increase in water use by urban and tourists activities.. The study showed that from the 1990s and until 2005, irrigated land continue to increase despite it being increasingly difficult to obtain water. This occurred after 2000, at a slower rate, and with enormous diversity and complexity. An example is the fact that The irrigation of vineyards has increased (a trend typical of all basins in the southern Iberian, making grapes the third largest irrigated crop in Spain after olives and maize), particularly in the Altiplano area, where most of the vineyards in the basin are located, and where grape irrigation has increased by 35 percent in recent years. Likewise, new irrigation areas had been created in the edges of established irrigated districts. In some cases this happened, in parallel with a reduction in the surface of former irrigation districts, due to urbanization. The study showed that this process also occurred in the southern, coastal zones (irrigated land in the Campo de Cartagena) one of the finest examples of the fruit and vegetable industry, with cutting edge technological innovation, high profit margins and close connections with national and international markets. Irrigation have also expanded considerably in inland areas which, due to their altitude and climate, were not formerly considered appropriate for horticulture production, but that can now be cultivated thanks to the development of technology and the development of more adapted and different crop varieties.

The study showed that, surprisingly, the increase in irrigated land occurred at the same time as water shortages increased (i.e. when previously existing irrigated areas were unable to achieve their harvest potential due to a lack of sufficient water). These parallel processes may explain the great political pressures to obtain more resources and to increase the overexploitation of aquifers. In addition, new demands for water emerged in recent years, related not so much to agriculture as to urban developments and tourism. All of this was against the backdrop of the worst drought ever recorded in Spain in the basins of the Upper Tagus, Júcar, Eastern Andalusia and Segura, from 2005 to 2008 – a period when, despite avoiding water rationing to homes and industry throughout Spain, it was not possible to avoid restricting water supplies for agriculture.

In Murcia, a region where farming accounts for some 7 percent of the GDP, the exceptional measures implemented (drought wells, desalinization plants producing 4 hm^3 of desalinated water in 2003 and up to 96 hm^3 by 2008, and water rationing, etc.) even produced surpluses of some types of crops (particularly citrus) that in some instances were not even harvested.

The former analysis of the basic characteristics of water use and availability in the Segura basin (the basin receiving water from the transfer) is necessary to understand the existing conflicts in this area. There are, however, other factors related to the ceding basin (the Tagus Basin) that need to be considered to understand the present and potential future conflicts.

The Tagus basin has strategic importance in Spain because it is the basin where the functional region of Madrid – the capital of Spain – is located. Madrid has a population of 6 million. There has been a process of increasing economic and spatial integration of Madrid with the surrounding cities such as Guadalajara and Toledo at the present time, and others such as

Cuenca, Ávila and Segovia having increasingly close links. The water demands of this active economic area are likely to increase in the near future. This may mean that the water available to be transferred from the Upper Tagus may be reduced. The Mid and Upper Tagus have been classified as ecologically sensitive areas in the new plans, and this means that the river must have larger flows for ecological reasons. This will also affect the volume of water available for the Transfer.

Another bone of contention concerns the implementation of a water transfer from the Tagus to the Upper Guadiana, whose aquifers 23 and 24 are over-exploited. Indeed, despite being in the same region (Castilla la Mancha), this transfer between different basins touches upon another of the most complex areas of conflict in the medium term in Spain. In fact the problem in the upper basin of the Guadiana is similar to the situation in the Segura basin. Its basic characteristics are outlined in Table 10.1.

In the Upper Guadiana Basin the River Basin Authority has prepared the "Special Upper Guadiana Plan" which should have been approved in 2002, was only prepared in 2004–8. The plan was intended to pave the way for new planning regulations for the affected region, working on the basis of the greatest possible social consensus, incorporating modifications in land use, reorganizing existing irrigated land, and with some of the water use rights being reallocated by the Guadiana Basin Authority.

The plan was expected to be executed by a consortium created for this purpose consisting of the Spanish authorities at the central and Autonomous Community levels, together with representatives of civil society and the local authorities. The results obtained and the progress made as regards both earmarked investment and new regulations governing land use, irrigated land and water allocation are now quite different from the aims pursued and overall, the tension in the areas of conflict (the Tablas de Daimiel national park, the quality of the potable water supply for certain towns, aquifer over-exploitation and pollution, etc.) has been reduced thanks to the higher rainfall recorded since May 2008. However, since no solutions to structural factors have yet been found, conflicts will reappear with the next droughts, and possibly be even worse due to the forecast effects of climate change.

The effects of climate change in the Upper Tagus, the Guadiana Basin and the Segura Basin are important, and will probably escalate the conflicts. This can be summarized as follows:

- A reduction in the volumes available and a reduction envisaged in the volume transferable from the Tagus, which is already some 270 hm^3/year. This will necessarily cause social tension because the irrigated land in the Segura alone was established on the basis of resources almost four times those that could be attained.
- Extreme vulnerability to the inevitable periodic droughts that affect this area, with even tenser social conflicts if no joint, stable, long-term

Table 10.1 Evolution of water use and challenges of the Upper Guadiana Basin

Period	*Basic aspects*
1970–87	The gradual replacement of waterwheels by hydraulic pumps and expanding irrigated land cause extractions from aquifer 23 to substantially outstrip its replenishment capacity.
4 February 1987	Temporary declaration of overexploited aquifer (art. 171 RDPH)
1988	Overexploitation peaks with extractions of almost 600 hm^3
1992–4	Severe drought deteriorates the state of the aquifer. Measures taken in the farming industry to mitigate its disastrous effects.
15 December 1994	Definitive declaration of overexploited aquifer (Western Mancha Aquifer). Extraction and Exploitation Plan. The Board of the Guadiana Basin Authority must approve the exploitation regime every year.
1996–2004	Between 1996 and April 2004, the Spanish authorities sanctioned 434 irrigators for unauthorised, illegal extractions or for breach of licence (an average of 55 sanctions per year).
2005	2,082 sanction proceedings begun and not resolved. 19 preventive measures issued for water extractions posing great risk to the aquifer. 68 notifications sent out for the subordinate execution of well closures in fulfilment of final court sentences, pending execution. 15 well closures (resolutions) Approximately 2,600 cases awaiting subordinate execution being carried out.
2006–April 2008	Due to the drought, illegal extractions persist. The Guadiana Basin Authority takes action, resulting in serious social tension and a new Basin Authority Chairman. Annual Aquifer Exploitation Regulations are gradually passed; serious problems in reaching a consensus. Fewer problems after Basin Authority Chairman replaced. There are reasonable doubts about the extent to which the Exploitation Regulations are complied with. Compulsory to install water meters: Act 11/2005 amending the Water Act. Emergency measures passed in Castilla la Mancha to ensure potable water supply although in fact it is not possible to distinguish it from irrigation water because of the link with aquifer use. Action taken by consensus under the Castilla la Mancha AGUA Scheme (Water, Agriculture, Biodiversity, Town and Country Planning Joint Committee).

solutions are reached by consensus (consensus that is impossible today, and likely to remain so for a long time).

- An inevitable reduction in the guaranteed access to the water resources that could be known to be available with hindsight.
- Serious environmental problems in ecosystems of interest, including Red Natura 2000 and the Tablas de Daimiel national park.
- Serious impact on the cultural heritage of sites of world interest such as the Huerta de Aranjuez (irrigated area of Aranjuez).

The role of market instruments and their impact on water policy effectiveness and efficiency

The fact that water is considered to be a production resource in Spain means that the tools used to charge for costs and allocate resources can play a fundamental and essential role in water policies. This is particularly so because we are in a scenario of rapid change in society and natural heritage that is characterized not only by the new dynamics stemming from global change, with lower rainfall *inter alia*, but also an international economic context apparently obliging us to reconsider the role of agriculture in developed countries and to open up new avenues of action in the production of crops such as biofuels which may give rise to new demands for irrigated land – the effectiveness and efficiency of which must be analyzed and assured beforehand.

For conflict resolution is it important to establish measures in Water Plans for facilitating the reallocation of existing water use rights via direct trading among users or through public water banks. This would need to be duly combined with the internalization of the external effects and involve the competent authorities in the different regions. It is important to recognize that the Spanish water legislation does not contemplate in a strict sense a water "market". The current legislation allows voluntary agreements among users, but insures that there is a considerable degree of public intervention and control to safeguard the public nature of the resource and ensure that the general interest in its use will prevail in the long term. In this sense trading of water use rights is highly regulated and it can only operate in the framework of defined objectives and with the conditions established by public authorities. In Spain, in spite of limited experience, they have proved to be efficient mechanisms for dealing with mismatches between demand and available resources, once the essential resources for supplying the population and sustaining natural heritage are guaranteed. The fact that, for example, water banks (Water Exchange Centres) operate under public control and that can be managed by the Autonomous Communities (for transactions within their territory) ensure that these mechanisms are suitably integrated in the regional spatial policies and strategies for economic development.

Water plans consider the specifications of the regional spatial plans in relation to existing and projected land use. They may also need to specify the maximum amount of water that can be used in each Autonomous Community in times of drought, and take this maximum into account when allocating/redeeming concessions. Future scenarios must consider what would be the water use if there is an improved allocation of resources considering uncertainty and risk, rather than considering always that the solution to shortages would be building new infrastructures.

The experience gained from cases in the Ebro Delta and areas affected by the transfer has shown that only on very few occasions is the construction of new infrastructures the most efficient solution when the associated costs and impacts are assessed. It has also shown that only on very few occasions are the new users of these infrastructures able to cover their costs. It would seem essential to incorporate, pursuant to current legislation, an assessment of the economic and social considerations and opportunity costs of the new investments envisaged, and also carry out feasibility analysis of the intended use of water if all the costs are to be passed on. Similarly, it is necessary to stipulate who is to bear the related investment, maintenance and operating costs, and how, when and why.

Passing on the costs of water services and charging them to users, and taking into account the opportunity costs of activities competing for the use of water are all key factors when analyzing mechanisms for the efficient allocation of water resources. Furthermore, the exceptions to paying for the costs of water services (this is the criterion for establishing and allocating subsidies) would need to be clearly spelled out in relation to political, environmental, and spatial development goals, and incorporated clearly in the River Basin Management Plans. They need to be made known and explained in public information and participation processes.

Major changes were made to the Spanish legislation to respond to the drought suffered by a good part of Spain in the 2005–8 period. This is when water trading was developed. Ten transactions involving water use rights took place between different demarcations, and five offers to acquire rights were made by the Exchange Centres of the Guadiana, Segura and Júcar Basin Authorities, successfully, in widely varying circumstances. But their results are not all perfectly clear. Otherwise, there are some lessons from the experience that may make it advisable to take the following fundamental factors into account in the future:

1. It is advisable to promote public water banks as tools to promote reallocation and more efficient use and as a tool for the prevention and solution of conflicts. In this respect, in the case of the Special Upper Guadiana Plan, trading of water use rights, duly complemented by the internalization of the associated external costs, could be efficient mechanisms for dealing with droughts or structural scarcity, and also to relieve the pressure on ecosystems.
2. The experience of trading and reallocation of water use rights made

through Exchange Centres (Water Banks) showed that prices for water have attained far higher levels than the estimated loss of profits in the farming activity, hence it may be advisable to evaluate the mechanisms used to improve the process. It may be essential that water prices are in keeping with the real social benefits and associated costs. Public subsidies that facilitate transactions cannot be considered the benchmark for water management by public authorities.

3. The total public resources devoted to the operations carried out in Spain in recent years have been considerable and not always in proportion to the intended aims and the results obtained. Sufficient thought must be given to socio-economic and environmental objectives before contracts are signed to ensure that the exchanges do not have external effects (of an economic or environmental nature, for example) or that there are abusive prices in operations (due to a lack of competition, transparency or flexibility) and that the overall balance is positive, demonstrating and ensuring that the process is socially acceptable. To provide another guarantee for this acceptance, public hearings must be contemplated to enable people who feel prejudiced or affected, including trade unions, businesses and ecological organizations, to appear in person.
4. Some Autonomous Communities in Spain (Community of Castilla La Mancha for examples) have considered that the transfer of resources could have a significant impact on zones declared to be places of special interest (LIC) and birdlife sanctuaries (ZEPA). Although strictly speaking the law does not require, in the opinion of the water authorities, the dossier to undergo a formal environmental impact evaluation, it is nonetheless obvious that a detailed appraisal of the foreseeable environmental impacts and any necessary remedial measures should be carried out in order to comply with the provisions of the Water Framework Directive, the Habitat Directive and the present-day Spanish Law on Natural Heritage and Biodiversity.
5. The impacts of the transactions would need to be monitored to incorporate improvements in the functioning of trading. Water use rights in Spain are given by public authorities on the basis of administrative decisions and are given for very long periods of time. Water right holders need to fulfill certain conditions but renewals do not allow to gradually adapt the rights to suit changes in society or available resources. Given the experience of implementation of trading in Spain, is advisable that public authorities insure that current concession conditions are complied with. This should prevent resources from being sold by license holders that are not using the water, and all the more so if they have not used it for three years and the cancellation of said concession is warranted. Furthermore, the fact that the existing system in Spain means that the concession license does not guarantee the availability of the allocated resources indicates that the concession

concept itself should be quickly reviewed and brought into line with what can and must be expected of a service supplying water for an economic, environmental or social activity in terms similar to other public services. This may mean a payment of a tariff for water consumption measured by the respective meter and guaranteed by the authorities following the recognition of the availability of the resource at the total applicable costs.

Conclusion

The overall conclusion is that Spain has made great progress in addressing water problems and has implemented new tools with varying results. Water planning and management has become more complex due to the new organization of the regional administrations with increasing responsibilities and wider legal mandates. This poses challenges to insure that actions benefit the general interest and to guarantee an effective or efficient use of water resources.

Mention must also be made of a certain factors of great importance in understanding the conflicts outlined in the preceding paragraphs:

- The debate about water policies in the political and social arena of the affected Autonomous Communities has acquired a disproportionate weight and has become the focus of clashes among political parties to win votes. Water demands have become the cornerstone of a new (or renewed) regional identity in some Autonomous Communities.
- The farming industry has become a solid, corporate block, despite its declining weight in the economy and employment, anchored and supported by a defensive response to the potential threats of reductions in the region's own or anticipated water resources.
- After heavy political and media campaigns, most of the population has accepted certain principles (not all very accurate) of great social and political importance:
  - the vital nature of water for development;
  - the concept of the permanent water shortage;
  - the importance of traditional, regional agriculture;
  - the awareness that farming is a core industry in regional development;
  - the exemplary way water is managed by regional farmers; and
  - the link between sustained development and increased water supply to satisfy current and envisaged demand.

Faced with this situation, it may still be essential to recognize all water management costs (including environmental conservation costs and opportunity costs) and charge these costs to those who benefit from its use in order to move towards solving existing problems. At the same time, this requires to consider exceptions justified by other goals (economic, social or territorial cohesion) to be publically stated and likewise the public subsidies associated with said exceptions. This measure may be crucial for solving inter-territorial conflicts and must be based on an in-depth economic analysis by a qualified team of experts providing a guarantee for civil society as a whole and reinstating the credibility of technical studies, to determine:

- all the costs related to the use of water in each part of the territory, including the cost of the environmental conservation of the affected ecosystems;
- the subsidies that would be necessary to make it feasible for this water to be used by every user anywhere in the territory, and who or how they would be paid for;
- the real alternative measures available to tackle the demand-supply question in each territory; and
- the water availability guarantees that each alternative offers to each user.

It is only in the framework of the consensus mentioned earlier and a "scientific" technical approach not conditioned by politics that it may be possible to advance towards an accepted and acceptable solution, while helping the technical and scientific studies in this sort of issue recover their essential roles of providing information and helping decision-making.

Public water banks controlled and managed by public authorities may be another way for achieving a more rational use of water in over-exploited areas and those with temporary shortfalls. The shift from concessions to public contracts in which users pay for the availability and use of the water they consume and the competent authorities guarantee the flow stipulated in the contract and undertake to pay compensation for damages arising from non-fulfillment, may be a new, necessary approach that would need to be assessed in the move towards greater rationality and the need to cope with future uncertainty as regards available water and the environmental demands to be dealt with. It would not be a good solution for the new water plans to repeat the approaches adopted in the plans of the 1990s.

References

Caja de Ahorros Provincial de Murcia (1984) *El agua en la Región de Murcia.* Murcia, Spain: Caja de Ahorros Provincial de Murcia.

Consejo Económico y Social de la Región de Murcia (2000) *Recursos Hídricos y su Importancia en el desarrollo de la Región de Murcia.* Murcia, Spain: Consejo Económico y Social de la Región de Murcia.

EC (2007) *Communication from the Commission to the European Parliament and the Council, Addressing the Challenge of Water Scarcity and Droughts in the European Union.* Brussels, Belgium: European Commission.

EC (2007) *The Green Paper. Adapting to Climate Change in Europe – Options for EU Action.* Brussels, Belgium: European Commission.
Fundación Nueva Cultura del Agua (2007) *Informe de Seguimiento de la Aplicación de la DMA en España.* Seville, Spain: FNCA.
Guadiana Basin Authority (1995) *Plan Hidrológico de la Cuenca del Guadiana.* Madrid, Spain: Spanish Ministry of the Environment (MMA).
Guadiana Basin Authority (2008) *Plan Especial del Alto Guadiana.* Ciudad Real, Spain: Spanish Ministry of the Environment.
IPCC (2007) *Climate Change 2007: Summary Report.* Geneva, Switzerland: IPCC.
MMA (2000) *Libro Blanco del Agua en España.* Madrid, Spain: Spanish Ministry of the Environment.
MMA (2000) *Análisis del Plan Hidrológico Nacional.* Madrid, Spain: Spanish Ministry of the Environment.
MMA (2006) *Plan Nacional de Adaptación al Cambio Climático.* Madrid, Spain: Spanish Ministry of the Environment Publications.
MMA (2007) *Precios y Costes de los Servicios del Agua en España. Informe Integrado de Recuperación de Costes en los Servicios del Agua en España. (Artículo 5 y Anejo III de la Directiva Marco del Agua).* Madrid, Spain: Spanish Ministry of the Environment.
MMA (2007) *El Agua en la Economía Española: Situación y Perspectivas. Economic Analysis Group.* Madrid, Spain: Spanish Ministry of the Environment.
MMA (2008) *Elaboración de la Estrategia de Financiación del Plan de Cuenca: Pasos, Criterios e Información.* November. Madrid, Spain: Economic Analysis Group, Spanish Ministry of the Land, Rural and Marine Environment.
OSE (2008) *Agua y Sostenibilidad: Funcionalidad de las Cuencas.* Madrid, Spain: Spanish Sustainability Observatory (OSE).
Segura Basin Authority (1997) *Plan Hidrológico de la Cuenca del Segura.* Madrid, Spain: Spanish Ministry of the Environment.
Tagus Basin Authority (1997) *Plan Hidrológico de la Cuenca del Tajo.* Madrid, Spain: Spanish Ministry of the Environment.
Tagus Basin Authority (2005) *Informe Resumen de los Artículos 5 y 6 de la Directiva Marco del Agua en la Demarcación Hidrográfica del Tajo.* Madrid, Spain: Tagus Basin Authority.

11 Voluntary water trading in Spain

A mixed approach of public and private initiatives

Alberto Garrido, Josefina Maestu, Almudena Gómez-Ramos, Teodoro Estrela, Jesús Yagüe, Ricardo Segura, Javier Calatrava, Pedro Arrojo and Francisco Cubillo

Introduction and scope

This chapter reviews the experience of water trading in Spain. Trading has had a limited role so far, but may become an especially important instrument in water management, given the severity of water scarcity, recurrent droughts and the expected climate change impacts in Spain. This chapter reviews existing and predicted problems of scarcity. It analyzes trading since it became common practice in 2005 and summarizes the water trading that occurred during the four-year drought period from 2005 to 2008, when it was most active. The chapter considers whether trading has been so far effective to cope with scarcity and if it has promoted more efficient water use. The chapter analyses asks whether market allocation is perceived as equitable by all potential participants, and if there is evidence showing that environment has been improved. Improvements in production efficiency cannot compensate the costs of environmental deterioration or more inequitable outcomes, in the eyes of stakeholders or water users, even if a cost-benefit analysis may indicate that it does. Finally, it identifies the major strengths and weaknesses of the existing regulation of trading.

Scarcity and climate change impacts in Spain

In the case of Spain, there is a clear unbalance of water availability between northern, central and southeastern areas. An important area of the country falls under very low amount of water availability (between 0 and 100 mm per year). Precipitation ranges from close to 2,000 mm per year in northwestern areas to less than 200 mm in the southeastern ones (Figure 11.1). Similarly, runoff values range from more than 1,000 mm to practically zero in the most arid areas (Figure 11.2).

According to the existing climate change scenarios, water resources will be severely affected by climate change. In a recent study, CEDEX (2011) did not find clear evidence of climate change impacts on runoff. The

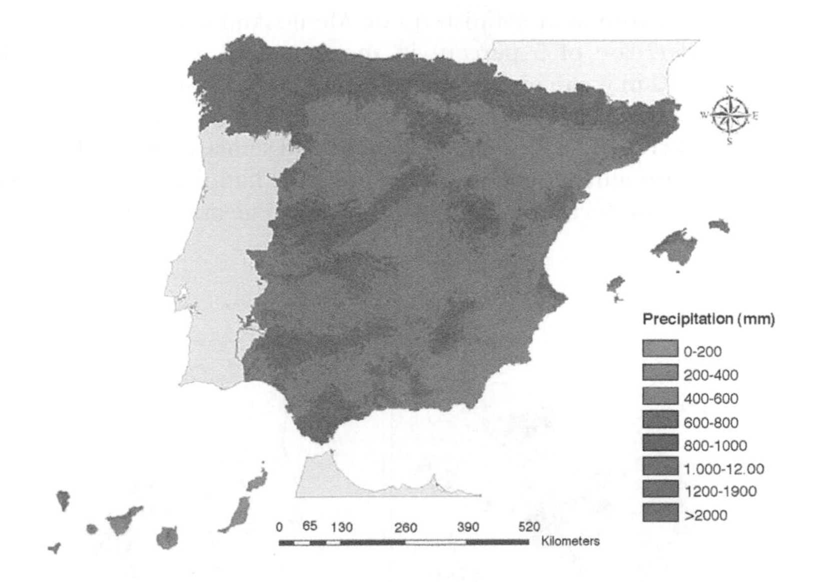

Figure 11.1 Average annual precipitation (mm/year)

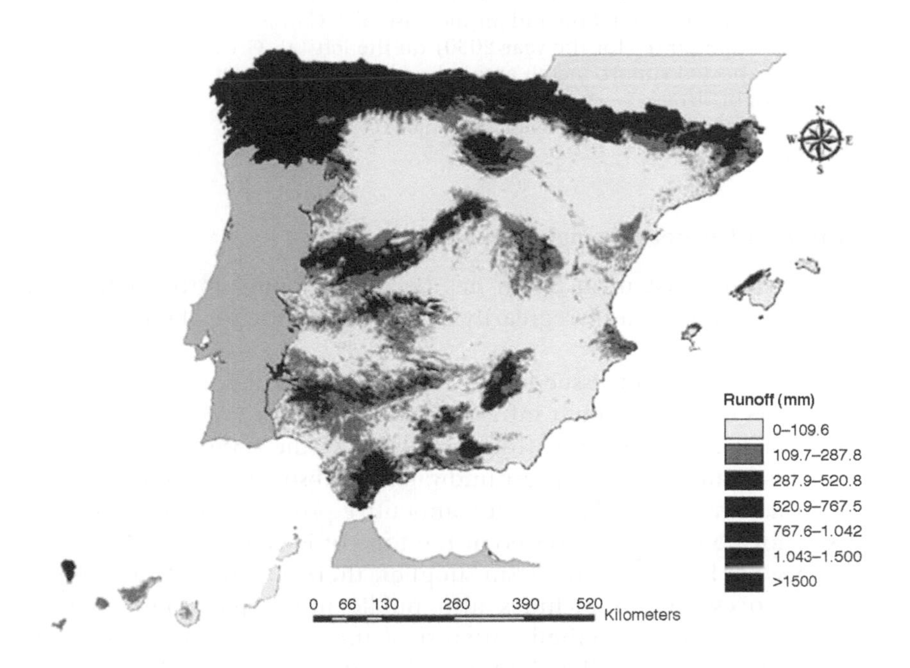

Figure 11.2 Average annual runoff (mm/year)

Ministry of Environment (Ministerio de Medio Ambiente 2000) estimated that for a decrease of 5 percent in mean annual precipitation and an increase of 1ºC in mean annual temperature, a decrease between 9 and 25 percent is expected for 2030 in runoff, depending on the river basin district. Most critical Spanish areas are arid and semiarid ones, where water scarcity and drought problems are greater, as it happens in the Guadiana, Canarias, Segura, Júcar, Guadalquivir, Sur and Balearic Island river basins (Figure 11.3).

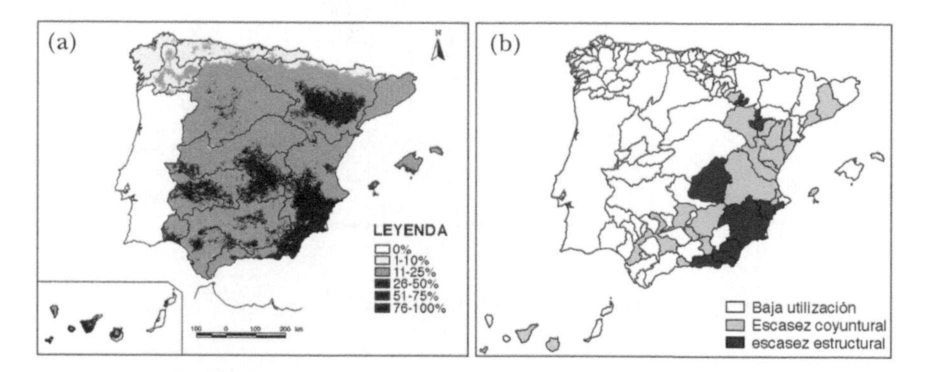

Figure 11.3 Impact on runoff reduction for a decrease of 5 percent in mean annual precipitation and an increase of 1°C in mean annual temperature (for the year 2030) on the left (a). Greater values of expected runoff, match areas already presenting water scarcity on the right (b).
Source: Ministerio de Medio Ambiente (2000)

Definition of water use rights in Spain

Water use rights system in Spain has been developing through time to match water scarcity and irregularity as well as provide flexibility to water managers.

Typically, water rights issued by the Water Authorities[1] are made available to users of publicly built infrastructures (dams or water transfers) or privately built infrastructures with permission of the state (hydroelectricity). Rights granted to pump groundwater or abstract resources directly from surface waterways. There is a competitive process (public tender for licenses) for potentially interested at the time of issuance for hydropower applicants. For irrigators and urban suppliers there is technical and administrative process, which includes also public information and aims at establishing the socio-economic interest of the request and its technical and environmental feasibility. Rights expire after 75-year use, but in practice they are renewed automatically if use continues.

The right is restricted to the "use" for which it was issued. Changing the type of use, location, abstraction or return points is not permitted without an explicit approval by the river basin agency (RBA). Rights can be transmitted but those appurtenant to a plot or farmland must be transmitted with the land. Rights then are carefully defined by at least name of the holder, abstraction point, type of use, calendar, the involved engineering, plots and crops to be irrigated and irrigation technologies, usable volume or flow, return flows. Rights can be forfeited by the RBA if water is not used by the holder.

Average entitlements are subject to technical efficiency standards where the RBAs can unilaterally reduce the volume to which a right-holder is entitled, based on technical, environmental or even economic grounds, and this can be done without compensating the holders of water use rights. Average entitlements of rights are also subject to water availability (RBA evaluates), and rights differ in the priority of their access to water depending on the type of use (domestic, environmental, agricultural, hydropower or industrial), and hence users could be seen a reduction in their entitlements considering the order of priority.

Opportunities and limits to trading with water use rights

The 1999 Water Law reform identifies two ways to exchange water use rights:

- One that involves two right-holders that voluntarily agree on specific terms of trade and jointly file a request in the River Basin Agency to exchange a number of water rights. The RBA screens, reviews and approves an exchange request in 30 days.
- The second involves publicly run and administered water banks (or *water exchange centers*, as they called in the Water Law). Initiated by the RBAs, water banks issue public tenders for potentially interested right-holders who would be willing to relinquish their water rights temporally or for the remaining maturity period. Water banks, in some cases instead of purchasing or leasing out the offered water rights, have purchased the land to which the water is appurtenant.

The Spanish regulatory framework is such that makes it unique compared to others in the world. It is best defined by reviewing the limitations to the type of exchanges that can be filed by right-holders to the RBAs. We identify legal, environmental and institutional limitations to trade water use rights.

Legal limitations

There are two kinds of legal barriers: public administration barriers, and barriers related to the way water rights are defined. The exchange of water use rights are subject to obtaining approval from the public water administration. They use criteria related to the use of publicly-managed water works, to arbitrage competing and equal right-supported uses, to considering community interests (mosquito larvae control in reservoirs, for example), the service needs of cities (including aesthetic, urban amenities), navigation, or even to meet international responsibilities of the public agencies (treaties involving quantities or flows across state boundaries). These barriers can be erected and made explicit in the regulation or can be customarily applied on the spot, with the result of granting or denying permission to certain market exchange requests. These result from public agencies' responsibilities and service. Without them, the market would not only be environmentally harmful but also poorly enforceable.

Limitations to trading also stem from the way water rights are defined. In the case of Spain, Law specialists differ in interpreting whether the rights definition hamper unnecessarily the market (Ariño Ortiz and Sastre Beceiro, 2009) or simply enforce the very Water Law tenets (Embid Irujo, 2009). Ariño Ortiz and Sastre Beceiro (2009, pp.100-101) identified the following legal constraints for the case of Spain:

- **Time limit**: maximum duration until the right-holder's use right expires.
- **Rights to consumptive use** cannot be sold to holders for non-consumptive users (hydropower), and vice versa.
- **Destination limit**: assignment to another holder or entitled organization of an equivalent or higher category as per the order of preference established by river basin planning or in accordance with the Water Act (Art. 60, 2001 Consolidated Water Act). Under exceptional circumstances and provided it is in the interests of the general public, the Ministry of the Environment may authorize licenses in a different order of preference (Article 67.2).
- **Quantitative limit**: "the annual volume subject to the license may not exceed what is actually used by the holder," where "regulations shall govern the calculation of this annual volume, based on the volume used over a specified number of years, adjusted as the case may be to the target allowance established under the river basin plan and proper use of water, and may never exceed the assigned volume" (Article 69). Subject to prior discretionary administrative authorization by the RBAs, there are four options: (a) authorization of the license; (b) refusal of the license, based on a reasoned decision in the following cases: "when it has a negative effect on the river basin resource use system, third party rights or environmental reserves, or does not comply with the requirements specified above, where stakeholders have no right to compensation whatsoever." The generic term of reference "negative" can potentially lead to arbitrary refusals that are difficult to control. (c) RBAs can also exercise preferential acquisition rights over the volume to be licensed during this period of time, thus

putting an end to all types of private use. (d) Administrative silence then applies: "Considered as authorized, without effects between the parties thereto until such time, after one month as of notification to the RBAs, provided they do not object when licenses are granted to members of the same consumer association, and after two months in all other cases" (Article 68.2).

- **Formal limits**: the following formal requirements exist: (a) "Exchange requests must be written and a copy of the agreement sent to the RBAs and consumer associations to which the licensor and licensee belong within fifteen days of signature" (article 68.1 2001 Consolidated Water Act). (b) "Should the license be granted to and by consumers for irrigation purposes, the contract must specify the land subject to non-irrigation by the licensor for the term of the contract, as well as the land to be irrigated by the licensee and the assigned volume" (Article 68.1). In addition to formal implications, this prohibits dealers or traders. As opposed to gas or electricity, water trading or re-sale (water licenses acquired under agreement) is forbidden. (c) RBAs must register water licenses in the Water Registry (Art. 68.4, 2001 Consolidated Water Act).
- **Geographical limits**: licenses for the use of infrastructure connecting areas subject to different river basin plans may only be authorized if they come under the National Hydrological Plan or other specific laws relating to a particular diversion scheme.
- **Limits on open pricing**: according to Art. 69.3 of the 2001 Consolidated Water Act, "regulations may determine maximum price limits" for water licenses. Competitive pricing is substituted by administrative intervention.
- **Market organization limits**: the Act provides for setting up Water Exchange Centers at the initiative of the Cabinet and at the request of the Minister of the Environment (in the cases foreseen in Articles 55, 56 and 58, 2001 Consolidated Water Act). In such cases, RBAs are allowed to buy and sell water rights. However, these Water Banks are subject to public contract procedures ("acquisition and disposal of water rights carried out under this section must comply with the principles of publicity and open bidding in accordance with the procedures and selection criteria established under applicable regulations").

Institutional limitations

In many areas there are institutional limitations that are erected by the water area-of-origin institutional representatives. Exchanges can serve to facilitate moving water across regions even across basins. Once third-party impacts are identified and evaluated, requested exchanges face more obstacles. Governments of the areas of origin erect institutional barriers to prevent users located in their territories from selling water to others in adjacent political boundaries. So in practical terms, the institutional barriers that are being erected for out-of-basin sales could mimic an environmental tax, which increases the price of the trade in a way that would help internalize the potential damage in the area-of-origin basins.

Inter-sectoral barriers occur when representatives of one sector fights collectively exchanges that go against its political standing within the hierarchy of water rights and political priorities. This is generally the case of irrigators. The literature shows how farmers are initially reluctant to sell water out of the sector (see Easter *et al.* 1998). There are long-term strategic reasons for combating out-of-sector water sales, chief among them the fear that the economic forces go against them and eventually their tradable rights will be questioned and perhaps irrigators will be deprived of them (Howitt 1994; Albiac *et al.* 2006).

The last set of institutional barriers stem from stakeholders' pressure. In general, green activists contest water markets because of the way water is treated as a commodity.

Environmental limitations

Environmentally based limitations are those enforced by public agencies responsible of the stewardship of the ecological quality of rivers and water bodies. In general, environmentally-based limitations are based on rigorous evidence. Occasionally, the agency in charge of approving a request to exchange water rights imposes reductions in transferable volume/flow, so that a variable proportion of the agreed exchanged volume should be left in the water courses of the area of origin.

A review of the 2005–8 exchanges

The experience with water markets in Spain is patchy, heterogeneous and discontinuous. In reviewing then, we consider the following characteristics: (a) water prices; (b) whether buyer and seller are in different basins; (c) which of the exchanging formats were used; (d) the traded amounts; (e) the location of the trading partners; (f) the role of RBAs or the Ministry; and (g) the potential environmental impacts.

A tentative grouping of exchanges can be made based on the consideration of these characteristics.

Case 1: Operations of the exchange centers of the Guadiana and the Jucar River basins

The Guadiana Exchange Center has made three public offers, on 25 October 2006, on 30 March 2007, and in September 2007. The goals were to buy rights to reduce pumping rates to raise aquifers' water tables, totaling 250 million m^3 to ensure the conservation of the Tablas de Daimiel

Wetlands (with UNESCO's qualification). They were targeted to the irrigators in the upper Guadiana basin. The exchange center acquired land property rights with appurtenant water rights. There were some tender's conditions: publicity and competitive process; preference to those licensees closer to the natural parks, river banks or protected areas; lowest bidders. The sellers were required to remove pumping infrastructures, seal the wells and install meters in remaining wells. The maximum prices of the offer were 10,000 euros/ha for land without permanent crops and 6,000 euros/ha for land with permanent crops; minimum price was 3,000 euros/ha. The results of the three public offers are presented in Table 11.1.

Table 11.1 Public offering of water and land rights in the Guadiana Basin

Offer	*Budget*	*Number of bids*	*Actual purchases*
25 October 2006	600,000 euros	228 bids, representing 1.96 million euros	55 hectares at 8,611 euros/ha
30 March 2007	10 million euros	41 bids	1,060 hectares
September 2007	30 million euros	79 bids, 12.3 million euros	1,268 hectares

The experience in the high Guadiana Basin should be judged against the complex institutional and environmental setting prevailing in the basin since the late 1980s. A number of authors have analyzed the remarkable continuity and importance of the abstraction level from aquifers' in some cases without appropriate permits or rights (Molina *et al.* 2009); the long-evolving and relentless trend of increasing groundwater use in the basin (Garrido *et al.* 2006); the succession of failed attempts to curtail extraction rates; the need to transfer from the Tagus basin to ensure the conservation of the Tablas de Daimiel Wetland, without which it may have lost its UNESCO's qualification and all its flooded area. The Guadiana public offerings were planned to continue in 2008 and following years, but the economic crises in the Spanish economy brought the so-called High Guadiana Special Plan to a sudden stop.

In the Júcar River Basin there has been one public call in 15 December 2006. The public goals were: to raise water tables in the upper Jucar's aquifer (reduce extractions by 100 million m^3) and to enhance flows to the lower Júcar basin. The exchange center targeted the irrigators in the upper Júcar basin (Province of Albacete) to reduce water pumping in the province. The allocated public Budget was 12 million euros and the range of prices for the public acquisitions were between 0.195 euros/m^3 and 0.13 euros/m^3. The River Basin acquired use rights for the entire 2007 irrigation season giving preference to those sellers with greater impact on the river and lowest bids with the condition of sealing the wells and adequate enforcement. The final acquired volumes represented 20 percent of actual right-holders' total water use.

The Júcar exchange center found that there were not enough bidders to cover the entire budget and target volume. The reasons behind it may have been (a) lack of information, (b) increased of cereal prices by the time farmers had to sign up for the program; and (c) the fact that many farmers could not afford to sell because they have fruit crops, which cannot survive without summer water applications.

Case 2: Voluntary water exchanges

There has been a short number (albeit important in volume) of water exchanges voluntarily agreed by right-holders, in the Guadalquivir and Segura (in this case totaling 70 million m^3). In the Guadalquivir, some exchanges were between the same right-holder permuting his rights from the lower basin (with more salinity concentration) with his rights in the upper basin. As a result, more water is used in the upper section at the expense of uses of more saline water in the lower sections.

Most voluntary exchanges among right holders have involved right-holders located in different river basins (jurisdictions) and have required the explicit approval of the Ministry of the Environment. Some exchanges were agreed (contracted) in 2006 (six in number, totaling 75.5 million m^3), 2007 (17 in number, representing 102 million m^3), and 2008 (two, with 68 million m^3). In all of them farmers in the area of origin (Tagus and upper Guadalquivir basins) leased out their water rights to farmers and urban users in the recipient basins of Segura (Sindicato Central de Regantes del Acueducto Tajo-Segura and Mancomunidad de los Canales del Taibilla) and the Andalusian Mediterrananean Basins (Aguas del Almanzora, which mainly service irrigators).

The exchanging format in Andalusia, which enabled transfers of water from the Upper Guadalquivir to the Andalusian Mediterranean Basins), involved water purchasers in the recipient basin (by the company Aguas del Almanzora, S.A.) purchasing or leasing irrigated land with appurtenant water rights in the lower Guadalquivir, and transferring the water rights across the basin to irrigate more land in the Mediterranean Basin. There are three geographical sites involved in the arrangement: (i) water rights linked to land in the lower Guadalquivir basin were transferred to the (ii) Mediterranean Basin using the (iii) Aqueduct Negratín-Cuevas de Almanzora, whose abstraction point is in the Upper Guadalquivir. Since this aqueduct's abstraction point is in the Upper Guadalquivir basin, the transfer effectively involved water taken 200 km upstream the location it would have been used under normal conditions. To reduce the environmental and third-party impacts of the reduced flows from the Aqueduct's headwaters to the lower Guadalquivir, a volumetric tax of 50 percent was enacted by the Ministry of the Environment, which implied that the contractor was given permission to transfer only 50 percent of the water rights attached to the land purchased.

The agreed prices in the exchanges ranged between 0.15 euros/m^2 and 0.28 euros/m^2 (Garrido and Calatrava 2009). In the case of the irrigation districts of Canal de Estremera and Canal de Aves, farmers received a payment of 2,400 euros/ha for fallowing their irrigated land (exceeding the value of the crops (maize) they would have grown under normal conditions). The Ministry of Environment exempted the exchanging parties from paying the fees applicable to all regular aqueduct's beneficiaries. In the case of the inter-basin Tagus–Segura Aqueduct the fees ranged from 0.09 euros/m^3 for irrigators to 0.12 euros/m^3 for water agencies supplying municipalities in the recipient region. In the case of Andalucia since the exchange involved only one partner (i.e. company Aguas del Almanzora, S.A.), there was no price or economic compensation.

Strengths and weaknesses of water trading in Spain

The number of exchanges in Spain is still small. In fact, voluntary market exchanges have moved less water than those brokered through the water exchanges centers, which required RBAs' budgetary allocations to run them. As reviewed above, part of the acquired water rights were used to meet environmental standards. The low level of transactions may be due to the fact that obtaining a permit to exchange water rights is still a cumbersome process, for which administrations have not worked out a rapid procedure. The problem is that the lack of completion and the thinness of the markets may have offered ample room for both monopsony and monopoly behaviors. The consequence is that some sellers sold at prices well above the willingness to accept a compensation for relinquishing their rights, if farm productivity in their region and bid prices in the banks offer some indication of what those compensations might have been requested under a more competitive regime.

Overall the buyers used the water in more valuable uses than the sellers, contributing to economic efficiency of water use. Farm productivity differences among area-of-origin and recipient areas provide an indication of the gains-from-trade. Figure 11.4 plots the water apparent productivities in the provinces from which water was taken and the province these supplies were used (Garrido *et al.* 2010). It represents the water productivity differences in provinces of the participating farmers in the Tagus-Segura exchange through in the inter-basin aqueduct, with Murcia being the recipient province, and Toledo, Madrid and Guadalajara, the area of origin.

The exchanges (voluntary and through the exchange centers) also had environmental benefits, because they allowed the Basin Agencies to acquire flows and volumes which were used to enhance environmental flows and to raise the water tables of overexploited aquifers.

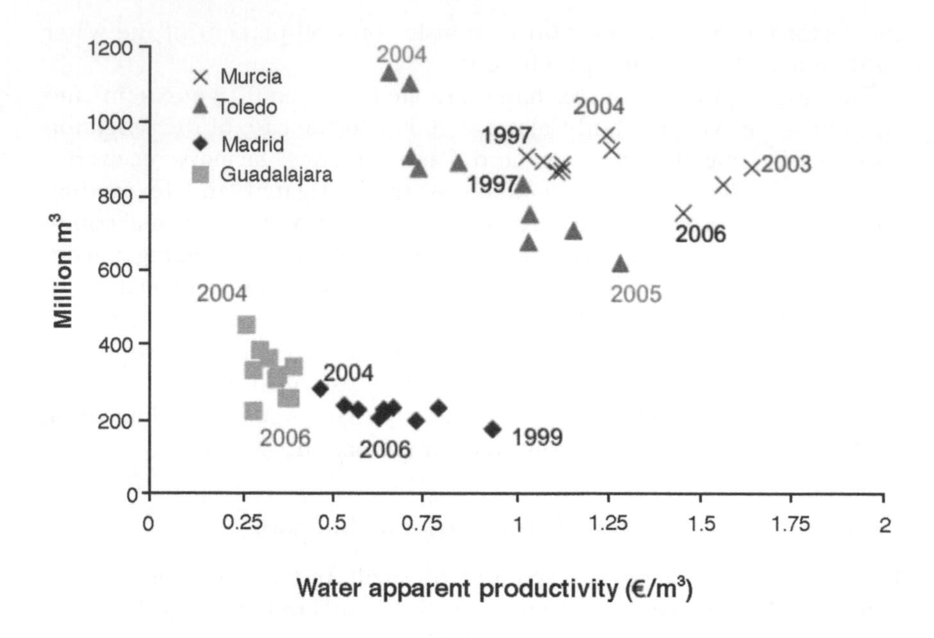

Figure 11.4 Water productivity of irrigated agriculture in the provinces from which water was transferred and to which was supplied in the Tagus–Segura Exchange
Source: Garrido *et al.* (2010)

There were some issues raised by the Government of Castille La Mancha, the area of origin, that contested the exchanges based on the hypothetical direct and indirect employment impacts in the rural areas where water use holders were selling rights. However, there seems to be little evidence of such third-party effects, as few farm labor is used in the area-of-origin from which resources were sold.

The most controversial issues

Water markets have not been free of controversy in Spain. The major issues of the debate have been: (i) whether water trading means water privatization; (ii) whether the existence of input subsidies to water services distorts the operations of the markets; (iii) whether what there has been is really trading of "paper"; and (iv) whether there are environmental impacts of water use rights exchanges.

There are some voices that argue that if water can be freely sold at an agreed price this may represents a hidden privatization process. Since profits are made out of selling rights that were given for free, an illegitimate rent is created which falls in the hands of private right-holders. There is thus the risk that the market may engender "water owners" who profit from water marketing at the expense of the public interests. They advocate that if "water planning" was more efficient, market reallocation would not be needed. Others consider that this may not be possible in Spain, where water resources fall unambiguously within the public domain. Water use rights are granted according to strict and clearly defined conditions for a limited period of time. The use rights are granted by the state after a long, transparent and cumbersome procedure, and they are subject to changes and contingent of water availability.

Some argue that input subsidies may have distorted prices. The fact is that private agents participating in inter-basin exchanges were exempted from the fees charged for using aqueducts. The gains from trade of such agreements were thus augmented by a significant amount, which may represent up to 50 percent of the paid price. Subsidies of these fees may have increased artificially the exchanged amounts and the profits made by both parties. Furthermore, as farmers do not pay the full cost of the water services they benefit from (Calatrava and Garrido 2010), market exchanges are distorted by an inflated willingness-to-pay which in any case should be netted of the full cost at the time it is sold in a market exchange. They argue that full cost recovery of water services should be a precondition for obtaining permission to become voluntary water sellers.

There has been some voices arguing that there has been paper "trading" and that this may have had environmental impacts. This is based on the fact that sellers have sold the entire volume which corresponds to that stated in the "entitlement," and not the actual water use regularly made by its holder under the non-trading situation (as the Law indicates this would be the effective right). As the buyer withdraws the entire volume written in the "entitlement," the exchange may have resulted in an increased water use, worsening the scarcity conditions of the basin. As most of the sellers have been irrigators using flooding irrigation techniques, return flows augment ground and superficial flows downstream the abstraction point for the original irrigator right holder.

Although the public authorities argue that there are enough institutional arrangements in place to avoid environmental impacts (the Water Law assigns the role of stewardship to the basin agencies, or to the Ministry of Environment), there are some that argued that this may not be the case. As yet no document, analysis or report has been made public evaluating the impact of the exchanges. In the case, for example, of the exchange with potentially more harmful effects – the referred above which involved three geographical points in the Guadalquivir and the Mediterranean Basins – it was mandated that one out of two water units sold should be left in the Guadalquivir basin. In this particular case, as the exchange involved just one agent, who owned the land from which the water rights were to be transferred and the land onto which the transferred water would be applied, the only way to tax the exchange was instituting a "water tax." No study has been made so far to analyze the environmental impacts of the recorded exchanges.

Conclusions

Based on past trends and climate change models, Spain is among the countries whose water resources will become most severely affected (Bates *et al.* 2008). Most models project lower runoff for most basins and more extreme events, droughts and floods, exacerbating many of the already stressed basins. More flexible allocation and reallocation mechanisms are considered necessary to enable water managers to cope with increasing competition for the resource and recurrent droughts. Obtaining efficiency gains are unavoidable if Spain wishes to adapt to climate change impacts. The functioning and effects of water trading has been limited, uneven and discontinuous (mainly during the last drought) with few buyers and sellers and strong public presence in promoting and facilitating the exchanges. There are important limitations, and observed patterns of trading in Spain can not yet be associated with any type of permanent water marketing. Still, it is an issue of much debate and controversy in a country where all water is public domain. Surprisingly for their detractors, water trading has been an important means for the river basins to improve ecological flows, raise water tables in aquifers supplying wetlands or downstream users, and improve the pollution dilution capacity of water bodies. To be able to realize their full potential and avoid its limitations, trading should be better regulated and more transparent, avoiding monopolistic practices and limiting input subsidies to those engaged in trading.

Note

1. In this chapter "Water Authorities in Spain" means the River Basin Agencies or, in basins entirely contained in a single Autonomous Community, the regional water agency. See Garrido and Llamas (2009) for a detailed description of the institutional framework in Spain.

References

Albiac, J., Hanemann, M., Calatrava, J., Uche, J. and Tapia, J. (2006) "The Rise and Fall of the Ebro Water Transfer." *Natural Resources Journal* 46(3): 727–57.

Ariño Ortiz, G. and Sastre Beceiro, M. (2009) "Water Sector Regulation and Liberalization." In A. Garrido and M. R. Llamas (eds), *Water Policy in Spain*, 95–106. Leiden, The Netherlands: CRC Press/Balkema.

Bates, B. C., Kundzewicz, Z. W., Wu, S. and Palutikof, J. P. (eds) (2008) *Climate Change and Water.* Technical Paper of the Intergovernmental Panel on Climate Change. Geneva, Switzerland: IPCC Secretariat.

Calatrava, J. and Garrido, A. (2010) *Measuring Irrigation Subsidies in Spain: An Application of the GSI Method for Quantifying Subsidies.* Global Subsidies Initiative

(GSI) of the International Institute for Sustainable Development (IISD). Available at http://ssrn.com/abstract=1656825 (accessed 12 July 2012).

CEDEX (2011) *Evaluación del Impacto del Cambio Climático en los Recursos Hídricos en Régimen Natural.* Madrid, Spain: CEDEX.

Easter, K. W., Rosegrant, M. and Dinar, A. (eds) (1998) *Markets for Water: Potential and Performance.* New York: Kluwer Academic Publishers.

Embid Irujo, A. (2009) "The Foundations and Principles of Modern Water Law." In A. Garrido and M. R. Llamas (eds), *Water Policy in Spain,* 107–14. Leiden, The Netherlands: CRC Press/Balkema.

Garrido, A. (2007) "Water Markets Design and Evidence from Experimental Economics." *Environmental and Resource Economics* 38: 311–30.

Garrido, A. and Calatrava, J. (2009) "Trends in Water Pricing and Markets." In A. Garrido and M. R. Llamas (eds), *Water Policy in Spain,* 129–42. Leiden, The Netherlands: CRC Press/Balkema.

Garrido, A. and Llamas, M. R. (eds) (2009) *Water Policy in Spain.* Leiden, The Netherlands: CRC Press/Balkema.

Garrido, A., Martinez-Santos, P. and Llamas, M. R. (2006) "Groundwater Irrigation and its Implications for Water Policy in Semiarid Countries: The Spanish Experience." *Hydrogeology Journal* 14: 340–49.

Garrido, A., Llamas, M. R., Varela-Ortega, C., Novo, P., Rodríguez-Casado, R. and Aldaya, M. M. (2010) *Water Footprint and Virtual Water Trade in Spain.* New York: Springer.

Howitt, R. E. (1994) "Effects of Water Marketing on the Farm Economy." In H. O. Carter, H. R. Vaux Jr. and A. F. Scheuring (eds), *Sharing Scarcity: Gainers and Losers in Water* Marketing, 97–133. Davis, CA: University of California Agricultural Issues Center.

Ministerio de Medio Ambiente (2000) *Water in Spain.* Madrid, Spain: Centro de Publicaciones del Ministerio de Medio Ambiente.

Molina, J. L., García-Aróstegui, J. L., Benavente, J., Varela, C., de la Hera, A. and López-Geta, J. A. (2009) "Aquifers Overexploitation in SE Spain: A Proposal for the Integrated Analysis of Water Management." *Water Resources Management* 23(13): 2737–60.

Part II

Concerns about water trading and how we are dealing with them

Introduction: concerns about water trading

Carlos Mario Gómez, Eduard Interwies, and Stefan Görlitz

Evidence from experience and case studies presented in the previous chapters demonstrate that allocating water to its most valuable uses provides opportunities to increase economic and social welfare. Allowing for water trading is a way to leave water users the way to spontaneously agree on a better allocation of a pre-existing set of water property rights.

In a properly working water trading scheme both the current and the would-be owners of water property rights are able to find a mutually beneficial agreement serving two purposes simultaneously: on one side, the increase in the value obtained by the economy with the amount of water available, and, on the other, sharing the financial surplus thus obtained. By definition the water transactions that take place are those that are mutually beneficial for all the individuals directly involved in the trade and, at the same time, those that increase the market value of the concerned water-using activities such as agriculture, manufacturing, hydropower, or drinking water production. There is little doubt that allowing for water trading is an alternative to foster some economic development goals in areas such as expansion of agriculture, urban development, or hydroelectricity, and it is particularly so in the face of water scarcity.

However, the main concerns are whether what is good for the individuals engaged in water trading can also be judged as equally desirable for the entire society. At the same time, we can also ask whether the changes induced by the trade in the water-using economic activities are compatible with the goals of sustainable development. The different concerns surrounding water markets that are discussed in this part of the book are connected to these social goals of water governance that definitively cannot be taken for granted in water trading schemes: the equity and/or the ecologic sustainability of the water allocation resulting from water trading.

In fact, the apparently strong evidence in favor of water trading may become weak when the scope of the analysis is extended beyond the local short-term efficiency gains resulting from reallocating water. The chapters in this section of the book illustrate how water-trading advantages become more uncertain when, for example, we add to the picture the third parties not directly involved in water bargaining. The same may happen when we consider the water resource rather than the particular amounts of water used for individual purposes. In a similar way, the evident short-term welfare gains can become uncertain when we consider that in the medium and the long term water uses cannot exceed the ability of water ecosystems to continue providing the economy with the critical water services it depends on.

Trading is a clear mechanism to foster economic efficiency but markets do not necessarily do it in an equitable manner. Not only are markets not a valid means to provide access to water and sanitation to the poor; in the particular case of water, markets also do not include in the bargaining process all the parties and individuals to whom water is important. Among those who risk to be excluded there are the so-called third parties. They include all the water users that can be affected because of the effect transactions may have over the amount of water available at any point in the basin: those affected by, for example, changes in river flows, water returns or groundwater recharge along the river basin. In addition to that, water resources are not only valuable because of the services provided to different economic activities to which water contributes as an essential input. Water resources also support many environmental services, such as recreation, health, and biodiversity, that cannot possibly be part of water trading deals. In fact water transactions involve only a rather limited set of water users while the interest of all others including those affected by the non-use environmental services provided by water connected ecosystems can only be protected by properly working water governance institutions. In mature water markets, governance includes institutions with expertise, consolidated water administration, comments and approval processes, and regulation of social, environmental, and third party effects, including the tax base of areas of origin. The most important regulation is limitation of trade to historical use and consumed volumes, to minimize third party and environmental impacts.

In addition, the social perception on the purpose and the means of water policy has radically changed since the first modern experiments of water trading were implemented in the early 1980s. The initial priorities – to support agricultural, industrial, and urban development, and to overcome the constraints resulting from a limited supply of water resources – have slowly shifted towards a new vision focused mostly on preserving water assets rather than administering water flows and also on balancing economic progress with the other important goals of attaining social fairness and ecological sustainability. This recent and still-in-progress change in water policy affects the perception about the performance of water trading. Some advantages are now less evident partly because most of the environmental outcomes of water trading remain simply unobservable (for example, a proper monitoring of water transactions in Chile is still pending), or have not been given political importance so far. An example of the latter is the recent recognition that trading opportunities can increase water scarcity as they may put into use water rights that would remain otherwise unused. Water markets allowing the transaction of hitherto non used waters and of nominal-as opposed to effectively used volumes-entitlements have raised concerns regarding sustainability and social equity.

In fact there are many individual opportunities and actions that are connected to each other through the water environment (involving users up- and downstream, those who benefit from water use and those who appreciate the preserving non-use services more, the current importance of water uses versus the future generations options, etc.). The recognition of that is but one of the elements below the many concerns about the potential advantages and disadvantages of allowing water trading.

Regarding water trading, the chapters in this part of the book draw attention to the importance of social and environmental concerns and values, on the one hand, and to the subsequent importance of the scheme's design on the other. Water markets need to be assessed not only by their local short term efficiency gains but by their contribution to foster economic progress, social justice and ecological sustainability. The difficulty of this task demonstrates clearly the importance of design and institutional arrangements in shaping – and judging – water trading systems. Naturally, there is no generally accepted approach to the framework and design of water markets. On the contrary, questions regarding the amount of economic liberty granted in a water trading scheme, the level of regulation by public authorities, the necessary legal security concerning private property rights, the inclusion of externalities, the issuing of permits or water use rights etc. are hotly debated topics. Generally, however, it has to be stated that there can be no generalized approach to forming water markets, as each trading scheme´s design has to take into account the scheme-specific circumstances, both spatial (e.g. hydro-geological background of watershed/river basin; specifics in water use etc.) and factual (e.g. economic, ecologic and social objectives).

In Chapter 12, Joseph W. Dellapenna firmly criticizes the temptation, supported by some economists, of pushing water markets as the single most effective solution to water allocation challenges. He states that the exigent conditions required for a water market to work efficiently, which are present in few existing markets, but are especially unlikely regarding water – exclusive, secure, well-defined, non-attenuated, transferable and enforceable private property rights, absence of market distortions among buyers and sellers, and absence of uncovered external benefits and costs – are nearly impossible to install in the case of water, and that the yet existing water markets are only functioning through heavy intervention of public authorities. Moreover, he argues that advocates of water trading schemes often fail to recognize the public good nature of water and the considerable transaction costs inherent in attempts to treat it as a private good, thus neglecting a further restraint to the proper functioning of a non-regulated water market. Dellapenna calls in his article for a careful consideration of water markets as a tool for reallocating water use rights. Because of the importance of water as a social good, the public authorities need to take an important role as regulator. Therefore, water markets are, according to the author, a potentially effective (and efficient) economic instrument among others that, in any case, can only work in a proper regulatory environment.

In Chapter 13, David Katz explores the kind of reforms that have the potential to make water markets work also for the environment (e.g. for guaranteeing and going beyond minimum environmental flows and for protecting crucial ecosystem services). The author provides a review of the theory and empirical literature concerning water markets and environmental flows, as well as possible third party impacts incurred by transferring great volumes of water along great distances. Katz provides compelling evidence on the economic value the public assign to the maintenance of environmental flows and shows how water markets are a potentially significant option for securing these flows. He recognizes, however, the various challenges that impede a proper market allocation of environmental flows – gaps in information regarding economic and ecologic benefits of environmental flows, regulatory constraints, and the nature of water as a non-stationary, both public and private good.

In the third and final chapter in this section, Robert J. Rose (Chapter 14) focuses on the potential of transferable emissions schemes to reduce widespread water pollution loads The critical importance of the institutional set-up is underlined by the analysis of its more necessary components, such as the enforcement of strict caps, and also through the discussion of the role of the command-and-control instruments in place that need to be adapted, when not removed, in order to guarantee an effective functioning of such a trading alternative. The developed theoretical framework is then applied hypothetically to the Po River Basin in order to highlight the type of analysis necessary to create the least constraining policy that sufficiently addresses existing hot spots and issues regarding fate and transport of pollutants. The author concludes by highlighting the importance of a progressive strategy during the transition. The best option, according to Katz, is to start with simple options, and only allow more complicated alternatives as the advantages and disadvantages become more evident and social acceptability increases.

12 The myth of markets for water

Joseph W. Dellapenna

Introduction

The emerging climate disruption will change precipitation patterns around the world (IPCC/TEAP 2007). The changes will have far reaching effects on innumerable aspects of the lives of humans and other living things (UNECE 2009; Roy *et al.* 2010; Wilcove 2008). The resulting challenges to water management institutions and to water law regimes occur in a world in which those systems are already under stress because of growing populations and growing per capita demand (Dellapenna 2010). The resulting stresses, moreover, arise in a world dominated by the "Washington consensus." Pro-market dogmas – the view that markets are the best way to manage resources and the economy and should be used both to allocate resources and to distribute wealth within society (Stiglitz 2002) – formed the "Washington consensus" embraced by the US Treasury Department, the World Bank group, and the International Monetary Fund. From the 1980s onward, the three institutions promoted market systems for water, including privatizing water utilities, treating water as a commodity, and relying on markets as the primary management tool (Dellapenna 2008; Dinar 2000). The resulting controversy raises serious questions about the utility of markets as a tool for addressing the growing global water crisis or for managing water generally. This section considers whether markets can work for water resources, drawing upon legal and economic theory, on the actual effects of the privatization of water utilities, and the consequences of treating water simply as a commodity.

The effort to rely on markets as the primary tool for responding to the growing water crisis has generated intense controversy (Naegele 2004; Rothfeder 2001). While I side with those who are critical of such efforts, I do not suggest that markets generally are a bad idea. My point is modest: markets need to be carefully considered before being adopted as a mechanism for social ordering for a particular field of activity rather than reflexively instituted in the belief that markets always work best. Markets are not always a satisfactory way to manage certain aspects of economic or social activity. This should hardly be surprising to anyone who actually examines the empirical evidence – something the economists of the Washington consensus seemed unable to do (Bauer 2004). Ronald Coase, winner of the Nobel Prize in Economics, demonstrated that a private-property/market system is a most efficient mechanism for allocating resources for particular uses when it works and that the particular rules of law applied to disputes over resources will not affect how those resources are allocated to particular uses so long as markets work (Coase 1960). Coase, however, stressed that markets fail when there are significant barriers to their functioning, particularly transaction costs.

Are markets the best model for managing water resources?

Many economists argue that markets are the best method for achieving environmental goals (Jaeger 2005; Keohane and Olmstead 2007). Markets have been particularly stressed for managing water resources. Systems of public allocation of water can result in significant inefficiencies and waste, particularly if there is no mechanism for shifting water from old uses to new ones. The problem is compounded by the tendency of major regulated industries to capture the regulators because of the need to rely on information provided by those industries and the rotation of high-level administrators between the public agency and the industries (Morgan and Chapman 1995). Market incentives, we are told, also lead people to actions that are desirable from a social perspective" (Heal 2000). Can markets in fact solve these problems?

The asserted benefits of markets for water

Economists tend to argue that water can be managed more productively and more efficiently when treated as a tradable standardized commodity rather than as a product of engineering or an integral part of nature (Gomez and Loh 1996; Robbins 2003; Thobani 1996). According to this view, the most efficient use of water can be made only through private, for-profit markets, primarily because of the theoretical benefits of competition (Griffin 2006; Lee 1999; Merritt 1997). They argue that private markets, with individuals buying and selling water according to economic rules, rather than management by government regulators, can determine the true economic value of water, ensuring that water is allocated to its highest and best use. It is important to stress that the concern here is exchanges involving substantial amounts of water over wide distances and (often) widely differing uses, not mere temporary changes of limited quantities of water between neighboring users. Yet market transactions on this scale have been vanishingly rare, unlike small scale transfers between similar users in close proximity to each other (Dellapenna 2000).

For large, long-distance transactions, we are told, the maximizing of the economic efficiency of water allocation arises from the maximizing of the financial return gained from the allocation (Rosegrant and Gazmuri 1995). This is a simple tautology because economists measure the value of water by society's willingness to pay for the increment of production resulting from the additional allocation of water (Dasgupta *et al.* 1995). Yet some economists distinguish between the true "economic" value of goods and the "financial" value of goods. Proponents of this view recognize that there are important ecological, environmental, aesthetic, and spiritual benefits, in addition to the benefit of human survival, which should be met outside the market for the good for society as a whole. The obvious difficulty is quantifying such social benefits outside the market pricing mechanism.

Leaving aside for the moment such non-marketizable benefits, a number of potential benefits would seem to arise from private markets for water. Ownership of clearly defined water rights is said to provide security to water users and to allow them to be more responsive than under the centralized allocation of water. Tradable water rights provide an inherent value or opportunity cost that creates built-in incentives to conserve water and to put it to the most productive use, allowing the movement of water to higher-value uses in a way that is cheaper and fairer than the alternatives. Such rights encourage investment and growth in activities that require assured supplies of large quantities of water and in water-saving technologies because the investors will benefit from the savings (Rosegrant and Binswanger 1994; Thobani 1996). Markets supposedly avoid the "tragedy of the commons" by assigning prices to scarce commodities, creating pressure to economize on such scarce resources (Perelman 2003). Such a system also might provide an incentive for industry to locate to areas where water costs less – often rural areas that otherwise have few employment opportunities.

Economists tell us there are several requirements for a well-functioning competitive water market to exist. There must be exclusive, secure, well-defined, non-attenuated, transferable and enforceable private property, rights (Hart 2008; Winpenny 2003). There should be an absence of collusion or market power among buyers or sellers and an absence of unpaid-for benefits (positive externalities) and uncompensated costs (negative externalities). Finally, each transaction should operate in a competitive market under the motivation to maximize profits. Under such conditions, demand and supply forces will determine the quantities to be traded and the unit price for the water, with water moving from low value uses to higher value uses. These conditions, it turns out, are nearly impossible to create for water.

Why markets for water fail

It is notable that markets for water – particularly water in a natural water body ("raw water") – are rare. As Terence Lee noted, "The idea of treating water as an economic good... is so novel that using markets, rather than bureaucratic decision, for water allocation makes almost everyone responsible for water policy very nervous" (Lee 1999: 78). Water markets have seldom produced significant changes in the ways water is used; such markets as do exist involve relatively small amounts of water sold among similar users in close proximity to each other, often among shareholders in a mutual ditch company or the like (Howe and Goemans 2003; Zaman *et al.* 2005). When there have been so-called markets intended to bring about major changes in the time, place, or manner of use, they functioned only through heavy-handed intervention of the state (Dellapenna 2000). The process in reality is not a market. So, if markets for raw water are so good, why are they so seldom used? Market advocates, however, seldom address such questions, preferring to denigrate their critics for holding cultural, religious, even mystical prejudices about water that prevent water from being treated as it should – just like any other commodity (Anderson and Snyder 1997: 17–29, 114–15). But water is not like other resources.

Despite the theoretical advantages of water markets, even a "perfect" market has disadvantages, not least of which is that their proper functioning requires transactions by rational individuals, with perfect knowledge, deliberately choosing to maximize utility (Dellapenna 2007b). One need think no further than the several recent "market bubbles" to doubt the rationality of some economic decisions (Perelman 2003). Rationality problems have become the focus of cognitive psychologists and by certain economists working in the fields of "behavioral economics" or "socio-economics" (Jones and Goldsmith 2005; Kelman 2003; Korobkin and Ulen 2000; Lubet 1996). They show that irrationality is built into how people live and make decisions, irrationality that prevents the market models from working in the way that economists assume. Of course, an economist's notion of irrationality often is just another person's idea of taking into account different values than those favored by economists – values that are impossible to price and therefore impossible to appraise or manage through a market. These might well be the cultural, religious, even mystical prejudices about water that economists love to denigrate. In addition to rationality problems, the complexity of the modern economy means that decision-makers not only often lack perfect knowledge, they often have only rudimentary knowledge.

Effective water management must do more than provide economic efficiency; it must also provide efficient decision-making, clearly communicate its rules and regulations, and furnish prompt enforcement (Coase 1960). Nor will efficiency principles always coincide with equity and political considerations. In fact, there is some evidence that equity is more important than efficiency for stable water management systems (Godden 2005; Howe 1996). Apart from these generic difficulties with markets (which are not so different from the generic difficulties with public management of water resources), the special problem of water that makes markets unsuitable is the inability of bilateral water transactions between sellers and buyers to reflect the effects of the transaction on third party, public, or environmental interests (Lund and Israel 1995). Furthermore, a system of tradable property rights impedes the development of effective river basin planning and environmental and ecological protection. Finally, there is the potential loss of the authority and control of the internal water resources of a nation or state to foreign buyers.

The core problem with markets – particularly the inability of markets to factor in adequately the external effects of market transactions – is transaction costs. Transaction costs can severely distort the economic efficiency of private water transactions, rendering theoretical market models that ignore transaction costs completely irrelevant (Dellapenna 2000). Nearly always a change in the time, location, or manner of use of water affects other water users. If the core idea that drives markets as useful tools for society is that a person's interests will only be affected if that person consents, then the effects of the proposed change on such third parties must be accounted for in determining whether the change in the ownership or use of a water right should take place (Dellapenna 2007b). Transaction costs, in short, are the costs of determining the costs to third parties and then compensating them for those costs.

While it is easy enough for someone to own and manage water unilaterally in small amounts (for example, bottled water), withdrawing water on a large scale necessarily affects many others, making it difficult to contract for the consent of all significantly affected persons. But, unless the agreement of all affected parties is obtained, third parties may be effectively deprived of their right to use water without their consent and wealth is transferred from those who formerly used water to those who thereafter would use water. Typically those who lose out are small users without capital resources or alternative sources of supply of water. And this doesn't even begin to consider the impact on persons or ecosystems that do not have legally recognized rights to use water but that nonetheless enjoy the fruits of water's location or use. Thus allowing market transactions causes the transfer of wealth from the general public, particularly its poorer members, to the privileged few. The law usually protects against such externalities by the rule, found in nearly all legal systems, that one cannot alter the time, place, or manner of the use of water without the consent of other holders of water rights (Thobani 1996). Yet, if the rights of third parties and the public are protected, transaction costs with respect to all but the smallest water bodies quickly become prohibitive (Dellapenna 2000).

The reality of transaction costs should give even the most free-market oriented economist cause to consider whether true markets can function effectively for water resources (Dellapenna 2000, 2007b). One of the best examples of the problem was a proposed water trade between the City of Denver and the Coors Beer Co. (Colorado Supreme Court 1972). Coors, then known for the high quality of the water used in its brewing, could not produce enough beer to satisfy demand for its product without finding more water. Denver, a fast growing city, also sought new sources of water. Denver and Coors agreed to exchange Coors' "clear mountain stream" for the right to use unlimited quantities of Denver sewage water in its brewery. The transaction failed not because of possible outrage by beer drinkers, but because farmers downstream from Denver (organized as the Fulton Irrigating Ditch Co.) obtained an injunction against the trade. The transaction would have deprived the farmers of the return flow on which they relied, even though, 30 years earlier, they had contractually recognized the seniority of Denver's rights over their own. Coors or Denver could have paid the farmers to surrender their right to block the deal, but that would have left others, further downstream, free to challenge the transaction. The number of potential claimants was many thousands, at least, and some of them could become "holdouts" – persons who either were wholly unwilling to deal, or were willing to deal only at prohibitive prices wholly out of line with the market values at stake. Denver and Coors abandoned the transaction. Coors went on to establish satellite breweries in other parts of the country because of its inability to obtain more water for its original brewery (Dellapenna 2000, 2007b).

Advocates for water markets sometimes insist that the protection of third-party rights represents an overly rigid legal regime. If only such requirements were removed, markets would flourish. This mischaracterizes the situation. Protection of third-party rights operate to prevent market-generated externalities from destroying the property rights of the third parties. Rather than representing government intervention that prevents or distorts markets, such protections are the minimum that is necessary to ensure that property rights – each person's property rights – are transferred only through markets. Richard Posner, a champion of using markets for water, has described why such third-party rights must be protected if society is to ensure that water is used efficiently:

> If the effects of return-flow were ignored, many water transfers would reduce overall value. Suppose A's water right is worth $100 to him and $125 to X, a municipality; but whereas A returns one half of the water he diverted to the stream, where it is used by B, X will return only one fourth of the water it obtains from A, and at a point far below B, where it will be appropriated by D. And suppose B would not sell his right to A's return flow for less than $50, while D would sell his right to the municipality's return flow for $10. Given these facts, to let A sell his water right to X because it is worth more to X than to A would be inefficient, for the total value of the water in its new uses (X and D's) – $135 – is less than in its old uses (A and B's) – $150. The law deals with this problem by requiring the parties to show that the transfer will not injure other users. In practice this means that A and X in our example, in order to complete their transaction, would have to compensate B for the loss of A's return flow; they would not do so; and the transaction would fall through, as under our assumptions it should.
>
> (Posner 2007: §3.11)

Things get even more complex where the transfer increases return flows. Because of these complexities, transaction costs will prevent the functioning of markets so long as third party rights are recognized except on a small-scale without major changes in where or how water is used.

The law generally does not protect persons or natural entities that do not hold water rights, yet the effects on those persons and entities can be equally dramatic. For example, an irrigating farmer supports various businesses and industries that provide seed, farm equipment, and other daily or seasonal needs. Both a permanent and a seasonal labour market develop to support the farmer. When the farmer's water withdrawal permit is transferred to another location or manner of use, an entire business community and the tax base of local governments are adversely affected (Mann and Mecon 2002). Even in a relatively humid area like Georgia (in the United States), values for agricultural land with a permit to withdraw water are almost double those for land without a permit (Demeo 2003). This might not be a problem if only a few farmers trade their permits, but as more farmers do so, the drain on the community mounts up. Should there be a limit? Who decides when the limit is reached? How should the law determine which farmer can trade and which cannot? These are serious problems that have a significant economic impact on individuals and local governments. These are problems that market mechanisms simply do not address, suggesting a need to consider social equity, as well the protection of other private rights, in any transaction in water rights.

Economists argue that such effects can be eliminated if the transfer is restricted to the consumptive use of the water withdrawn. This presupposes that one can readily determine what the consumptive use is. For municipalities and industries that return water to one or another water source through a "point source" outlet, precise measurements are possible. And for such activities, the percentage of the water withdrawn that is consumed often is very low – although the return flow often will be heavily polluted if it is not treated properly before its return. For agricultural and other uses that do not involve a point source discharge, determining the amount of return flow can be difficult or impossible. Absent proof of the amount of consumptive use, the transaction will be barred (Dellapenna 2000).

Complicating the problem of transaction costs and public values is that the value of water often extends beyond its financial benefits. To many individuals and societies, water symbolizes security, opportunity, and self-determination (Draper 2005; Shiva 2002). In areas where water scarcity is normal, water is associated with life, power, and status (de Haan 1997). Certain water uses are associated with quality-of-life issues or have social or civic purposes that cannot be quantified in a market valuation. Thus minimum flow patterns established for environmental and ecological purposes, as well as the water needed to meet other public purposes, must be established by political processes and not through markets.

Finally, beginning with the Dublin Principles of 1992, there has been recognition that water management should be based on four guiding principles; economic, sociopolitical, gender, and economic issues (Dublin 1992). These principles have developed into a doctrine of integrated water resources management, the primary goal of which is to develop a comprehensive understanding of the available water resources and human and ecological water resource needs within a given basin, and then to manage those resources in an equitable and sustainable basis. Integrated water resources management stands on a foundation of four legs: economic efficiency and growth; social equity; environmental and ecological protection; and effective governance. Water management processes are required to integrate the assessment, management, protection, and use of all water resources within a basin to meet human needs on a sustainable basis while still protecting the integrity of the resource and its associated ecosystems. Tradable property rights to water prevent integrated water resources management because the market's focus is exclusively on economic efficiency, neglecting or negating the other three legs.

Box 13.1 Supposed real-world examples of successful markets

Market advocates often use certain real-world examples in an effort to prove that markets work successfully, including the Chilean water law of 1981, the California Water Bank, and the "sale" of water from the Imperial Valley Irrigation District to San Diego. These examples are illusory.

Market advocates claim that Chile provides an example of how successful market institutions can be in allocating water (Mohanty and Gupta 2002; Rosegrant and Binswanger 1994; World Bank 1994). The Chilean system of tradable water rights, however, does not confront externalities, such as third party effects or environmental and ecological issues (Bauer 2004; Vergara 1997). Moreover, market advocates base their conclusions on the theoretical effects of water markets without examining the actual functioning of the Chilean law (Bauer 2004). Little market activity has happened in Chile despite the grand claims for that law. Almost no trades occurred in any river basin other than the Limarí, and even there trades largely stopped because of intense public opposition. Carl Bauer concluded that the Chilean system seems "incapable of handling the complex problems of river basin management, water conflicts, and environmental protection" (Bauer 2004: 86–89).

The two California examples, beloved by some champions of markets (Huffman 2004), are not markets at all, but rather examples of government management masquerading as a market (Dellapenna 2007b; Gray 1994b). The California Water Bank, created in 1992 in response to an intense drought, moved water out of agriculture to certain northern California cities (Gray 1994a; Israel and Lund 1995).

California allowed the water bank to take water rights without protection for third parties rights. The water bank, moreover, was the only legal buyer for the 350 persons willing to sell water rights, and was the only legal seller for the 20 municipalities willing (and allowed) to buy water rights. The Bank's prices were set administratively, not from bidding in a market. The Bank also selected buyers and sellers by administrative fiat. This was not a market in any meaningful sense of the term, but the implementation of government policy through economic incentives, with at least a veiled hint of coercive power.

In 2003, another five-year drought provoked the transfer water from several large irrigation districts in southern California to San Diego (Dellapenna 2007b). The city asked the Imperial Valley Irrigation District to sell about 11 percent of its allocation from the Colorado River, but the District board voted 3–2 in December 2002 to reject the offer. Federal Secretary of the Interior Gail Norton then cut the District's allocation of water from the Colorado River by 15 percent, indicating she would restore it only if it was sold under the terms of the rejected contract. In the end, the District board "accepted" the contract by another 3–2 vote. This was not a market at all given the heavy government involvement in selecting the buyer and the seller, in setting the terms of the transaction, and in coercing "agreement." The transaction did provide cash to the owners of the farms served by the district, but it provided nothing but unemployment for the farm workers idled in order to free up water for San Diego and disaster for ecosystems dependent on runoff from the farms. Even the landowners considered themselves short-changed.

Public management as an alternative to markets

Although individuals and organizations have tried for years to quantify social, environmental, and other non-production costs and values (e.g. valuing social, religious or aesthetic needs and desires or valuing ecological sustainability and ecosystem services), our inability to fully understand and to fully quantify the total impacts of our economic decisions and our inability to assign generally accepted economic values to those actions severely limits our ability to evaluate and balance the overall costs and benefits of such actions in economic terms. Thus, ecosystem services are not priced in the market place, but they are real enough and inure to the benefit of all, including many who may not be participants in the withdrawal and use of water for a given economic activity (Draper 2005). Preserving the opportunities for future generations (intergenerational justice) is also an important goal, although future generations cannot participate in the market. Moreover, environmental issues affecting ecological sustainability often occur over long time horizons; for example, climate disruption issues have developed over a century or more and will take a similar period to play out. Species extinction, biodiversity loss, and the storage of nuclear waste involve similar time horizons, time horizons that are completely outside the normal range in economic decision making (Heal 2000).

The externalities and other difficulties demonstrate that there cannot be market for water without public oversight and intervention (Heal 2000). As a result, historically public water allocation has been the norm, in part because water has been considered a public asset, not a marketable commodity (Le Moigne *et al.* 1997). For effective water management, stakeholder involvement and public participation are essential (Gleick 1998; World Commission on Dams 2000). Water management restricted to private transactions between willing buyers and sellers precludes such a governance structure. Public management can promote equity objectives and enable decision-makers to deal with some of the unusual aspects of water resources, while the public sector, so long as the political will can be found, is able to fund large-scale water development that is too expensive for the private sector. Public management can also protect the poor and consider environmental and ecological needs. The case for public allocation is particularly strong in inter-sectoral allocation and reallocation because the state is often the only institution that includes all users of water resources and has jurisdiction over all sectors of water use (Dinar *et al.* 1997). The resulting public management, if done effectively, involves a multi-disciplinary endeavor, involving the natural and social sciences, engineering, politics, and law, as well as economics. The preference for public management does not preclude recourse to economic instruments as public managerial tools (Seroa da Motta *et al.* 2005; Wichelns 2006). Economic incentives can take any number of forms, from taxes, to "water banks," to assigning operation and maintenance to farmer organizations or water districts, to user charges, to volumetric or quasi-volumetric pricing. One simply must not confuse such tools with markets. The question is how to structure economic incentives so as to ensure a reasonable modicum of efficient use coupled with adequate protection of public values (Anderson and Snyder 1997; Dinar and Subramanian 1997; Freeman 2002; Heal 2000; Oates 1996).

Concluding remarks

Talk of privatization and markets reflects a simple demand to an end to the treatment of water as a free good. Pricing water, or other economic incentives, forces water users to evaluate the social consequences of their conduct more realistically (Damania 1996; Dinar and Subramanian 1997; Merritt 1997). To deny that water is also a public good, however, is simply wrong. Consider that even the strongest advocates of free-market economics use water metaphors to describe the few public goods that they will recognize: "common pool resource," "spillover effects," and so on. Yet advocates of markets for water hardly mention the public nature of water and barely consider the transaction costs inherent in any attempt to treat water as a private good. While some economists do acknowledge that the inherently public nature of water precludes true markets, they often end up advocating "transferable allocation permits" as the best method for allocating water to particular uses (Rosegrant and Binswanger 1994; Rosegrant and Gazmuri 1995; van Egteren and Weber 1996). They seem unable, however, to explain how such tradable permits would differ from markets.

The paradigm of property remains the ownership of land. Land can be marked off and considered for most purposes as the exclusive domain of a particular owner with, despite certain restrictions on the rights of landowners, limited regard for the effects of the owner's conduct on other persons or property. Land, for the most part, stays put within its boundaries. Flowing water, an ambulatory resource, simply does not fit very easily into such a paradigm. Concepts of property in water can be broadly divided into three types: common property, private property, and public property (Rose 1994). The three types each correspond closely to the three real-world models of water law found today in the United States (Dellapenna 2000). Riparian rights is a near perfect embodiment of the model of common property – each riparian owner decides for herself when, where, how, and how much water to use and outside decision makers become involved only if two riparian owners directly interfere with each other. Appropriative rights are as close as one finds to a private property model to water rights. The right to use water is defined as to the timing, location, purpose, and amount of use, as well as according to a strictly enforced temporal priority ranking ("first in time, first in right"). And increasingly states in the United States are turning to regulated riparianism, an application of a public property model to the right to use water. The right to use water depends upon a time-limited permit allowing the state collectively to determine, and periodically to re-determine, the socially best use of the water.

The correspondence between the forms of American water allocation law and the several basic models of property rights is important because it enables us to predict with some certainty whether existing forms are adaptable to changing circumstances, or whether an entirely new form must be substituted when circumstances of water demand or supply change dramatically. Treating water as common property leads into the tragedy of the commons as soon as water becomes a scarce commodity in a particular region, and thus state after state in the eastern United States has abandoned traditional riparian rights (the common property model) for regulated riparianism (the public property model) – and not, as some economists and others predict, for a private property model (Dellapenna 2000). There are reasons, some highly specific to the situation of the eastern states of the United States, why those states did not adopt a private property model. At bottom, however, the problem is that markets have simply failed to emerge even under appropriate rights – the private property model – for the reasons already discussed. Because of the utter failure of true markets, states have been left to use the imperfect public property model as the best available.

Attempts to commodify water generate the inequities that follow from markets without bestowing the benefits that markets, when functioning at their best, can provide – the benefits of rational management and efficient use that to some extent justify the inequities generated by the use of markets. Indeed, the utter unsuitability of markets for managing raw water – water in bulk in its natural sources – raises questions of why anyone, including economists, would insist on treating water solely or even primarily as a market commodity. Blind faith seems a better explanation than the rational application of well-founded economic theory to yet another natural resource. True markets must remain marginal to the management of large quantities of raw water for numerous diverse users, and economic incentives must be seen as supplemental to public management. This is not to deny that economics is relevant, merely that it is not the only relevant mode of analysis. Today, resistance to markets for raw water is stiffening and has achieved some real successes, both nationally and internationally. Rather than viewing this as a failure of policy makers to enact the necessary market reforms, efforts to reform water law need to consider alternatives to markets – alternatives that should include economic incentives, even if not markets – as means for adapting to the global climate disruption. It is time to put the Washington consensus into the past and to move forward without such crippling preconceptions – not to eliminate the market under all circumstances (remember bottled water), but to recognize it as an option, often not a very good option, for raw water.

References

Anderson, T. and Snyder, P. (1997) *Water and Water Markets: Priming the Invisible Pump*. Washington, DC: Cato Institute.

Bauer, C. J. (2004) *Siren Song: Chilean Water Law as a Model for International Reform.* New York: Routledge.

Coase, R. H. (1960) "The Problem of Social Cost." *Journal of Law and Economics* 3: 1–44.

Colorado Supreme Court (1972) "City and County of Denver Versus Fulton Irrigating Ditch Company." *Pacific Reporter* 506: 144–53.

Damania, O. (1996) "Pollution Taxes and Pollution Abatement in an Oligopoly Supergame." *Journal of Environmental Economics and Management* 30: 323–36.

Dasgupta, P., Bengt, K. and Maler, K. G. (1995) "Current Issues in Resource accounting." In P.-O. Johansson, B. Kristrom, and K.-G. Maler (eds.), *Current Issues in Environmental Economics*, 117–52. Manchester, UK: Manchester University Press.

de Haan, E. J. (1997) "Balancing Free Trade in Water and the Protection of Water Resources in GATT." In E. H. P. Brans, E. J. de Haan, A. Nollkaemper, and J.

Rinzema, (eds.), *The Scarcity of Water: Emerging Legal and Policy Responses*, 245–59. Berlin, Germany: Springer.

Dellapenna, J. W. (2000) The Importance of Getting Names Right: The Myth of Markets for Water." *William and Mary Environmental Law and Policy Review* 25: 317–77.

Dellapenna, J. W. (2007a) "Transboundary Water Sharing and the Need for Public Management." *Journal of Water Resources Planning and Management* 133: 397–404.

Dellapenna, J. W. (2007b) "Introduction." In Amy K. Kelly (ed.), *Waters and Water Rights*, ch. 6. Dayton, OH: LexisNexis.

Dellapenna, J. W. (2008) "Climate Disruption, the Washington Consensus, and Water Law Reform." *Temple Law Review* 81: 383–432.

Dellapenna, J. W. (2010) "Global Climate Disruption and Water Law Reform." *Widener Law Review* 15: 409–45.

DeMeo, T. A. (2003) *Rural and Urban Water Tours Foster Invaluable Information Exchange on Georgia's Regional Water Resource Issues.* Atlanta, GA: Association of County Commissioners of Georgia.

Dinar, A. (ed.) (2000) *The Political Economy of Water Pricing Reforms.* Washington, DC: World Bank.

Dinar, A. and Subramanian, A. (eds) (1997) *Water Pricing Experiences: An International Perspective.* World Bank Policy Paper no. 386. Washington, DC: World Bank.

Draper, S. E. (2005) "The Unintended Consequences of Tradable Property Rights to Water." *Natural Resources and Environment* 20(1): 49–55.

Dublin (1992) "Dublin Statement on Water and Sustainable Development." Dublin International Conference on Water and the Environment. Available at www.gdrc.org/uem/water/dublin-statement.html.

Freeman, A. M. III (2002) *The Measurement of Environmental and Resource Values: Theory and Methods.* 2nd edition. Washington, DC: RFF Press.

Gleick, P. H. (1998) "Water in Crisis: Paths to Sustainable Water Use." *Ecological Applications* 8: 571–9.

Godden, L. (2005) "Water Law Reform in Australia and South Africa: Sustainability, Efficiency, and Social Justice." *Journal of Environmental Law* 17: 181–205.

Gomez, S. and Loh, P. (1996) "Communities and Water Markets: A Review of the Model Water Transfer Act." *Hastings West-Northwest Journal of Environmental Law and Policy* 4: 63–73.

Gray, B. (1994a) "The Market and the Community: Lessons from California's Drought Water Bank." *Hastings West-Northwest Journal of Environmental Law and Policy* 1: 17–47.

Gray, B. (1994b) "The Modern Era in California Water Law." *Hastings Law Journal* 45: 249–308.

Griffin, R. C. (2006) *Water Resource Economics: The Analysis of Scarcity, Policies, and Projects.* Cambridge, MA: MIT Press.

Hart, K. J. (2008) "The Mojave Desert as Grounds for Change: Clarifying Property Rights in California's Groundwater to Make Extraction Sustainable Statewide." *Hastings West-Northwest Journal of Environmental Law and Policy* 14: 1213–39.

Heal, G. (2000) *Nature and the Marketplace: Capturing the Value of Ecosystem Services.* Washington, DC: Island Press.

Howe, C. W. (1996) "Water Resource Planning in a Federation of States: Equity Versus Efficiency." *Natural Resources Journal* 36: 29–36.

Howe, C. W. and Goemans, C. (2003) "Water Transfers and their Impacts: Lessons from Three Colorado Water Markets." *Journal of the American Water Resources Association* 39: 1055–65.

Huffman, J. L. (2004) "Water Marketing in Western Prior Appropriation States: A Model for the East." *Georgia State University Law Review* 21: 429–48.

IPCC/TEAP Special Report (2007) *Working Group II, Fourth Assessment Report: Impacts, Adaptation, and Vulnerability.* Available at http://ipcc-wg1.ucar.edu/wg1/wg2-report.htm.

Israel, M. and Lund, J. R. (1995) "Recent California Water Transfers: Implications for Water Management." *Natural Resources Journal* 35: 1–32.

Jaeger, W. K. (2005) *Environmental Economics for Tree Huggers or Other Skeptics.* Washington, DC: Island Press.

Jones, O. D. and Goldsmith, T. H. (2005) "Law and Behavioral Biology." *Columbia Law Review* 105: 405–502.

Kelman, M. (2003) "Law and Behavioral Science: Conceptual Overviews." *Northwestern University Law Review* 97: 1347–92.

Keohane, N. O. and Olmstead, S. M. (2007) *Markets and the Environment.* Washington, DC: Island Press.

Korobkin R. B. and Ulen, T. S. (2000) "Law and Behavioral Science: Removing the Rationality Assumption from Law and Economics." *California Law Review* 88: 1,051–1,144.

Le Moigne, G., Dinar, A. and Giltner, S. (1997) *Principles and Examples for the Allocation of Scarce Resources among Economic Sectors.* Options Méditerranéennes Séries A, no 31. Available at http://ressources.ciheam.org/om/pdf/a31/CI971533.pdf.

Lee, T. R. (1999) *Water Management for the 21st Century: The Allocation Imperative.* Cheltenham, UK: Edward Elgar.

Lubet, S. (1996) "Notes on the Bedouin Horse Trade, or Why Won't the Market Clear, Daddy?" *Texas Law Review* 74: 1,039–57.

Lund, J. R. and Israel, M. (1995) "Water Transfers in Water Resource Systems." *Journal of Water Resources Planning and Management* 121: 193–205.

Mann, R. and Mecon, R. (2002) *Economic Effects of Land Idling for Temporary Water Transfers.* Report, California Department of Water Resources. Available at http://www.watertransfers.water.ca.gov/docs/econ_effects.pdf.

Merritt, S. (1997) *Introduction to the Economics of Water Resources: An International Perspective.* New York: Routledge.

Mohanty, N. and Gupta, S. (2002) *Breaking the Gridlock in Water Reforms through Water Markets: International Experience and Implementation Issues for India.* Available at http://www.libertyindia.org/policy_reports/water_markets_2002.pdf.

Morgan, S. P. and Chapman, J. I. (1995) *Special District Privatization: A Report Prepared for the Association of California Water Agencies.* Available at http://www.acwa.com/mediazone/research/privat.asp.

Naegele, J. (2004) "What is Wrong with Full-Fledged Water Privatization?" *Journal of Law and Social Challenges* 6: 99–130.

Oates, W. E. (1996) *The Economics of Environmental Regulation.* Cheltenham, UK: Edward Elgar.

Perelman, M. (2003) *The Perverse Economy: The Impact of Markets on People and the Environment.* New York: Palgrave Macmillan.

Posner, R. (2007) *Economic Analysis of Law.* 7th ediiton. New York: Little, Brown.

Robbins, E. (2003) "Winning the Water Wars: In the West, They Say that Water Flows Uphill to Money." *Planning* 69(6): 69.

Rose, C. M. (1994) *Property and Persuasion: Essays on the History, Theory, and Rhetoric of Ownership.* Boulder, CO: Westview Press.

Rosegrant, M. W. and Binswanger, H. P. (1994) "Markets in Tradable Water Rights: Potential for Efficiency Gains in Developing-Country Water Resource Allocation." *World Development* 22: 1613–25.

Rosegrant, M. W. and Gazmuri, R. S. (1995) "Reforming Water Allocation Policy through Markets in Tradable Water Rights: Lessons from Chile, Mexico, and California." *Caudernos de Economía* 32: 291–316.

Rothfeder, J. (2001) *Every Drop for Sale: Our Desperate Battle over Water in a World about to Run Out.* New York: Tarcher.

Roy, S. B., Chen, L. and Girvetz, E. (2010) *Evaluating Sustainability of Projected Water Demands under Furture Climate Change Scenarios.* Lafayette, CA: Tetra Tech.

Seroa Da Motta, R., Alban, T., Saade Hazin, L., Feres, J. G., Nauges, XC. and Saade Hazin, A. (2005) *Economic Instruments for Water Management: The Cases of France, Mexico and Brazil.* Cheltenham, UK: Edward Elgar.

Shiva, V. (2002) *Water Wars: Privatization, Pollution and Profit.* Cambridge, MA: South End Press.

Stiglitz, J. E. (2002) *Globalization and Its Discontents.* New York: W. W. Norton.

Thobani, M. (1996) *Tradable Water Rights: A Private Property Approach to Resolving Water Shortages and Promoting Investment.* FDP Note no. 34. Washongton, DC: World Bank.

UNECE (United Nations Economic Commission for Europe) (2009) *Guidance on Water and Adaptation to Climate Change.* Geneva, Switzerland: UNECE.

van Egteren, H. and Weber, M. (1996) "Marketable Permits, Market Power, and Cheating." *Journal of Environmental Economics and Management* 30: 161–73.

Vergara, A. (1997) "Perfeccionamiento Legal del Mercado de Derech de Aprovechamiento de Aguas." In *Convencion Nacionale de Usarios del Agua, 17–18 October, 1997, Arica Chile, Santiago,* 83–96. Santiago, Chile: Confederacion de Canalistas de Chile.

Wichelns, D. (2006) "Economic Incentives Encourage Farmers to Improve Water Management in California." *Water Policy* 8: 269–85.

Wilcove, D. (2008) *No Way Home: The Decline of the World's Great Animal Migrations.* Washington, DC: Island Press.

Winpenny, J. (2003) *World Panel on Financing Water Infrastructure: Financing Water for All.* Available at www.gwpforum.org/gwp/library/ExecSum030703.pdf.

World Bank (1994) *Peru: A User-Based Approach to Water Management and Irrigation Development.* Report No. 13642-PE. Washington, DC: World Bank Available at www-wds.worldbank.org/external/default/WDSContentServer/WDSP/IB/1995/04/11/000009265_3961214160422/Rendered/PDF/multi_page.pdf.

World Commission on Dams (2000) *Dams and Development: A New Framework for Decision-Making.* Available at www.dams.org/report.

Zaman, A. S., Davidson, B. and Malano, H. M. (2005) "Temporary Water Trading Trends in Northern Victoria, Australia." *Water Policy* 7: 429–42.

13 Cash flows

Markets for environmental flow allocations

David Katz

Introduction

For much of the last century, in standard water allocation practices ecological water needs were often disregarded and ecosystems were left to make do with residual amounts of water left over after all offstream uses received allocations. Water left instream was often considered a waste of a resource. Such practices resulted in severe degradation of freshwater ecosystems, including the loss of over half of the world's wetlands (Finlayson and Davidson 1999) and nearly 30 percent of freshwater species (WWF 2006). An index of freshwater ecosystem health based on populations of freshwater species showed a decline of 35 percent between 1970 and 2007 (WWF 2010). However, with a growing understanding of the importance of water flow as a "master variable" in determining aquatic ecosystem functioning (Poff *et al.* 1997), scientists, economists, and policy-makers are increasingly focusing attention on securing sufficient flows for ecological objectives (Petts 1996; Baron *et al.* 2002; Postel and Richter 2003; Hirji and Davis 2009; Poff *et al.* 2010).

The growing recognition of the value of allocation of water for environmental purposes is reflected in local and national laws, in international agreements, and even in the protocols and guidelines of international financial organizations.[1] Several countries have standards and even laws specifying minimum flow quantities for rivers. South Africa's national water law goes even further, requiring that basic human needs and ecological needs receive priority allocations, designating that only water remaining after these two objectives are met be available for allocation for other purposes. Some of the most effective examples of legal protection of water for ecological purposes are not found in water laws. In the United States, for example, the most useful legal and regulatory instruments for environmental flow protection have been the Endangered Species Act, which requires protection of habitat for endangered and threatened species, and the Federal Power Act, which authorizes the Federal Energy Regulatory Commission to require supply of flows for environmental objectives as a condition for dam licensing or relicensing.

In terms of international law, the United Nations Convention on the Law of the Non-Navigational Uses of International Watercourses (1997) lists environmental conservation as a criterion to be considered in the allocation of transboundary waters, and the Berlin Conference Report of 2004 – a summary of international water law – acknowledges the importance of environmental flows for ecological purposes. Numerous other international conventions and agreements also provide support for allocation of water for environmental objectives. International lending institutions such as the World Bank have also begun to consider aquatic ecosystem needs as part of their project evaluation criteria (Hirji and Panella 2003; Hirji and Davis 2009), and the World Commission on Dams (WCD) emphasized taking into account environmental needs in its recommended best practices for dam design and operation (King *et al.* 1999).

One reason for the increased acceptance of allocation of water for environmental purposes is the realization that ecosystem services and other instream uses of water can provide significant economic benefits.[2] A study in the western United States found the marginal value of water for recreational fishing to be higher than that for irrigation in 52 out of 67 watersheds observed (Hansen and Hallam 1991). Another study found that the ecological benefits of protecting Mono Lake in California exceeded the replacement cost of water from alternative sources by a factor of 50 (Loomis 1998, 2007). Ecological and social benefits from water in the Hadejia Jama'are floodplain in Nigeria were valued at US$9,600 to US$14,500 per cubic meter (in 1989–90 dollars), compared to just US$26–40 per cubic meter for upstream diversion of the water for irrigation (Barbier and Thompson 1998). The public has also demonstrated a strong willingness to pay for aquatic ecosystem services, even in developing countries (e.g. Becker and Katz 2006; Ilija Odeja *et al.* 2008).

Given these potential and actual economic benefits and given the market's potential to allocate goods to their highest value, many economists have long advocated for use of markets in provision of water for ecosystem services (e.g. Anderson and Johnson 1988; Colby 1990). Over the past two decades, water markets have indeed begun to play such a role, and though this role is still relatively modest, it can be expected to increase as water markets become more common and more sophisticated (Brookshire *et al.* 2004; Garrick *et al.* 2009; Grafton *et al.* 2010). While water markets provide for direct acquisition of water for environmental flows, they may also have unintentional environmental impacts as water transactions between offstream users move water within and out of basins. This chapter presents a review of the theory and empirical literature addressing water markets and the environment. It addresses both the use of markets to acquire flows for environmental purposes as well as the potential impact on the environment of water market transactions between other sectors, so called third party impacts. The study highlights opportunities for using the market to increase environmental water allocations, but also the numerous challenges involved.

Theoretical rationale for market instruments and environmental flows

Economic approaches to allocating environmental flows

Much of the existing economic literature fails to differentiate between instream flows and environmental flows, a distinction increasingly made in the scientific literature. Instream flows are all waters left in a stream, river, or wetland. Environmental flows are waters left in a stream, river, or wetland that provide for ecosystem functions. Thus, environmental flows represent only a subset of instream flows. Instream flows for hydropower production, navigation, and recreation may or may not benefit aquatic ecosystems, which thrive under a natural flow regime and may not be able to function under a highly altered one. Thus, not only is the amount of water left instream important for provision of ecosystem services, but also the timing and quality of this water. This review addresses environmental flows, recognizing that instream flows for other purposes may also contribute to ecological functioning.

In terms of environmental flows, basic economic models of quantitative water allocation can be summarized as being of one or a combination of the following three types described below.[3] The type of objective function used will depend in large part on the perceptions of the modeler and on the regulatory regime in place.

Model 1: environmental flow as an externality

Maximization of utility from allocation of water to non-environmental sectors only. The environment receives whatever water is left over, if any, after allocation to other sectors. Damage to the environment affects overall utility, but is not included in the objective function, as represented in Equation 1.

$$\text{Max } U(w_1,w_2, \ldots, w_i) \text{ s.t. } \Sigma\, w_i \leq W_T\text{; but actual } U(w_1,w_2, \ldots, w_i, w_e) \qquad [1]$$

Where U represents utility, w is quantity of water allocated to sector I, i is water consuming sectors other than the environment, W_T is total amount of water available for allocation to all purposes, and w_e is quantity of water allocated to the environment.

In this case, overall utility is a function of water allocations among competing uses, including environmental purposes, however, for the purposes of the maximization problem the model does not explicitly incorporate environmental values in the objective function and water allocation is restricted to non-environmental sectors. Such models were the predominant water allocation model in the past, as nature was considered simply a reservoir for supplying water to other uses, and leaving water instream was considered a waste of resources, or more commonly, was simply not considered at all. The typical output from such models is that the environment (streams, wetlands, lakes, etc.) receives only residual amounts of water after other uses have exhausted their shares. The result is an inefficient under-provision of freshwater for environmental flows and high levels of environmental degradation.

Model 2: minimum flows

Maximization of utility from allocation of water to non-environmental sectors, subject to an overall water constraint and an environmental constraint. Such a method is represented by Equation 2.

$$\text{Max } U(w_1, w_2, \ldots, w_i) \text{ s.t. } \Sigma\, w_i + w_e \leq W_T \text{ and s.t. } w_e \geq E \qquad [2]$$

Where the notation is the same as for Equation 1, with the addition of E representing some minimum flow requirement for environmental purposes.

This second type of model differs from the first in that the environment is explicitly incorporated into the objective function. The optimization question is still how to maximize overall utility, and the choice variables are still restricted to non-environmental uses. However, the model guarantees that at least a certain specified amount of water will remain instream. In regions in which water fully appropriated (and sometimes even in cases in which it is not), the result of such a model is that the environmental flow constraint is binding, and thus, the environment receives only the minimum flow required by law.

This model is used most often when a minimum flow volume is required by law or statute, or when such mandates are being contemplated. It is also useful for parties that are interested in restoration of numerous sites, but who recognize a budget constraint. Such a model and such regulation beg the question of how one determines minimum flow requirements. The answer depends largely on the intended instream use. Calculations may be relatively easy for some uses, such as boating and navigation, however, they have proved quite difficult in the case of flows for maintenance of ecosystem services.

Early scientific attempts to define minimum water requirements for ecological purposes used rules of thumb that related to channel size or average flow rates.[4] One the most popular methods in practice has been the Tennant Method, which proposes a percentage of natural flow necessary to preserve different levels of quality in streams (Tennant 1976). The popularity of these types of methods is due in no small part to the ease and low cost of calculation. Such simplistic methods, however, are now widely recognized as insufficient to provide ecological functions (IFC 2002). Other methods, such as the instream flow incremental methodology (IFIM), determine actual water needs of specific species (Stalnaker *et al.* 1995: 45).

Such methods generally take years of research and significant budgets to complete, making them less popular as policy tools. Even when used, they are generally appropriate when the status of a specific species is important, such as preserving a threatened or endangered species, or promoting a valuable game species. Optimizing flows for one species, however, does not optimize flows for the ecosystem as a whole, and so such flow recommendations also may be ecologically sub-optimal.

Ecosystems depend on variation in streamflows in order, for instance, to scour channels, provide access to habitat and protected areas for various species, indicate spawning times, and disperse seeds along shore banks. Richter *et al.* (1996, 1997) highlighted five hydrological characteristics important for ecosystem functioning: magnitude, frequency, timing, duration, and rate of change of hydrologic conditions. Most minimum flow regulations address only the magnitude of water, and some the timing (for instance requiring minimum flows only in the dry season). Rarely, however, do they address the full range of ecosystem water needs. As such, minimum flows tend to be far from optimal from an ecological perspective. As a burgeoning literature on the economic value of aquatic ecosystems is proving, minimum flow regulations are also often not economically efficient. Efficient allocations for environmental flows are often orders of magnitude larger than those required by minimum flow requirements (Loomis 1987, 1998).

Model 3: competitive market

Maximizing utility from allocation of water to all sectors including the environment, as represented by Equation 3.

$$\text{Max } U(w_1, w_2, \ldots, w_i, w_e) \text{ s.t. } \Sigma w_i + w_e \leq W_T \qquad [3]$$

This method for allocating water allocates economically efficient quantities for all sectors, including the environment. Thus, more than other models, competitive market allocation has the theoretical potential to maximize social welfare by providing environmental flows. According to economic theory, efficient allocation can be accomplished by administrative or regulatory fiat or through markets. However, several obstacles to such a reallocation by central water authorities exist. For one, as priority water rights in many regions have been claimed by offstream uses, allocating water for environmental flows often requires reallocation. Reallocation is likely to be opposed by existing users unless they receive adequate compensation, and strong lobbies may form in opposition (Gillilan and Brown 1997). Furthermore, administrators often lack knowledge regarding marginal benefits of environmental flows, and thus, cannot determine the shadow price of water, and thus, the appropriate quantities for allocation to environmental purposes.

Markets can potentially address both concerns. Because they depend on voluntary transactions, sellers are compensated for their loss of water, reducing opposition. In addition, use of a decentralized market mechanism to reallocate water obviates the need for a central allocator to know marginal values of water in various sectors. In a fully functioning market, the shadow price will be revealed as the going market price. Thus, markets represent a potentially powerful tool for acquiring environmental flows.

Theoretical and practical challenges to market allocation of environmental flows

Numerous challenges to efficient market allocation of environmental flows exist, however, some stemming from regulatory constraints, some from lack of scientific knowledge, and others from water's peculiar nature as a non-stationary, public and private good, with variable stochastic supply. In terms of regulatory constraints, not all legal regimes allow for market transfers of water rights. Observers have noted at least three conditions necessary for successful water markets:

1. clear and defined rights to and limits on freshwater extraction;
2. designation of environmental objectives as legitimate beneficial uses of water; and
3. establishment of mechanisms to allow transfer of water rights (Aylward 2008; Garrick *et al.* 2009).

In many countries and regions, water rights are tied to land rights. Efforts to unbundle water rights have been critical to advancing the market in some places (Garrick *et al.* 2009), but can be difficult in practice (Johnson 2007; Lave *et al.* 2008; Bretsen and Hill 2009), In other countries, water is fully owned by the government and/or cannot be sold or leased. In legal systems in which only the government is allowed to hold instream water rights, this monopsony position limits the effectiveness of the market. Also, in many areas instream or other environmental uses are not recognized as beneficial uses. In fact, in many cases establishment of rights is dependent on extraction. Thus, even if other conditions exist allowing for market transfer of water, the environment is unlikely to benefit from it.

In some cases, even when Pareto improving transactions are allowed, they do not take place (Young 1986; Wilkensen 2008). One explanation for a relative lack of market activity is that owners may fear that lease or sale of some water rights would jeopardize rights to water in the future (Howitt and Hansen 2005). For instance, if farmers leased water in a drought year, they may fear that this would indicate to government officials that their initial allocation of water rights was inefficient, and thus, put those rights at risk in the long term. Similarly, water banks, one form of market based water reallocation, like regular banks, function based on the principle that deposits are secure redeemable assets. Water rights holders who value aquatic ecosystems may be unwilling to sell water rights for environmental flow purposes, if they fear that the water will not actually be kept instream, but will simply be extracted by downstream parties. In short, water markets are unlikely to be active if property rights are poorly defined or enforced.

Another significant obstacle to efficient markets for environmental flows stems from the fact that they are primarily public goods. Thus, economic theory predicts that they will be under-provided, as some beneficiaries free-ride off the provision of others. In legal systems in which private parties can purchase water rights, coordination among multiple beneficiaries can result in high transaction costs that limit market effectiveness.

A practical limitation that stems from environmental flows public good nature is that their economic value does not necessarily translate into financial value. Because the consumer surplus provided by environmental flows does not generate revenue streams the way that most offstream uses do, parties acting on behalf of the environment can find it difficult to compete successfully in the marketplace. Even in cases in which official water management agencies are involved in market acquisitions of water, there is evidence that some such bodies view "economic values without cash transactions... as 'second-class' values" (Loomis *et al.* 2003: 22).

In order to determine the efficient amount of water to allocate to environmental flows, it is necessary to tie flow levels to ecosystem services, which are can then be valued in the market place. While the level of scientific understanding regarding the benefits of environmental flows has been increasing rapidly over the past years, it is still in its infancy. Several knowledge gaps exist in this realm. Most research on flows and ecosystem functioning, for instance, has looked at flows necessary for *maintaining* ecosystem services. These may differ substantially from flows necessary to *restore* ecosystem services, and it is often in arid areas with degraded streams, that the marginal value of water for the environment is highest (Gillilan and Brown 1997), and in which environmental flows are most able to compete in a competitive marketplace.

For these reasons, many environmental flow recommendations are for allocating flows experimentally and allowing for readjustment, according to an adaptive management framework. Large scale experiments with water releases for environmental purposes along the Colorado River in the United States, for instance, have only partially achieved their designated goals, and have been twice revised since the first release. Without reliable information regarding the ecological and environmental benefits of environmental flows, economists cannot produce reliable estimates of economic values of such flows. In this sense, incorporating environmental flows into economic models is much more problematic than incorporating other instream flow goals such as promoting recreation or navigation, which can be estimated more easily and reliably.

The goal of environmental flow allocations is generally to maintain or restore ecosystem services. However, quantitative allocations alone may be insufficient to achieve such objectives. In addition to adequate quantities of water at the right time, water needs to be of sufficient quality and other criteria such as wetland soils or adequate riverbank geomorphology need to be in place. Reviews of the effectiveness of environmental flow transactions have indicated that other such limiting factors may be preventing anticipated environmental gains (Hardner and Gullison 2007). Institutions such as wetland banking and stream restoration banking, which allow for development or damage in one area in exchange for improvement in others, have been criticized for failing to deliver the ecosystem services lost. The theoretical benefits of water quality trading have been promoted by many (e.g. Weinberg *et al.* 1993), but such markets have been of limited success in practice. Attempts to identify and market specific ecosystem functions are problematic for a number of reasons (Lave *et al.* 2008). What is certain, however, is that it is difficult to predict the ecological benefits from environmental flows, and therefore all the more so the economic value that would be necessary for efficient environmental flow allocation in a marketplace.

The above three criteria, while necessary, are insufficient to ensure functional markets for environmental flows. Garrick *et al.* (2009) list several additional other factors that they have identified as crucial, including ensuring adequate funding to overcome public goods problems and development of institutional capacity to plan and coordinate transactions and monitor and enforce environmental flow allocations.

In addition to the direct role markets can play in supplying environmental flows, they can also affect environmental flow levels via third party impacts resulting from trade between other right holders. These impacts may be positive or negative, depending on whether the transfers are upstream or downstream, and on the manner and timing of the transfers. The current transfer of water from Imperial Valley to San Diego and Coachella in California, for instance, is likely to reduce flow to the Salton Sea by up to 40 percent, threatening numerous species that use the inland lake for habitat and food (Cohen 2007). Models of efficient market allocation of the Colorado River, however, show that the result would be less water taken by upstream states and more transferred downstream, which would be potentially beneficial for the environment (Booker and Young 1994). Empirical studies documenting which trend is actually more prevalent in practice are lacking and would be a welcome contribution.

The provision of environmental flows through markets can itself have both positive and negative third party impacts. The provision of instream flows can benefit not only the environment, but also other downstream users. Ward and Booker (2003) show that the provision of water for preservation of an endangered fish species in the Rio Grande River, would result in net economic benefits for both the United States and Mexico. Market acquisition of environmental flows can also have negative social impacts on community members in the selling region, however; and can even have negative environmental impacts, if, for instance, selling of surface water rights results in increased groundwater pumping in the selling region. This is likely to occur in areas in which surface and groundwater rights are administered separately, a not uncommon situation in many countries. Furthermore, the public good environmental benefits from agriculture, such as terrestrial habitat provision and preservation of open spaces, may be harmed as water is traded away. Again, little empirical work has been done to validate whether and to what extent such impacts occur in actuality.

In sum, markets offer an important option for provision of water for environmental objectives. They also allow for moving beyond minimum flows, which are unlikely to be ecologically or economically efficient. Several factors, however, serve to limit and/or detract from the effectiveness of markets for these purposes. These obstacles include regulatory and institutional factors, public goods problems, information limitations, and third party impacts. Furthermore, as will be noted in the following section on practical impacts, water markets can have significant social impacts (Chong and Sundig 2006).

Markets and environmental flows in practice: empirical results

Direct market environmental flow transactions

Growing experience with both water markets and environmental flow acquisition over the past two decades provides an opportunity to compare empirical evidence to the theoretical issues outlined above. It should be noted that the majority of such evidence reported in the English language literature focuses on experiences from a limited number of regions. The overwhelming amount of literature is limited to the western United States and Australia. In addition to the language bias, the primary reason that these two examples dominate the published literature is that water markets are not widely used. Even in other regions which do have water markets, such as Chile, Mexico, South Africa, and Spain, there seems to be little use of the markets for acquisition of environmental flows, and the overall environmental impacts of these water markets have not been well studied. In Chile, for instance, given the secure status of private water rights, the market may be the only legal means of obtaining water for the environment (Hodgson 2006), however, such activity does not appear to be occurring and environmental concerns are generally an afterthought in market activity there (e.g. Bauer 2004a, b). In the case of South Africa, because the national Water Act ensures that water for environmental and basic human needs are protected and cannot be harmed by transfers (Nieuwoudt and Armitage 2004), water markets there may be less needed.

Landry (1998) reported that between 1990 and 1997, 2 million acre-feet of water was leased through markets and 132,000 acre-feet purchased for environmental protection in the western United States. Several more recent studies that have examined the functioning of water markets in the American west show that the market for environmental flows has grown significantly in recent years (e.g. Loomis *et al.* 2003; Brookshire *et al.* 2004; Howitt and Hansen 2005; Brown 2006; Scarborough and Lund 2007; Brewer *et al.* 2008; Scarborough 2010). According to Brewer *et al.* (2008), between 1987 and 2005 the total the amount of water acquired for environmental purposes through markets in the United States was over 7 million acre-feet in terms of annual flow. The amount of activity varied greatly by state. For instance, according to one database, between 1987 and 2008, California accounted for 113 of the 447 transactions that took place among states with water markets, while Utah and Wyoming accounted for only 2 each.[5] The variation is a function of differences in regulatory mechanisms, government funding, and activity among non-governmental organizations (NGOs).

In Australia, use of water markets for environmental purposes has increased significantly in recent years. Initial programs for government buybacks of water rights have been very successful, with the government having little problem finding enough willing sellers to obtain the designated amounts. Recently, government programs to obtain water for environmental flows have been greatly expanded. In 2008, out of a AU$12.9 billion water budget, the Australian government allocated AU$3.1 billion to buy water from irrigators in the Murray–Darling Basin for environmental purposes over the next 10 years, and an additional AU$5.8 to increase water efficiency in farmlands (Qureshi *et al.* 2010; Grafton 2010). The markets have faced challenges however. Early experience with water markets, for instance, at times resulted in sale of so-called sleeper rights, rights that in any case were not being utilized, or dozer-rights, those which were seldom used, and so failed to actually augment stream flows (Chong and Sundig 2006; Grafton *et al.* 2010). Such experience has emphasized the need to evaluate the quality or seniority of rights being purchased. Modeling of the Australian policy demonstrated that efficiency improvements may reduce return flow, thereby mitigating the expected environmental benefits (Qureshi *et al.* 2010). Thus, scholars have calculated that even with the impressive budgets dedicated to securing environmental flows – budgets that more than outweigh the potential foregone profits from irrigated farm income in the basin – the policy is unlikely to secure the necessary quantity of water (Grafton 2010).

Environmental purchases are a minority, though not a negligible share, of water market activity. Environmental purchases accounted for 9 percent of overall United States water market transactions between 1987 and 2005,[6] but for 23 percent of the volume of the annual flow and 22 percent of the water committed (Brewer *et al.* 2008). Agriculture was the primary source of water sold for environmental objectives, accounting for 80–85 percent of such transactions, with the rest coming primarily from urban transfers. A small number of market transfers between environmental purposes were also documented. While data on overall market transfers for environmental purposes in Australia were not available, agriculture appears to be overwhelmingly the most significant seller there too.

Options for market-based water acquisition include purchase of water rights, short- and long-term leases, dry-year options, according to which obligate right holders are paid to provide water in low rainfall years only, and forbearances which the water user agrees to stop irrigating as of a certain date and to leave the water instream in exchange for a cash payment (Neuman 2004; Garrick *et al.* 2009). In addition to transferring entitlements, arrangements have also been reached in which entitlement holders are paid for implementing conservation and/or efficiency improvements, with the resulting saved water being dedicated to environmental flow purposes. Leases account for the bulk of the US environmental flow transactions (Brown 2006; Scarborough 2010).[7] In addition to providing flexibility, leases often face lower bureaucratic hurdles and lower transaction costs, and can be approved more quickly than permanent sales (Scarborough 2010). Because water is often most needed for environmental purposes during dry years, when the lease prices are high, many advocates for market acquisitions of environmental flows have pointed to option contracts as a method for lowering costs (e.g. Hafi *et al.* 2005). Bjornlund and Rossini (2010) note, however, that purchasing options for water during high flow periods when prices are low, for instance in order to allow for flooding of wetlands, may also represent a cost-effective way of achieving environmental objectives.

In both the US and Australia, government agencies of various levels, from municipal to federal, account for the overwhelming majority of all environmental water purchases. In many cases such purchases are required by law or as a matter of policy decision. In such cases, the market is not functioning to reallocate water according to its highest valued use (as depicted in Model 3 above), but rather, as a least cost means for achieving some minimum flow level (Model 2 above). The value of water markets should not be dismissed, however, as they allow governments to gain information about the marginal value of water among sellers, identify less cost options for achieving their self-designated targets, and minimize political opposition which can severely delay administrative reallocation.

In the US, NGOs such as water trusts – organizations specifically dedicated to raising money for purchase of water for environmental purposes – and other environmental organizations have been actively participating in water markets in recent years. NGOs have not been active in direct purchases in Australian water markets, but have played a role in identifying sellers for environmental flows and in coordinating willing donations of water rights for environmental purposes (Garrick *et al.* 2009). Despite accounting for a minority of environmental transactions, NGOs can still play an important role in water markets. They identify water needs and opportunities that are not addressed by government programs. NGOs have also shown that they can often work more quickly than cumbersome government programs. As contracts can take up to years to be approved (Grafton *et al.* 2010), expediting transaction times can be truly valuable, especially when responding to temporary drought conditions. Moreover, NGOs are often capable of attracting sellers who are resistant to selling or leasing rights to the government (Garrick *et al.* 2009; Scarborough 2010).

While important, one should not overestimate the role of private individuals and organizations in obtaining environmental flows in a competitive market, however. A closer inspection of the funding sources of some of the leading water trusts in the US reveals that much of their funding comes from government sources. For instance, according to the annual report of the Oregon Water Trust, one of the first and largest water trusts,[8] over half of the organization's budget came from government funds. Thus, rather than a true expression of the ability of private funds ability to secure environmental flows, the main contribution of these organizations may be in being better able and quicker to target opportunities for market acquisition of instream flow than governments directly.

Prices paid for water for environmental flows in the US tended to be lower than those paid by municipalities and irrigation (Brown 2006). There are several possible explanations for such a phenomenon. It may reflect that water for environmental purposes has a lower marginal value than other uses and/or is purchased in less competitive markets, or it may reflect the dominant market power of the government as primary buyer. Also, numerous studies have found that sellers in water markets are motivated by factors other than purely financial gain (Bjornlund 2004; Chong and Sundig 2006), and so lower prices for environmental flow purchases may also reflect that some people are more willing to sell or lease rights for environmental purposes than for other purposes. The lower price may also support the contention discussed above that the public good status of environmental flows has resulted in an under-provision by open markets. Overall, however, the prices paid for environmental water in sales and leases in the US markets did tend to conform with the estimates of the value of such services using various nonmarket economic valuation methods (Chow and Sundig 2006), thus adding credibility to the use of such methods in policy planning.

Some observers have worried that water markets may result in strategic speculative purchasing of instream flow, only to be sold later when the price is higher (Gillilan and Brown 1997). There is little evidence that such practices are occurring in practice. According to the UCSB database, between 1987 and 2008 no transactions have been recorded in which water was transferred from environmental to non-environmental purposes. Furthermore, Scarborough and Lund (2007) note both that in the US inflation adjusted prices for water have risen little over the past two decades and that governments are the largest purchaser of instream water rights and are unlikely to engage in speculative buying.

Indirect impacts of market transfers on environmental flows

As noted in the previous section, transactions for environmental purposes represent a minority of overall water market activity. Environmental externalities in terms of flow changes due to other non-environmentally oriented market transactions have not yet been well documented or studied in a systematic manner. Howitt and Hansen (2005: 61) note that such "externalities and third-party damages are likely to become more important as a greater volume is traded," however, to date most evidence is anecdotal. Because costs for pumping uphill are often prohibitive, many, perhaps most market transfers are from upstream to downstream users. This can have net benefits for the environment if the water is transferred instream and at appropriate times. Many transfers are conducted via constructed infrastructure, however, rather than via instream flow. Thus, even if they transfer water to downstream parties, environmental impacts can be substantial (Tisdell 2001).

Various regulatory mechanisms exist to limit or mitigate third party environmental damage from such transactions. In many areas, transactions are limited to within watershed trades. In the US, some markets have been limited to trade in consumptive use only, in order to protect return flows. As water rights are traditionally for withdrawals, however, calculation of consumptive use levels can add significantly to the transaction costs of market transfers (Brewer *et al.* 2008). In both cases, these limitations reduce environmental impact, but at the expense of restricting the cost-saving potential of the market and overall market activity.

Given both the potential third party impacts and the lack of direct financial income from many forms of environmental flows, some states have imposed a tax on water market transfers. This tax can be in the form of a financial transaction fee, or, as is the case in Oregon and parts of California, a requirement that a percentage of water transacted be dedicated to environmental flows (McKinney and Taylor 1988; Howitt and Hansen 2005). Such fees may have to be substantial, and thus, may face political resistance, should significant amounts of water be needed for environmental flows (Tisdell 2010).

Conclusion

The increasing recognition of the importance of environmental flows can be seen in the type and amount of regulations, legislation, and international agreements addressing the issue. Numerous economic valuation studies show that the public increasing values environmental flows and the amenities they provide. Given this economic value, water markets represent a potentially significant option for securing water for these flows. Existing water markets are already serving this purpose in the US and Australia, and this trend is on the increase there. Environmental flow acquisition has yet to take off in water markets in other countries, however.

The provision of water for environmental objectives via markets is still largely determined by government regulation, and often results only in provision of minimum flows, which is rarely either ecologically or economically optimal. While in theory markets allow for efficient allocation of water among competing users, including the environment, because environmental flows are public goods, markets are unlikely to provide efficient allocations. High transaction costs and scientific uncertainty regarding benefits from flow augmentation also contribute to under-provision of environmental flows in market situations. As such, even with active markets, governments play a primary role both in mandating and in provision of environmental flows. As such, markets serve more as mechanisms for reducing costs and political resistance to reallocation, rather than as means for efficient allocations. Acquisition of environmental flows by private parties, including NGOs, accounts for a minority of such flow provision, and is likely to continue to do so in the future. These parties can and do play important roles, however, in terms of reducing transaction costs, identifying and providing for flows that are not part of government mandates, and securing water from potential sellers skeptical of government programs.

While third party environmental impacts of water markets have long been recognized as potential and actual concerns, most of the literature on this topic has been anecdotal or case study in nature. As environmental flows grow in importance and water markets increase in popularity, it is imperative to have a broad body of empirical research on which to refine policy if such water markets are to truly benefit the environment.

Notes

1. For reviews of legal mechanisms and policies for provision of environmental flows, see Dyson *et al.* (2003: 118), Scanlon and Iza (2004), or Katz (2006).
2. For reviews of several such economic studies, see Wilson and Carpenter (1999); BoR (2001); NPS (1995, 2001); or Loomis (1998).
3. For simplicity's sake, the above typology of economic models addresses only water quantity, and ignores aspects such as water quality or timing of flows, which are known to affect economic value. The three models represent the bulk of approaches commonly found in economics literature. Other quantitative methods have been proposed, but are less used in practice. For instance, Tisdell (2001: 116) presents the option of "minimizing the average sum of squared differences between the actual and natural flow regimes, subject to the extractive use of water and available water for environmental use." Also, some game-theoretic models have been proposed (e.g. Supalla *et al.* 2002; Janmaat 2008). Harou *et al.* (2009) and Murphy *et al.* (2009) offer quantitative models for determining optimal flows.
4. For reviews of ecologically based methodologies for determining environmental flow needs, see IFC (2002) or Tharme (2003).
5. This database is maintained by the University of California Santa Barbara (UCSB) and published online at www.bren.ucsb.edu/news/water_transfers.htm. Scarborough (2010) claims to have documented over

2,800 transactions between 1987 and 2007, the value of which exceeded half a billion dollars. While the total number of transactions differed greatly from that contained in the UCSB database, the geographical variation was similar.

6. According to the UCSB database, transfers of water towards environmental purposes accounted for 447 out of 4,168 (10.7%) of total water market transactions occurring between 1987 and 2008. In certain states, for instance, in the Pacific Northwest with its commercially important salmon stocks, environmental purposes accounted for the majority of transfers (Howitt and Hansen 2005).
7. Leases accounted for over 70 percent of the US environmental flow transactions reported in the UCSB database.
8. As of 2009, after a merger with Oregon Trout, the organization is now called The Freshwater Trust.

References

Anderson, T. L. and Johnson, R. N. (1988) "The Problem of Instream Flow." *Economic Inquiry* 24: 535–54.

Aylward, B. (2008) *Water Markets: A Mechanism for Mainstreaming Ecosystem Services into Water Management?* Briefing paper, Water and Nature Initiative. Gland, Switzerland: IUCN.

Barbier, E. B. and Thompson, J. R. (1998) "The Value of Water: Floodplain Versus Large-Scale Irrigation Benefits in Northern Nigeria." *Ambio* 27: 434–40.

Baron, Jill S., Poff, N. L., Angermeier, P. L., Dahm, C. N., Gleick, P. H., Hairston, N. G., Jackson, R. B., Johnston, C. A., Richter, B. D. and Steinman, A. D. (2002) "Meeting Ecological and Societal Needs for Freshwater." *Ecological Applications* 12(5): 1247–60.

Bauer, C. J. (2004a) "Results of Chilean Water Markets: Empirical Research since 1990." *Water Resources Research* 40: W09S06.

Bauer, C. J. (2004b) *Siren's Song: Chilean Water Law as a Model for International Reform.* Washington, DC: Resources for the Future.

Becker, N. and Katz, D. (2006) "Economic Valuation of Resuscitating the Dead Sea." *Water Policy* 8(4): 351–70.

Bjornlund, H. (2004) "Formal and Informal Water Markets: Drivers of Sustainable Rural Communities?" *Water Resources Research* 40: W09S07.

Bjornlund, H. and Rossini, P. (2010) "Climate Change, Water Scarcity and Water Markets: Implications for Farmers' Wealth and Farm Succession." Paper delivered at the 16th Pacific Rim Real Estate Society Conference, Wellington, New Zealand, January.

Booker, J. F. and Young, R. A. (1994) "Modeling Intrastate and Interstate Markets for Colorado River Water Resources." *Journal of Environmental Economics and Management* 26(1): 66–87.

BoR (Bureau of Reclamation). (2001) *Economic Nonmarket Valuation of Instream Flows.* Denver, CO: Bureau of Reclamation (BoR).

Bretsen, S. and Hill, P. J. (2009) "Water Markets as a Tragedy of the Anticommons." *William and Mary Environmental Law and Policy Review* 33: 723–83.

Brewer J., Glennon, R., Ker, A. and Libecap, G. (2008) "Water Markets in the West: Prices, Trading, and Contractual Forms." *Economic Inquiry* 46(2): 91–112.

Brookshire, D. S., Colby, B., Ewers, M. and Ganderton, P. T. (2004) "Market Prices for Water in the Semiarid West of the United States." *Water Resources Research* 40: W09S04 (doi:10.1029/2003WR002846).

Brown, T. C. (2006) "Trends in Water Market Activity and Price in the Western United States." *Water Resources Research* 42: W09402.

Chong, H. and Sundig, D. (2006) "Water Markets and Trading." *Annual Review of Environment and Resources* 31(1): 239–64.

Cohen, M. (2007) "Salton Sea at a Crossroads." *GeoTimes* August. Available at www.geotimes.org/aug07/article.html?id=feature_saltonsea.html (accessed September 2007).

Colby, B. G. (1990) "Enhancing Instream Flows in an Era of Water Marketing." *Water Resources Research* 26(6): 1113–20.

Dyson, M., Bergkamp, G. and Scanlon, J. (2003) *Flow: The Essentials of Environmental Flows*. Gland, Switzerland, IUCN.

Finlayson, C. and Davidson, N. (1999) *Global Review of Wetland Resources and Priorities for Wetland Inventory: Summary Report*. Wetlands International and the Environmental Research Institute of the Supervising Scientist, Australia. Available at www.wetlands.org/RSIS/WKBASE/GRoWI/welcome.html (accessed 13 July 2012).

Garrick, D., M. Siebentritt, A., Aylward, B., Bauer, C. J. and Purkey, A. (2009) "Water Markets and Freshwater Ecosystem Services: Policy Reform and Implementation in the Columbia and Murray-Darling Basins." *Ecological Economics* 69(2): 366–79.

Gillilan, D. M. and Brown, T. C. (1997) *Instream Flow Protection: Seeking a Balance in Western Water Use*. Washington, DC: Island Press.

Grafton, R. (2010) "How to Increase the Cost-Effectiveness of Water Reform and Environmental Flows in the Murray–Darling Basin." *Agenda: A Journal of Policy Analysis and Reform*17(2): 17–40.

Grafton, R. Q., Landry, C., Libecap, G. D. and O'Brien, R. J. (2010) *Water Markets: Australia's Murray-Darling Basin and the US Southwest*. NBER Working Paper 15797. Cambridge, MA: National Bureau of Economic Research.

Hafi, A., Beare, S., Heaney, A. and Page, S. (2005) *Water Options for Environmental Flows*. Canberra, Australia: Australian Bureau of Agricultural and Resource Economics, Natural Resource Management Division, Australian Government Department of Agriculture, Fisheries and Forestry.

Hansen, L. and Hallam, A. (1991) "National Estimates of the Recreational Value of Stream Flow." *Water Resources Research* 27(2): 167–75.

Hardner, J. and Gullison, R. E. (2007) *Independent External Evaluation of the Columbia Basin Water Transactions Program (2003–2006)*. 7 October. Available at www.nwcouncil.org/fw/program/2008amend/cbwtp.pdf (accessed 25 July 2012).

Harou, J. J., Pulido-Velazquez, M., Rosenberg, D. E., Medellín-Azuara, J., Lund, J. R. and Howitt, R. E. (2009) "Hydro-economic Models: Concepts, Design, Applications, and Future Prospects." *Journal of Hydrology* 375(3–4): 627–43.

Hirji, R. and Davis, R. (2009) *Environmental Flows in Water Resources Policies, Plans, and Projects: Findings and Recommendations*. Washington, DC: The World Bank.

Hirji, R. and Panella, T. (2003) "Evolving Policy Reforms and Experiences for Addressing Downstream Impacts in World Bank Water Resources Projects." *River Research and Applications* 19(5–6): 667–81.

Hodgson, S. (2006) *Modern Water Rights: Theory and Practice*. FAO Legislative Study 92. Rome, Italy: Food and Agriculture Organization of the United Nations.

Howitt, R. and Hansen, K. (2005) "The Evolving Western Water Markets." *Choices* 20(1): 59–63.

IFC (Instream Flow Council). (2002) *Instream Flows for Riverine Resource Stewardship.* Lansing, MI: Instream Flow Council.

Ilija Ojeda, M., Mayer, A. S. and Solomon, B. D. (2008) "Economic Valuation of Environmental Services Sustained by Water Flows in the Yaqui River Delta." *Ecological Economics* 65(1): 155–66.

Janmaat, J. (2008) "Playing Monopoly in the Creek: Imperfect Competition, Development, and In-stream Flows." *Resource and Energy Economics* 30(3): 455–73.

Johnson, J. L. (2007) "Property Without Possession: Defining Private Instream Rights in Western Water Law." Yale Law School Student Prize Papers, Yale Law School, New Haven, CT.

Katz, D. (2006) "Going with the Flow: Preserving and restoring instream water allocations." In P. Gleick, H. Cooley, D. Katz and E. Lee (eds), *The World's Water: 2006–2007: The Biennial Report on Freshwater* Resources, 29–49. Washington, DC: Island Press.

King, J., Tharme, R. and Brown, C. A. (1999) *Definition and Implementation of Instream Flows.* Thematic Review II.1: Dams, Ecosystem Functions and Environmental Restoration. Cape Town, South Africa: World Commission on Dams.

Landry, C. J. (1998) "Market Transfers of Water for Environmental Protection in the Western United States." *Water Policy* 1(5): 457–69.

Lave, R., Robertson, M. M. and Doyle, M. W. (2008) "Why You Should Pay Attention to Stream Mitigation Banking." *Ecological Restoration* 26(4): 287–9.

Loomis, J. B. (1987) "The Economic Value of Instream Flow: Methodology, and Benefit Estimates of Optimum Flow." *Journal of Environmental Management* 24(1): 73–90.

Loomis, J. B. (1998) "Estimating the Public's Values for Instream Flow: Economic Techniques and Dollar Values." *Journal of the American Water Resources Association* 34(5): 1007–14.

Loomis, J. B., Quattlebaum, K., Brown, T. C. and Alexander, S. J. (2003) "Expanding Institutional Arrangements for Acquiring Water for Environmental Purposes: Transactions Evidence for the Western United States." *International Journal of Water Resources Development* 19: 21–8.

McKinney, M. and Taylor, J. (1988) *Western State Instream Flow Programs: A Comparative Assessment.* Fort Collins, CO: National Ecology Research Center, Fish and Wildlife Service, US Dept. of Interior.

Murphy, J.J., Dinar, A., Howitt, R. E., Rassenti, S. J., Smith, V. L. and Weinberg, M. (2009) "The Design of Water Markets when Instream Flows have Value." *Journal of Environmental Management* 90(2): 1089–96.

Neuman, J. C. (2004) "The Good, the Bad, the Ugly: The First Ten Years of the Oregon Water Trust." *Nebraska Law Review* 83: 432–84.

Nieuwoudt, W. L. and Armitage, R. M. (2004) "Water Market Transfers in South Africa: Two Case Studies." *Water Resources Research* 40: W09S05.

NPS (National Park Service) (1995) *Economic Impacts of Protecting Rivers, Trails and Greenway Corridors.* Washington, DC: National Park Service (NPS).

NPS (National Park Service) (2001) *Economic Benefits of Conserved Rivers: An Annotated Bibliography.* Washington, DC: National Park Service (NPS).

Petts, G. E. (1996) "Water Allocation to Protect River Ecosystems." *Regulated Rivers-Research and Management* 12(4–5): 353–65.

Poff, N. L., Allan, J. D., Bain, M. B., Karr, J. R., Prestegaard, K. L., Richter, B. D.,

Sparks, R. E. and Stromberg, J. C. (1997) "The Natural Flow Regime." *Bioscience* 47(11): 769–84.

Poff, N. Leroy, Richter, B. D., Arthington, A. H., Bunn, S. E., Naiman,R. J., Kendy, E., Acreman, M., Apse, C., Bledsoe, B. P., Freeman, M. C., Henriksen, J., Jacobson, R. B., Kennen, J. G., Merritt, D. M., O'Keeffe, J. H., Olden, J. D., Rogers, K., Tharme, R. E. and Warner, A. (2010) "The Ecological Limits of Hydrologic Alteration (ELOHA): A New Framework for Developing Regional Environmental Flow Standards." *Freshwater Biology* 55(1): 147–70.

Postel, S. and Richter, B. D. (2003) *Rivers for Life: Managing Water for People and Nature.* Washington, DC: Island Press.

Qureshi, M. E., Schwabe, K., Connor, J. and Kirby, M. (2010) "Environmental Water Incentive Policy and Return Flows." *WaterResourcesResearch* 46: W0451.

Richter, B. D., Baumgartner, J. V., Powell, J. and Braun, D. P. (1996) "A Method for Assessing Hydrologic Alteration within Ecosystems." *Conservation Biology* 10(4): 1163–74.

Richter, B. D., Baumgartner, J. V., Wigington, R. and Braun, D. P. (1997) "How Much Water Does a River Need?" *Freshwater Biology* 37(1): 231–49.

Scanlon, J. and Iza, A. (2004) "International Legal Foundations for Environmental Flows." *Yearbook of International Law* 2004: 81–100.

Scarborough, B. (2010) *Environmental Water Markets: Restoring Streams Through Trade.* PERC Policy Series 46–2010. Bozeman, MT: The Property and Environment Research Center (PERC).

Scarborough, B. and Lund, H. L. (2007) *Saving Our Streams: Harnessing Water Markets A Practical Guide.* Bozeman, MT: The Property and Environment Research Center (PERC).

Stalnaker, C .B., Lamb, B. L., Henriksen, J., Bovee, K. and Bartholow, J. (1995) *The Instream Flow Incremental Methodology: A Primer for IFIM.* Fort Collins, CO: National Biological Service.

Supalla, R., Klaus, B., Yeboah, O. and Bruins, R. (2002) "A Game Theory Approach To Deciding Who Will Supply Instream Flow Water." *Journal of the American Water Resources Association* 38(4): 959–66.

Tennant, D. L. (1976) "Instream Flow Regimes for Fish, Wildlife, Recreation and Related Environmental Resources." *Fisheries* 1(4): 6–10.

Tharme, R. (2003) "A Global Perspective on Environmental Flow Assessment: Emerging Trends in the Development and Application of Environmental Flow Methodologies for Rivers." *River Research and Applications* 19(5–6): 397–441.

Tisdell, J. G. (2001) "The Environmental Impact of Water Markets: An Australian Case-Study." *Journal of Environmental Management* 62(1): 113–20.

Tisdell, J. (2010) "Acquiring Water for Environmental Use in Australia: An Analysis of Policy Options." *Water Resources Management* 24(8): 1515–30.

Ward, F. A. and Booker, J. F. (2003) "Economic Costs and Benefits of Instream Flow Protection for Endangered Species in an International Basin." *Journal of The American Water Resources Association* 39(2): 427–40.

Weinberg, M., Kling, C. L. and Wilen, J. E. (1993) "Water Markets and Water Quality." *American Journal of Agricultural Economics* 75(2): 278–91.

Wilkensen, M. (2008) "Farmers Won't Go with flow". *Syndey Morning Herald* 19 May.

Wilson, M. A. and Carpenter, S. R. (1999) "Economic Valuation of Freshwater Ecosystem Services in the United States: 1971–1997." *Ecological Applications* 9(3): 772–83.

WWF. (2006) *Living Planet Report 2006.* Gland, Switzerland: Worldwide Fund for Nature.

WWF. (2010) *Living Planet Report 2010 – Biodiversity, Biocapacity and Development.* Gland, Switzerland: Worldwide Fund for Nature.

Young, R. A. (1986) "Why Are There so Few Transactions among Water Users?" *American Journal of Agricultural Economics* 68(5): 1143–51.

14 Experiences with water quality trading in the United States of America

Robert J. Rose

Introduction

It has been estimated that over 400 water bodies worldwide are experiencing problems with eutrophication (Selman and Greenhalgh 2009), which is an excess of nutrients (nitrogen and phosphorus) that cause harmful algae blooms, hypoxia (reduced levels of dissolved oxygen), and other water quality problems. Sources of excess nutrient pollution include in no particular order: burning of fossil fuels which transform atmospheric N_2 into bioavailable NO_x, animal waste manure, human waste, synthetic fertilizers used in agriculture and horticulture, soil and stream erosion that releases sediment bound nutrients, and the loss of healthy ecosystems capable of assimilating a portion of excess nutrients.

Because the sources of nutrient pollution include many economic activities spread across the landscape, and because the damaging effects often occur downstream from such sources, nutrient pollution is if often considered a candidate for water quality trading. Water quality trading is premised on many of the same goals as greenhouse gas cap-and-trade schemes (i.e. reduced pollution with lower compliance costs, greater flexibility, and incentive for innovation), but differs in many important ways. In the US the vast majority of water quality trading programs are a response to regulatory drivers under the federal Clean Water Act, which is the nation's law and regulations that protect human health and the nation's rivers, lakes, and waterways. Overwhelmingly, experiences with water quality trading are in the US, and are driven by the Clean Water Act. (Selman 2009) As such the experiences described here are from that perspective. Most examples involve nitrogen and phosphorus, although examples with sediment runoff pollution and stream temperature also exist.

Unlike greenhouse gas cap-and-trade schemes, as of 2011 water quality trading programs in the US are almost exclusively local, limited to individual states, and within each state's individual rivers or local watersheds. Interest in multi-state trading schemes exists but at the design stage such as the Ohio River Basin, which the Electric Power Research Institute and Ohio River Valley Water Sanitation Commission are exploring (Electric Power Research Institute 2010).

There are roughly 50 examples of water quality trading programs in the US (US Environmental Protection Agency 2012). Many of the examples cited are not large, market-based trading programs but very small, uniquely designed adaptations to fit a specific need or circumstance. Also, many examples cited have not resulted in any actual trades. If viewed through the lens of large programs in which many trades have occurred, the conclusion would be that water quality trading in the US is not successful. This is not the appropriate way to view the status of water quality trading in the US since despite not having any explicit, federal law that allows or creates water quality trading programs, individual states within the US have been willing to explore the possibilities.

The reader should keep in mind that although today cap and trade is a concept synonymous with controlling greenhouse gases, there was once a time when policy makers favored command and control approaches. Also, just as greenhouse gas cap-and-trade policies may have not yet resulted in substantial reductions, this does not preclude cap-and-trade policies from being effective in the future. In the case of water quality trading there are perhaps five or six large programs that most experts are familiar, most of which will be cited in this chapter. This chapter focuses on the larger examples in the US for discussion purposes and noted trends.

Most water quality trading programs are limited to individual rivers or watersheds. Typically a diverse group of sources upstream within a river contribute to an overall load (kg) of nitrogen or phosphorus, in which the cumulative negative effects are experienced downstream. It is also possible that negative impacts exist upstream as well. In such as case, water quality trading must be constrained to protect the upstream portion. Where the science does allow, the state of Connecticut's Nitrogen Credit Exchange (Nitrogen Credit Advisory Board of Connecticut 2011), for example, allows for trades to occur across local rivers within the state, since the area impacted by nitrogen pollution is downstream in the Long Island Sound. This is an example in which nearly the entire state drains into the same water body in which nitrogen tends to not negatively affect upstream freshwater streams and rivers. Thus, Connecticut's program is an exception to most examples in the US to date. In contrast, the state of Virginia established one trading program for all its rivers that drain into the Chesapeake Bay, but constrains the trades, with a few exceptions, to trading within each river but not across from one river to another. In this example each river has an upstream freshwater portion and a downstream tidal portion. The tidal portions tend to be negatively affected by nitrogen and phosphorus, thus provide rational for constraining where trading may occur according to the science. In reality, the science does suggest that some amount of trading can occur across each of the rivers that drain into the Chesapeake Bay (personal communications with Richard Batiuk, Chesapeake Bay Program Office, 2010). As a policy decision, Virginia chose to constrain trading to within each river, with a few exceptions. The state of Virginia exemplifies the efficiency of having one state-wide program, in which trading areas can be constrained as needed or as chosen. This is in contrast to North Carolina, which developed separate trading programs, for separate watersheds (Selman 2009). In the case of North Carolina, their programs were designed early in the learning curve, prior to many trading programs that have since been developed. Thus Virginia as well as other states benefit from the other state's experiences, and have greater confidence to create state-wide trading programs, tailored to local circumstances within the state.

These examples support the larger observation that the status of water quality trading experiences in the US are currently local to individual states, sometimes with separate trading programs within the same state, and when required by the science limited to individual rivers or local watersheds.

Legal construct of water quality trading in the US

Most experiences in the US are reactions to requirements under the Clean Water Act. Because the concepts of water quality trading were not contemplated at the time the Clean Water Act was written, the Clean Water Act is silent on water quality trading. This has not proven to be an absolute barrier, but does create a unique legal construct. In response to states beginning to experiment with water quality trading, in 2003 the US Environmental Protection Agency (US EPA) published its Water Quality Trading Policy (US Environmental Protection Agency 2003), which is a non-binding policy intended to clarify the conditions in which water quality trading may occur. This policy provides clarity, but does not fundamentally alter the implicit, underlying legal construct for water quality trading in the US under the Clean Water Act.

Specifically, where driven by requirements under the Clean Water Act, the "buyers" looking to purchase "credits" are rather exclusively individual entities (e.g. wastewater treatment plants, large urban areas with stormwater runoff) operating under a Clean Water Act permit. The role and purpose of these permits has not fundamentally changed in decades, namely the requirement that defined pollutants be limited to a defined amount (in kg or kg/l) specified within the permit or best available technology be used. Clean Water Act permits applicable to classes of entities (e.g. similarly sized industrial plants of similar type) are subject to the same "technology standard", which all entities of that class must meet. A technology standard is considered to be a minimum, across the board requirement usually expressed as concentration (kg/l), and is not applicable to water quality trading as per the 2003 Water Quality Trading Policy.

Clean Water Act permits may also contain a "water quality-based limit," which is a unique limit on an individual pollutant, for a unique entity, specific to that water body where the entity is located, as justified by local waters that are considered to be polluted based on human health or aquatic life. Permit limits that are water quality-based are considered to be applicable to water quality trading since they are not an across the board requirement for all entities.

The allowed use of water quality trading to meet a water quality-based permit limit is explained in US EPA's 2003 policy. However, even though trading may be used, it is not required that the permitting authority allow trading, only that the permitting authority may decide to do so. US EPA has delegated 45 of the 50 states to issue and oversee Clean Water Act permits. As a result the decision to allow for water quality trading is overwhelmingly a state decision, not a federal or national decision. This further explains why all trading programs to date are unique to each state. This also further exemplifies that a permit holder's right to access a water quality trading program is not a guarantee.

Absent any additional state laws or regulations that state might develop, under the federal Clean Water Act a "trade" as part of a state water quality trading program is, in reality, unique to that permittee (buyer), in which that permittee is able to demonstrate compliance with a particular water quality based limit, by demonstrating that another entity (seller) created an equivalent reduction of pollution so as to satisfy the first entity's permit limit.

This rather technical definition means that each trade carries with it a unique risk to the buyer that the particular credit they acquired may not satisfy their permit requirements. It also means is that each trade, from a legal standpoint, is to be assessed by the permitting authority (usually the state), in which the federal US EPA has oversight. Thus, unlike national laws in the US that established national air emissions markets to control power plant sulfur dioxide, the buyer cannot have full confidence that each credit is valid since each trade is technically a stand-alone, unique action. As a practical matter, states that have designed water quality trading programs have tried to create conditions necessary for all credits to be valid. But even in this case, because each permit itself is unique and subject to objection by citizens, credits from one seller used by Buyer A might be approved, but additional credits from the same seller used by Buyer B could be rejected due to the public's objection.

Much of the state's efforts in the US is focused on addressing this unique legal framework, which has not proven to be a barrier, but does greatly increase the cost and complexity to the states to establish trading programs in the first place. In the examples of Virginia, North Carolina, and Connecticut, the legal approach taken by each state varies. In Virginia for example the state chose to activate water quality trading using state legislation signed by the governor. This was not necessary since as mentioned trading is implicitly possible under the Clean Water Act through the permitting process. State legislation was used in part to clarify legal issues as doing so can provide greater confidence to potential buyers and sellers.

In the case of Connecticut, the state used a part of the Clean Water Act in which the state allocates numerical needs for reductions of nitrogen, under what is known as a total maximum daily load (TMDL; Connecticut Department of Environmental Protection 2005). Within the TMDL the state created a 14-year schedule for nitrogen reductions, in which wastewater treatment plants may use trading over the 14 years to meet their obligations. Strictly speaking the permits are active for just five years at a time, but Connecticut used its TMDL to effectively span water quality trading over three consecutive permit cycles. There are other details unique to Connecticut's experience, but the emphasis here is the use of a TMDL as a framework for water quality trading.

Regardless of the state's choice to activate a trading program (e.g. legislation, TMDL, permit, state policy), if the buyer acts under a Clean Water Act permit it remains the case that each trade is legally unique to that permit, and is subject to being individually assessed inclusive of the public's input.

Risks to buyers

Risk within any market results in less buyers and/or sellers willing to engage in that market, which is true of water quality trading as a type of environmental market. In the US, water quality trading programs that have seen the greatest number of trades have been programs that limit or emphasize trading between permitted entities, in particular permitted wastewater treatment plants. A likely reason for this is that managers of each plant at a personal level may well know each other; they each manage similar technologies; end-of-pipe measurements are well established; and all are subject to the same federal Clean Water Act penalties for non-compliance. As result, they tend to trust each other, and have greater confidence that credits bought and sold between two permitted entities will be valid. In Connecticut, the state itself conducts the actual trades providing an added level of confidence.

A challenging example that is the subject of most discussion and research is where the seller is not a permitted entity under the Clean Water Act, which we use the term "non-point source." Because a non-point source such as farm or individual landowner is not required to have a permit, if they sell a credit and that credit is found to be invalid no penalty or legal action can be applied under the Clean Water Act. Whereas, if the buyer of that credit is permitted, under the Clean Water Act they are held liable by penalty for the fact the credit they acquired was invalid. In short, under the Clean Water Act there is no transfer of liability when a permitted entity purchases a credit from a non-point source.

This is not an insurmountable barrier. For example states have created banks where many credits exist (Pennsylvania Infrastructure Investment Authority 2012). If a particular credit is found to be invalid, in theory another credit will be available in the bank. This approach can help reduce the liability risk placed on the buyer, but not entirely. This is in contrast to air emissions trading emissions in the US governed by one national law, in which the seller is held liable if a particular credit is found to be invalid. Under air emissions trading in the US the buyer has complete confidence and is more willing to participate in trading.

Where a state establishes a bank to reduce a buyer's risk for water quality trading, doing so comes at a price, namely that an excess number of credits are needed to keep the bank solvent (i.e. enough credits to cover any needs), thus the price of credits to the buyer will be higher than otherwise necessary. Experiences in the US are too limited at this point to indicate if this added cost of having a bank would limit the number of trades, or if the bank sufficiently addresses the buyer's risk in the first place.

Another difference between a bank for water quality trading under the Clean Water Act and experiences with national air emissions trading is that credits under water quality trading are usually valid for just one year. The duration in which a create remains valid can be linked to either the retention time of the pollutant (e.g. nitrogen in Chesapeake Bay tidal waters can remain present for over a year before being swept into the sea), conditions of the TMDL, and/or conditions of the permit that the buyer is subject to. The exact duration of a credit and the exact definition is at this time subject to case-by-case interpretation on the part of the state and US EPA. Under air emissions trading for sulfur in the US, a credit is valid indefinitely even though the science would suggest the retention time of sulfur in the atmosphere is finite, and that sulfur's ecological and human health impacts are too finite in their duration.

It is beyond the scope of this discussion to define what the proper duration of a credit should be. The ramification of the decision whether it be science based, policy based, or a mix of both is that more or less risk is placed on the buyer as to the availability of credits. Water quality trading has to date tended to not have large surpluses of credits, which increases the buyer's risk. Where the credit's duration is indefinite as with the case of air emissions trading for sulfur, a large surplus of credits has been the norm since 1996 (US Environmental Protection Agency 2009: figure 2)

Where a non-point source creates credits for sale an additional risk to the seller and buyer is that the practices used to generate the credit may not reduce pollution runoff as effectively as assumed. Not knowing if the practice used were as effective is a major component of the risk described above, but also introduces in additional risk and additional costs. In order to compensate for the uncertainty, many states' trading programs may only value each kilogram of credit as half a kilogram. This helps to address the uncertainty, but in this case the effective price for each credit would be doubled.

Not knowing if the practice used is as effective places a risk on the general public as well, that a river or lake might not be as clean as should be. For this reason and many others, the public may not favor water quality trading. If trading occurs under one national law, such as air emissions trading for sulfur a single citizen objection likely cannot (and to date has not) prevent the trade from occurring. In contrast, since each trade under the Clean Water Act is legally unique to that particular trade for that particular permit, a single citizen objection is more likely to prevent any particular trade. This is additionally true if citizen ratepayer's money is being used by a publicly owned wastewater plant to purchase the credit. In this case a citizen ratepayer has a unique relationship to the buyer, not to mention the general public's right to express objection to any permit under the Clean Water Act.

There are other risks involved for the buyer and seller, but the risks described are somewhat unique to water quality trading in the context of the Clean Water Act. It should be noted that states have sufficient authority to create conditions of confidence that efficiently address these risks, but states are often hesitant since doing so necessary would place additional legal requirements (liability) on non-point sources beyond what is required under the federal Clean Water Act.

Future direction

In the US, interest in water quality trading has increased over the past decade, and in general more members of the environmental community are open to the concept, with the caveat that promised pollution reductions are real. Part of increased acceptance is likely familiarity. What is not clear is when and if large numbers of trades involving non-point sources occur, if the public's support will increase or decrease. To date larger numbers of trades involving permitted entities has proven to be reliable largely because each buyer and seller can be monitored at the end of a pipe. Legal liability unique to this situation (all the buyers and sellers are permitted), such as the experiences in Connecticut likely explain the high number of trades as mentioned, in which direct monitoring likely explains the public's acceptance.

For these reasons the future direction of water quality trading in the US is a focus on ensuring the level of confidence needed that credits are real and valid. As implied this could both increase the number of trades and the public's acceptance. Concurrently yet independently, both the science and activity under the Clean Water Act is gradually moving towards a need for greater control of nitrogen and phosphorus, both of which are generally applicable to trading. This should increase an interest for trading by potential buyers, namely permitted entities subject to water quality-based limits.

In the Chesapeake Bay region along the east coast, in 2010 US EPA issued its most explicit and demanding requirements for water quality trading, unique to the Chesapeake Bay watershed. This was done not as a law, nor as a regulation, but as expressed numerical needs for reductions of nitrogen, phosphorus, and sediment to restore tidal waters of the

Chesapeake Bay under a TMDL. One relevant aspect is that the inherent liability issues (i.e. a non-point source seller is not liable if credits sold are found to be invalid) are not addressed in the TMDL, since this can only be addressed by a state law or regulation, or changes to the Clean Water Act as directed by Congress.

However, the TMDL for the Chesapeake Bay does attempt to address many of the risks related to the need for buyers and sellers (as well as the states and EPA) to know if the credits are valid, such requirements as monitoring, reporting, and transparency. In theory, over the next decade the Chesapeake Bay may serve as an experiment. Assuming that conditions of trust are indeed established as to whether non-point source credits are valid, we will better understand if trust in the validity of credits is sufficient to see large numbers of trades involving non-point sources. Or, we may learn that even where trust in credits does exist, the inherent liability issues are indeed a barrier.

Complicating the matter will be the issue of credit prices. Even if credits have sufficient trust and if liability issues are not a significant deterrent, it may be the case that potential buyers are able to meet their permit limits at a cost reasonable enough to not engage in trades with non-point sources in large number.

What does seem clear is that where large numbers of permitted entities are allowed to trade among themselves and the state encourages such trades (e.g. Connecticut, North Carolina, Virginia), we can expect significant trading activity in which citizen ratepayers are accepting. We can also expect that such trades are not always driven by cost, but sometimes by greater planning flexibility. For example, the ability to postpone the installation of pollution control technology until more favorable financial loans are available, or a previous debt is fully paid, or to install technology while expanding the capacity of a wastewater treatment plant.

Although no conclusion can be drawn yet, within the past three years there has been greater interest at the regional level to begin a dialog for regional multi-state trading programs. All such discussion to date still rely on the Clean Water Act as the regulatory driver, which as discussed is not best equipped to address the many conditions necessary and risks.

No example is known where multiple state are exploring a regional water quality trading program, in which the states themselves create the legal need independent of the Clean Water Act. This is understandable, but also interesting since we have seen examples in the US where states have worked together to create regional Greenhouse Gas trading programs, driven by agreed upon pollution reductions determined by the states themselves (RGGI 2012).

In summary, water quality trading in the US has and is still evolving. In contrast to air emission examples the legal construct of the Clean Water Act is not best able to support such environmental markets. This is not, nor has been a barrier to creating such programs, albeit at the individual state level. Core issue of liability under the Clean Water Act will continue to pervade experiences with non-point sources. But where other sources of risk are addressed we will learn over the next decade if such liability issues are a core barrier, or just a unique circumstance that carries with it a higher than otherwise price. What is clear is that water quality trading among large numbers of permitted entities is likely to remain a permanent feature in the US particularly for nitrogen and phosphorus where applicable.

References

Connecticut Department of Environmental Protection (2005) *The Long Island Sound TMDL*. October. Available at www.ct.gov/dep/lib/dep/water/lis_water_quality/nitrogen_control_program/tmdlfs.pdf (accessed 13 July 2012).

Electric Power Research Institute (2010) *Ohio River Basin Water Quality Trading Project*. October. Available at http://my.epri.com/portal/server.pt?open=512&objID=423&mode=2&in_hi_userid=2&cached=true (accessed 13 July 2012).

Nitrogen Credit Advisory Board of Connecticut (2011) *Report of the Nitrogen Credit Advisory Board for Calendar Year 2010*. Available at www.ct.gov/dep/lib/dep/water/municipal_wastewater/nitrogen_report_2010.pdf (accessed 13 July 2012).

Pennsylvania Infrastructure Investment Authority (2012) *Nutrient Credit Trading Program*. Available at www.pennvest.state.pa.us/portal/server.pt/community/nutrient_credit_trading/19518 (accessed 13 July 2012).

RGGI (2012) "Welcome." Regional Greenhouse Gas Initiative. Available at www.rggi.org (accessed 13 July 2012).

Selman, M. (2009) *Water Quality Trading Programs: An International Overview*. Issue Brief. Washington, DC: World Resources Institute.

Selman, M. and Greenhalgh, S. (2009) *Eutrophication: Sources and Drivers of Nutrient Pollution*. Policy Note. Washington, DC: World Resources Institute.

US Environmental Protection Agency (2003) *Water Quality Trading Policy*. January Available at http://water.epa.gov/type/watersheds/trading/upload/2008_09_12_watershed_trading_finalpolicy2003.pdf (accessed 13 July 2012).

US Environmental Protection Agency (2009) "Acid Rain and Related Programs: 2009 Highlights." Available at www.epa.gov/airmarkt/progress/ARP09_4.html (accessed 13 July 2012).

US Environmental Protection Agency (2012) "Water Quality Trading." Available at http://water.epa.gov/type/watersheds/trading.cfm (accessed 13 July 2012).

Part III

Reforms to overcome legal and institutional barriers to trading

Introduction: considering institutional frameworks and legal reforms in trading

Carlos Mario Gómez

Contributions in this part of the book address the reforms required to enable the water trading schemes. The chapters deal with some of the general conclusions stressed in previous parts of the book. One of these is that water markets are not substitutes for good water governance. Water trading is not an alternative when the public sector lacks the will or the ability to define and enforce the norms limiting the use of water and protecting water resources.

Rather than that, synergy and complementarity between water markets and public regulations are pre-conditions for water trading to be able to make a real contribution to the purposes of water policy. Only when the political process comes out with a clear and enforceable decision over where, how much, and for what purpose water can be used can the market be used as an instrument to help the economy in doing the best within the constraints established by water regulations. Water trading is a means to an end. Thus the environmental objectives, the development goals, and the limits resulting from that over the use of water are all policy objectives that need to be decided in the policy arena and cannot be left to the spontaneous outcomes of water trading. In other words, water markets, when they exist, are only one functional mean of the water policy mix and of the complex array of water-governing institutions.

Once the decision to allow for water trading has been taken, finding the right place of water markets and adapting the existing institutional framework becomes the corner-stone of the political reform. This is illustrated in this section of the book by analysing this institutional transition both in Australia and Spain.

Part III is opened by Chapter 15, in which Miguel Solanes frames the way water trading works around the following central argument: the way water markets work depend both on external conditions of the overall economy and on how effective are existing water institutions.

Water trading can probably work only in properly working economic systems. In effect, water markets do not exist in isolation, and water is but one (although a very important one) among the many inputs used in the economy. In the same sense, we cannot expect to have an efficient use of water in an economy which inefficiency is pervasive. When, for instance, the exchange rate is overpriced, this will distort the economy and the use of all natural resources, including water. Interest rates lower than the opportunity cost of capital, artificially maintained by lax monetary policies and increasing external borrowing, will also boost the derived demand of water in the overall economy. Even in the presence of model water institutions, we cannot expect an efficient allocation of water if the agricultural sector is driven by price distorting subsidies. In all these cases, as Solanes shows, water trading can exacerbate the overall economic inefficiency.

On the other hand, the key institutional change to allow for water trading consists in the creation of a well-defined set of property rights. That is to say, a bundle of clearly specified privileges or obligations, belonging to an identifiable owner, protected from expropriation or damage and that can be exchanged with others through voluntary and enforceable agreements. The previous statement makes clear the minimal conditions required. Different from other markets, this can only be achieved by having a capable water administration, able to keep track of water rights and monitor transfer processes. In this sense, Solanes shows why, without the appropriate institutional set-up, markets may become ineffective or, even worse, create new opportunities for rent seeking and free riding leading to further deterioration of the many public goods involved in water management. This is illustrated with a variety of examples and experiences on water trading in different countries.

The complexities of the legal reforms required to allow for effective water trading are illustrated in two separate chapters revising the approach followed in Australia and Spain. The first is developed by Jennifer McKay (Chapter 16) while the second is by Antonio Embid Irujo (Chapter 17). Both chapters are rich in information presenting the options at hand, explaining the factors driving the alternatives chosen, and providing the whole picture on how water legal frameworks have evolved in two distant and different legal frameworks. However, with the benefit of hindsight, the reader is given the opportunity to make a rich comparative exercise from the two apparently opposite experiences.

Both McKay and Embid Irujo show how the emergence of water trading has been driven by similar factors in both cases. In some particular regions, such as the Murray–Darling Basin in Australia and the Mediterranean areas in Spain, water scarcity was perceived as a critical factor limiting economic progress particularly for agriculture and food security. At a certain point, in both cases, it became evident that going along with public works and water supply policies is not sufficient to keep pace with the increasing demand for water in the growing economy. Finally, in both countries, there is a growing recognition of the limits of the command and control system to cope with the regional and sectoral conflicts derived from water scarcity.

In spite of these similarities, the legal reforms chosen by both countries differ in some significant ways. One key factor explaining the different choices made lies in the evolving balance between the developmental objective of putting water policy at the service of the expansion of the economy, and the environmental objective of preserving and protecting the water sources on the other. Australia, for example, pioneered the adoption of water markets at a time when water was mainly driven by developmental purposes, without too much concern over environmental objectives. Obtaining efficiency gains from water reallocation was at the time at least as important as mobilizing additional water resources into the agricultural sector and the overall economy. At the same time, development objectives in Spain were pursued by pioneering water planning at a river basin scale, but relying mostly on command and control instruments and publicly driven allocation of water in the economy. In contrast to Australia, water markets emerge in Spain as a complement of a complex and well established institutional water management framework. Trading then was designed with strong requirements to consider environmental impacts and even as an instrument to achieve environmental goals.

Both countries differ in how far they go in the set-up of the property rights that can be traded. McKay explains, for instance, how unbundling water from land ownership was essential for the extension and the efficient work of water trading in Australia. In an opposite direction, Embid Irujo explains how attaching water use rights to land ownership, limiting trading to the right of using water and retaining the property of water in public hands is considered as a guarantee that water serves the purposes of water policy. The volume of water to be traded is another substantial difference: Spain limits transfers to actually effectively utilized water. In so doing it follows the American system of water trading, where only historically consumed water can be traded. Australia allows the transfer of non-used-waters, which may have so far resulted in important environmental and third party impacts.

The different story-line of water markets in Australia and Spain give some interesting clues to understand the different policy priorities and the institutional changes promoted in both countries. In this sense, McKay explains that the current political priority in Australia has shifted to a systematic effort to make water trading work for the environment. Apart from guaranteeing minimum flows it requires finding the way to enforce transparent and acceptable water consumption ceilings defined by a water authority. On the other hand, Embid Irujo explains why and how water trading needs to be given space to protect the public interest and environmental standards, with all provisions to continue to be defined and enforced by the institutions already in place. In other words, in Australia the strategy consists of adapting the water markets to work for the environment, while in Spain the priority consists of protecting the environmental objectives from the outcomes of water trading.

Legal reforms may be necessary for water trading to become not only a means to cope with water scarcity but also an instrument for adaptive risk management. Water trading makes it possible to adapt water allocation to a variable and uncertain water supply, limiting the negative impacts of droughts in the economy. Publicly defined water allocations can potentially be politically acceptable (and even efficient) in normal times, but lack the flexibility required to cope with droughts and with the climate change driven hydrological uncertainty. In Chapter 18 Almudena Gómez-Ramos shows how water trading has the potential to make a real contribution to enhance drought and climate change resilience. To do so she proposes to explore the complexity of improving water allocation in a context of uncertainty. Once the fundamental principles are set, Gómez-Ramos goes into the details of the water institutions in Spain in order to discuss the reforms that would be required in the legal system in order to allow for a better distribution of risk between water users. Apart from trading on current water, the author suggests trading on water options, and illustrates this proposal with an empirical simulation in southern Spain.

15 Water and development in Latin America

Rights, markets, economic context and institutional requirements

Miguel Solanes

Introduction

This chapter analyses the relevance of the macroeconomic context, the system of water rights, and the institutional organization for the functioning of water markets, so as to insure a sustainable insertion of water resources into the development processes. Both the definition and allocation of water use rights are assessed, together with the functioning of trading schemes. Evidence from all over the world shows that this is a critical factor, as are the overall macroeconomic context and the existence of economic incentives. This chapter emphasizes the role of institutions for sounder governance of water resources.

Water rights

In most countries water belongs in the public domain. However, water use rights granted to economic agents are protected under the property provisions of national and, in the case of federal countries, state or provincial constitutions. Water rights are a basic legal element of systems that have successfully promoted private investment in the development of the resource. A water right usually is a right to use, and ownership of a water right normally means a power of usufruct, and not ownership of the corpus of water itself.

A system of secure and stable water rights is an incentive to invest in the development and conservation of water resources and a precondition for implementing water trading. Furthermore, the stability and certainty of water rights, and the uses arising from them, provide recognition of existing economic activities and prevent the social unrest that would result from ignoring uses existing at times of change in water legislation.

An important regulatory element is the requirement that water rights are effectively utilized, and for beneficial use, in order to prevent monopolies. Effective water utilization usually demands investments of magnitude for the implementation of irrigation, electricity, or water supply and sanitation projects, among other possible water uses. Keeping idle water rights is therefore less costly for monopolizers than effectively using them. Controlling idle water rights is an effective manner to build barriers to market entry to competitors. Chile faced this situation concerning the electricity generation industry. A few companies held a substantial part of generation water rights. They did not use them, neither transferred them to competitors, therefore blocking competition and controlling electricity outputs. Most water laws control the hoarding of non-used water rights forfeiting or revoking non-used rights. Controlling monopolization was a major driver behind the amendments to the Chilean Water Law, in 2005.

In addition to this requirement, there is a general trend to condition the use of water in water use rights (Solanes and Gonzalez-Villarreal 1999). This conditioning includes formal (obtaining a permit) and substantive requirements (for example, no harm to third parties, environmental protection, reasonable efficiency, and payment of common control costs). It is usually possible, within certain limits, to impose new conditions after a water right, permit, or license has been granted.

Market instruments: private water transfers

If water allocation is important, even more fundamental is reallocation as resources become scarcer in relation to demand. Historically, new users could obtain water through appropriating water rights to which no previous claims had been established. At present, in the countries of Latin America and the Caribbean, although overall supply is plentiful, in many areas with concentrated economic development, all surface water and much of groundwater is appropriated. Alternatively, historically, new water supplies have been obtained through the construction of storage and conveyance facilities, projects usually undertaken with the help of substantial public investments. In many areas, as the best and the least expensive water sources have been developed and attention increasingly turns to the more expensive and poorer sites, the costs of new projects have begun to escalate.

On the other hand, due to population growth, urbanization and economic development, per capita water availability is decreasing. At the same time that demands for both instream and consumptive water uses expand with economic and population growth, water pollution diminishes the available quantities of good quality water and increases the costs of treatment. It can be expected that there will be increasing demands to reallocate supplies among uses and users.

In this situation, countries have to decide whether to use administrative mechanisms or water markets to achieve water reallocation. The choice between these alternatives has been the cause of numerous debates in the Latin American and the Caribbean (LAC) region. While acknowledging the importance of appropriate water management, the debate reflects, on the one hand, the poor performance of a conventional response (administrative reallocation) and, on the other, the difficulty of implementing a different alternative (water markets), which, on occasion, produces a profound contradiction to traditional practices and ideas. At a practical level, markets are difficult to implement, lacking appropriate registers and cadasters of both water sources and water rights.

The introduction of water markets in no way represents a blanket solution to every conflict. The market is but one of a series of management tools, and it needs to be appropriately designed, adequately regulated and correctly used. Its advantages will only be materialized insofar as the market approaches the competitive paradigm. The efficiency of competitive markets is based on many restrictive assumptions, and market failures (such as externalities, market power, public good characteristics, risk, uncertainty and imperfect information) raise the possibility that the transfer of water rights may be beneficial for the buyer and the seller, but inefficient from an overall economic, social, and environmental perspective.

In the situations of market failure, governments should intervene to correct distortions and restore the necessary conditions for economic efficiency. As water markets tend to diverge considerably from the competitive model, mainly because of the external effects (return flow, in stream flow and area-of-origin effects) that water right transfers usually cause, they must be regulated by the State. This has been confirmed by the experience of water markets in the western United States.

The market can indeed act as an effective means of water reallocation, but always provided it operates in an institutional framework capable of correcting the distortions that the nature of the resource inevitably generates, thereby fulfilling the function of the State and civil society in safeguarding the public interest and the resource itself.

It is then necessary to have an institutional and legal system that is compatible with water marketing. For water markets to operate, water rights must be secure, safe, and easy to identify. In turn, marketable entitlements must be based on the nature of the resource and its hydrological characteristics. The transfer of nominal entitlements leads to issues of equity and sustainability, as in Australia (Young 2010). In addition, market development is influenced by the economic and cultural context, where it takes place. Therefore, economic policies must also be appropriate for the development of water markets. Acquisition of water rights is an investment affected by the economic environment where it takes place.

Thus, a sustainable and efficient water market depends on the institutional and economic framework established by the state. Concomitantly, a water market without regulations to guarantee sustainable water supply, control of externalities, and prevention of monopolies, may not be an instrument of efficient water reallocation. Equally important, investments will not take place if the economic environment where water will be utilized does not offer opportunities for economic returns.

Experience has shown that the following rules are important for the operation of a water market:

- water must be put to beneficial use and continue to be used beneficially after reallocation;
- reallocation should not affect other users and must be in the public interest;
- in many jurisdictions, the transfer of water from one river basin to another can only occur with due consideration of local interests;
- in countries where the institutions for water management are not well developed, emphasis has to be placed instead on the implementation and improvement of appropriate systems for the allocation and registration of water rights; and
- only historically used waters can be transferred, non-used waters are not transferable.

(Anderson 1991)

As noted before, water markets and water rights do not perform in a vacuum. They are influenced by the macroeconomic context in which they operate. It is important, then, to consider macroeconomic policies and incentives, when assessing the potential contributions of water to development. Examples from Chile and Argentina illustrate the importance of economic context.

Macroeconomic environment and economic incentives

Argentina and Chile are good examples of the impact that general economic policies have on the productivity of water resources. Between 1985 and 1995, Chile enhanced the contribution of water to socioeconomic development. Irrigated agriculture, mining and other export-oriented activities, for which water is especially relevant, were key to the process (Peña and Brown 2004). In Argentina, over the same period, the surface area under irrigation declined, and the country's drinking water supply and sanitation services suffered to such an extent that some foreign investors withdrew from the country and filed claims with international arbitration tribunals (Diaz and Bertranou 2003). The reasons for the relative success of one system and failure of the other lie in the macroeconomic policies and the public policy decision-making criteria of the two countries.

Chile, with a different macroeconomic policy, consolidated its position has a world-class exporter of agricultural produce and other products for which water is an input. The Water Directorate was modernized and the quality of hydrological information improved.

The experiences of Argentina and Chile confirm that the decisions of water users are affected by general economic drivers, such as interest rates, uncertainty, prices, subsidies, exchange rates, property rights and taxes (Ciriacy-Wantrup 1951). High interest rates reduce investment in all areas, not only in water-related sectors, and they also reduce environmental conservation and protection efforts.

When capital is expensive, there is a natural tendency to overuse the resource or the environment. Users facing high interest rates are also likely to reduce their investment in improvements, works, and equipment. The same occurs when markets are erratic in terms of price and demand.

The effectiveness of water policies and associated investments, as well as the relevant legislation and organization, is thus dependent on macroeconomic policies, incentives, and the environment that they create. In the long term, they are so powerful and structurally significant that the best sectoral instruments and legislation cannot counteract their influence: "macroeconomic policy has a pervasive influence on the structure of incentives and performance in the entire water sector" (Donoso 2003; Donoso and Melo 2004). This has been evident in the provinces of western Argentina, where substantial subsidies for the use of groundwater were made a significant part of economic policy. There was no legal regulation that could prevent deterioration of the water situation when these policies offered such powerful incentives. When public policies are counterproductive, the unfavorable macroeconomic context erodes even the best institutional reforms.

So, at the very minimum, those in charge of water resources management and service provision must continuously analyze the impact of existing and proposed macroeconomic policies on the development of the water sector and maintain a fluid, frank, and active dialogue with those responsible for economic and social policies. This presupposes that water administration has a certain degree of autonomy and independence of political and sectoral interests, an active attitude, stability and resources (professional, financial, information, etc.) in line with its responsibilities, and maintains a close contact with water users and water-dependent sectors.

In many cases competition in global markets presupposes a high degree of technical skill in water management, not just because of the dynamics of water, but also because of competitive advantages. In Chile, for example, the sophistication of water management in vineyards, which is water efficient, aims at producing high-quality wines (of high value), not at saving water. Water savings are not relevant in taking investment decisions. What is relevant is that water-saving technology produces high-priced, high-quality wines for international markets. For this reason, producers frequently go far beyond national requirements. This is why clean production programs have been agreed to by some sectors (mining, agriculture, etc.).

Institutional context

Governance

Water management problems neither originate nor can be solved within the limits of water resources alone. They are rooted in social, economic and political factors. Sectoral lobbies often capture water management institutions and policies. They may be the outcome of the efforts by irrigators or by the hydroelectricity industry. Countries abound on examples of legislation and subsidies that end up in non-sustainable practices. For this reason, frameworks to analyze water resources governance should include drivers outside the water sector. Ignoring this fact may result in the proposal of greatly over-simplified, uniform, and generalized solutions that are simplistic, or ideological, and eventually counter-productive (Arbor and Giner 1996).

Governance refers to:

> the capability of a social system to mobilize energies, in a coherent manner, for the sustainable development of water resources. The notion includes: i) the ability to design public policies (and to mobilize social resources for their support) which are socially accepted, whose goal is the sustainable development and use of water resources; and: ii) to make their implementation effective by the different actors/ stakeholders involved in the process.
>
> (Rogers 2002: 1)

The management of a public good such as water is problematic and precarious when regulatory institutions do not adapt to the nature of the resource they regulate. The regulatory needs of a resource with the complexities of water (characteristics of public and private good, flowing, part of environmental processes, social need, economic good, affected by externalities and market failures) are much more complex than the regulation of simpler goods.

In this respect, the procedures for institutional change in Latin America and the Caribbean have often neglected the fact that markets need laws and regulations to function. Water markets do not function properly without free flows of information, competition, protection of property rights, and control of externalities:

> The libertarian utopia of complete laissez-faire does not give you prosperity... Markets... are human institutions, with human imperfections. They do not necessarily work well, and they are too important to be left to the ideologues.
>
> (Gewen 2002)

According to Stiglitz (quoted in Lloyd 1999), this ideology "sees the state as being irredeemably corrupt."[1] In some cases, the result of this distorted vision has been that administrative structures for water management and the regulation of water-related public services have been deliberately designed with *ex profeso* limitations on State power or with a distortion of their information base.[2]

The weakness of civil society

In developed countries with robust corporate structures representing various interest groups (such as industrial associations, public and user organizations, trade unions, and environmental groups), with a high degree of pluralism, with a relatively even balance of power among them and with effective support structures (such as adequate systems for the delivery of justice and education), consensus and self-regulation are tools that are increasingly used with the consequent reduction of transaction costs. However, this system, when transferred to societies in which there is both an imbalance of power and inequality of access among the various groups, leads to inequity. The group with the greater capacity of *de facto* political leverage ends up manipulating the political system for its own benefit.

This is a real issue in water trading. In this context, reference to civil society loses part of its *raison d'être*, because the *de facto* principles on which it functions are absent.[3] For example, in many countries, as in Chile, "due to their sheer economic size, utility companies have acquired an influence in the political system and in society as a whole, against which it is hard for regulators to contend" (Bitrán and Serra 1998).

Capture and corruption

Political and regulatory bodies can be captured.[4] For instance, in the privatization of drinking water supply and sanitation services in the city of Buenos Aires, the "poor information base, lack of transparency in regulatory decisions and *ad hoc* nature of executive branch interventions make it harder to reassure consumers that their welfare is being protected" (Alcázar *et al.* 2000: cover). "The regulation model (framework, rules of the game, and regulatory body) has been inefficient, fragile and weak. The capture of the regulator and/or the government has been mentioned as one of the main reasons for the governance problems of the concession" (Rogers 2002: 72). This would also be a problem for water trading.

Public policy decision-making

There is a need for the compulsory evaluation of the economic, social and environmental impacts of policies, subsidies and projects utilizing public funds.

There are many public policy decisions associated with water resources, such as the granting of public subsidies to promote water use, which can enhance or limit water contributions to national socio-economic development (Solanes 2005).

With regard to water as an agricultural input, its integration in the productive economy, in the case of Chile, has been strengthened by public policies that have considered not only water uses, but also improvements to the quality of agricultural produce, availability of external markets, and the design of appropriate marketing systems. Subsidies are only granted upon proof of their contribution to national development.

Water administration

Studies on water resources administration in Latin America and the Caribbean (LAC) have concluded that these systems have traditionally been, and in many cases still are, characterized by an essentially sectoral approach (ECLAC 1994). In the current conditions of growing water scarcity, rising externalities, and increasingly drastic and ruthless competition between water users; this approach is leading to ever-more disputes and inefficient water use. It also diminishes the authority of water administrations. The main limitations of sectoral approaches are:

- Lack of objective assessment of policies and projects, and often absence of technical criteria in the decisions-making process, as the sectoral administrator attempts to satisfy the demands of political constituencies, according to their functional interests, without considering the sustainability of the source of supply or the economic evaluation of the project.
- A separation of management functions (for example, issues related to the quantity and quality of water are often considered separately, as are matters related to surface and groundwater) that does not reflect the physical characteristics of water and its optimum use, thereby making it difficult to achieve an integrated vision of the resource.
- Uncoordinated, and often conflictive, use of common sources by different sectoral uses, such as electricity, irrigation, and water supply and sanitation.

In order to avoid such problems, many jurisdictions assign responsibilities for policy formulation, water allocation, and program and project evaluation to an agency that does not have responsibilities for the sectoral use of the resource (agriculture, energy, etc.). The technical specifications and the environmental and social functions of water make it inappropriate to place water management within the remit of purely economic ministries, or even environmental agencies, as in either case there is the risk of neglecting relevant considerations. It is for this reason that it is usually recommended that, when water administration is part of the general system of environmental or natural resource ministries, it should have some degree of functional autonomy to make it easier to carry out its functions (Solanes and Getches 1998). Such autonomy includes independent management of funds.

The most interesting experiences in the LAC region over the last few decades have been in Mexico, where water resources are managed by the National Water Commission (CONAGUA), and in Brazil, which recently set up the National Water Agency (ANA) with the principal objective of overcoming traditional conflicts and limitations imposed by a system in which, until recently, water had been under the responsibility of functional ministries. Other examples of non-user organizations, or at least of those that are not linked to specific water sectors, were the Ministry of Environment, Housing and Territorial Development in Colombia, the Water Resources Authority in Jamaica, the Ministry of Environment and Natural Resources in Venezuela, and the General Water Directorate of the Ministry of Public Works in Chile.

The World Bank (1993) emphasizes the need to separate policy-making, planning, and regulation functions from operational functions at each level of government. Thus, the World Bank agrees with the United States National Water Commission, which, back in 1972, already recommended that policy planning and sectoral planning must be separated from functional planning, design and construction, and operation by executing agencies (NWC 1972). This functional separation is fairly uncommon in Latin America and the Caribbean, yet its application in practice has proved successful. One case has been Chile, which since 1969 has maintained a clear separation of roles in the institutional structure of the state. This has avoided distortion of the regulatory and management functions and has generated a system that gives clear signals to the different economic agents, in both the public and the private sectors, about the relative scarcity of water.

Other important characteristics that are considered essential for a water authority, if it is to provide adequate governance to the sector, are that the authority should have a sufficiently high position within the government hierarchy. Another relevant consideration is that, given the technical complexity of water issues, and especially water trading, a number of countries respect administrative rulings on issues that require specific professional knowledge. This means that fact-finding must be determined in the first instance by the administration in charge of water resources management, and that they must be final unless they are unreasonable or arbitrary (Trelease 1974).

Conclusions

There are many cases of successful integration of water into sustainable development. They are based on a complex mix of economic context, legal institutions, use of economic instruments, and institutional arrangements, including:

1. appropriate macroeconomic context and economic incentives;
2. independent, non-sectoral, institutional arrangements for water managements;

3. a system of water rights conferring legal security and safe tenure;
4. unbiased and impartial assessment of the economic, social, and environmental impacts of water policies, projects, and fiscal subsidies; and
5. appropriate governance, not co-opted by special interest groups.

In addition, the insertion of water into development is enhanced by the availability and operation of water markets, since they increase the efficiency of water allocation and use. However, as water is not an ordinary commodity, markets and water rights must be designed according to the specificities of water resources. Paramount among them is the restriction of trade to effectively consumed and used water (e.g. US and Spanish systems), since trade of nominal rights may compromise sustainability and efficiency (as shown by the experience from Australia and Chile, allowing the transfer of water hitherto not used), and monitoring and control transfers to prevent externalities and impacts on third parties (through processes of inquiry and approval, as well as to regulation on behalf of the public interest under the principle of effective and beneficial use, as in the Spanish and US systems).

Notes

1. "Decisions were made on the basis of what seemed a curious blend of ideology and bad economics, dogma that sometimes seemed to be thinly veiling special interests. When crises hit, the... prescribed outmoded, inappropriate, if 'standard' solutions, without considering the effects they would have on the people in the countries told to follow these policies. Rarely did I see forecasts about what the policies would do to poverty. Rarely did I see thoughtful discussions and analyses of the consequences of alternative policies. There was a single prescription. Alternative opinions were not sought. Open, frank discussion was discouraged... Ideology guided policy prescription and countries were expected to follow the... guidelines without debate" (Stiglitz 2002: xiv).
2. For instance, Sappington (1993) suggested that, to facilitate private investment in public services, it might be advisable to make it more difficult to measure the true level of profits; for example, by developing accounting systems that reduce the visibility of profits or encouraging vertical integration of the regulated company in order that "creative" transfer prices may be used for reducing observable profits.
3. Laws are social products to which lawyers merely attach a certain technical content. What underlies the law and the institutional framework in this context is the result of a variety of social inputs and interests. Water laws in countries where there is a balance between institutional, social, economic, and environmental elements tend to reflect this structural equilibrium. In other countries, water laws reflect the basic imbalance of the underlying system, in water and in other natural resources (Solanes and Getches 1998).
4. Since regulatory decisions affect the regulated companies' welfare, there is an incentive for them to use resources at their disposal to try to influence regulators. The companies are almost always better organized, motivated, and financed than any other group, and so are often viewed as the single greatest threat to regulatory decisions being made in the public interest: "Their tenacity and creativity in pursuing favourable regulatory decisions is driven by their

knowledge of the substantial impact of regulatory decisions on their income and quality of life" (Zearfoss 1998: 65).

References

Works cited

Alcázar, L., Abdala, M. and Shirley, M. (2000) *The Buenos Aires Water Concession.* Policy Research Working Paper no. 2311. April. Washington, DC: World Bank.

Anderson, L. O. (1991) "Reallocation." In R. Beck, *Waters and Water Rights,* vol. 2. Charlottesville, VA: The Michie Company.

Arbor, X. and Giner, S. (1996) *La Gobernabilidad: Ciudadanía y Democracia en la Encrucijada Mundial.* Madrid, Spain: Siglo Veintiuno de España Editores.

Barry, G. (2002) "Beware of False Profits: an Economist has High Praise for Both Free Markets and Government Regulations." *New York Times Book Review,* 16 June.

Bitrán, E. and Serra, P. (1998) *Regulation of Privatized Utilities: The Chilean Experience.* Santiago, Chile: Universidad de Chile.

Ciriacy-Wantrup, S. (1951) *Dollars and Sense in Conservation.* Berkeley, CA: University of California, College of Agriculture, Agricultural Experiment Station.

Díaz, E. and Bertranou, A. (2003) *Investigación Sistémica Sobre Regímenes de Gestión del Agua: El Caso de Mendoza.* Santiago, Chile: South American Technical Advisory Committee (SAMTAC) Global Water Partnership (GWP).

Donoso, G. (2003) *Mercados de Agua: Estudio de Caso del Código de Aguas de Chile de 1981.* July. Santiago, Chile: Pontificia Universidad Católica de Chile.

Donoso, G. and Melo, O. (2004) *Water Institutional Reform: Its Relationship with the Institutional and Macroeconomic Environment.* Santiago, Chile: South American Technical Advisory Committee (SAMTAC) Global Water Partnership (GWP).

ECLAC (1994) *Agenda 21 and Integrated Water Resources Management in Latin America and the Caribbean.* LC/G.1830. 12 April. Santiago, Chile: ECLAC.

Gewen, B. (2002) "*Reinventing the Bazaar: A Natural History of Markets* by John MacMillan." *New York Times Book Review* 16 June. Available at www.nytimes.com/2002/06/16/books/beware-of-false-profits.html?pagewanted=all&src=pm (accessed 25 July 2012).

Lloyd, J. (1999) "The Russian Devolution." *New York Times Magazine* 15 August. Available at www.nytimes.com/1999/08/15/magazine/the-russian-devolution.html?pagewanted=all&src=pm (accessed 25 July 2012).

NWC (1972) *Water Resources Planning.* Springfield, VA: NWC, US National Water Commission, US Department of Commerce,.

Peña, H. and Brown, E. (2004) *Investigación Sistémica Sobre Regímenes de Gestión del Agua: El Caso de Chile.* Santiago, Chile: South American Technical Advisory Committee (SAMTAC) Global Water Partnership (GWP).

Rogers, P. (2002) *Water Governance in Latin America and the Caribbean.* Washington, DC: Inter-American Development Bank (IDB).

Sappington, D. (1993) "Comment on 'Regulation, Institutions, and Commitment in Telecommunications', by Levy and Spiller." In Michael Bruno and Boris Pleskovic (eds), *Proceedings of the World Bank Annual Conference on Development Economics 1993.* Washington, DC: World Bank.

Solanes, M. (2005) "Editorial Remarks." *Circular of the Network for Cooperation in Integrated Water Resource Management for Sustainable Development in Latin America and the Caribbean* 21 (February). Santiago, Chile: Economic Commission for Latin America and the Caribbean.

Solanes, M. and Getches, D. (1998) *Prácticas Recomendables para la Elaboración de Leyes y Regulaciones Relacionadas con el Recurso Hídrico.* Washington, DC: Inter-American Development Bank (IDB).

Solanes, M. and Gonzalez-Villarreal, F. (1999) *The Dublin Principles for Water as Reflected in a Comparative Assessment of Institutional and Legal Arrangements for Integrated Water Resources Management.* Stockholm, Sweden: Global Water Partnership (GWP), Technical Advisory Committee (TAC).

Stiglitz, J. (2002) *Globalization and its Discontents.* New York: W. W. Norton & Company.

Trelease, F. (1974) *Water Law, Resource Use and Environmental Protection.* St Paul, MN: West Publishing Corporation.

World Bank (1993) *Water Resources Management.* Washington, DC: World Bank.

Young, M. (2010) *Environmental Effectiveness and Economic Efficiency of Water Use in Agriculture: The Experience of and Lessons from the Australian Water Reform Programme.* Paris, France: OECD.

Zearfoss, N. (1998) *The Structure of State Utility Commissions and Protection of the Captive Ratepayer: Is There a Connection?* June. Silver Spring, MD: National Regulatory Research Institute (NRRI).

Additional references

Colby, B. (1995) "Regulation, Imperfect Markets, and Transaction Costs: The Elusive Quest for Efficiency in Water Allocation." In Daniel Bromley (ed.), *Handbook of Environmental Economics.* Oxford, UK: Basil Blackwell Ltd.

Colby, B. and Bush, D. (1987) *Water Markets in Theory and Practice: Market Transfers, Water Values, and Public Policy.* Studies in Water Policy and Management no 12. Boulder, CO: Westview Press.

Ingram, H. and Oggins, C. R. (1992) "The Public Trust Doctrine and Community Values in Water." *Natural Resources Journal* 32: 515–37.

Solanes, M. (2010) "Water Resources Legislation: a Search for Common Principles." *Investment Treaty News* 20 January. Available at www.iisd.org/itn/?view=archives.

Lee, T. and Jouravlev, A. (1998) *Prices, Property and Markets in Water Allocation.* Medio Ambiente y Desarrollo series no. 6. LC/L.1097. February. Santiago, Chile: Economic Commission for Latin America and the Caribbean (ECLAC).

Wells, L. (1999) "Private Foreign Investment in Infrastructure: Managing Non-commercial Risk." *Private Infrastructure for Development: Confronting Political and Regulatory Risks (8–10 September 1999, Rome).*

16 The theory and practice of Australian institutional reforms to incorporate water markets in integrated water resources management

Jennifer McKay

Introduction

Australia has a rich legal history in water law, as it moved from common law rules to elaborate water licensing systems very early in its history. This achieved a broader use of water to more users than just the riparian owners of land. Land and water remained bundled together until 1983, when South Australia created a limited water market system (Department of Water, Land and Biodiversity Conservation 2012). In 1992 the federal government provided financial incentives to all the State governments to start the processes to unbundle water from land to create water markets (Council of Australian Governments 1994; National Water Market 2012).

Water Markets were considered in order to promote the shifting of water to higher value uses (Davis and Swinton 2011). As one commentator says:

> Given the high level of commitment of the resource in some major irrigation areas, growth opportunities can be realised only if markets exist to deliver water to the most productive activities. Water trading is an equitable way of achieving structural adjustment in irrigation regions. Trading exposes the opportunity cost of water, which is its value in alternative uses to all farmers, whether or not they trade. It enables marginal producers who hold significant water entitlements to realise an asset that was previously valueless unless used or sold with the land. The study shows that there is enough water in the low-to-marginal value end of the irrigation market to supply all the likely long-term growth needs of higher-value intensive irrigation activities in the Murray–Darling Basin.
>
> (Australian Academy of Science and Technology 2006)

Clearly the policy was seen to have the potential to enable these outcomes but the evidence from the practice of the markets is not always clear and several impediments exist. This paper will discuss these and present a solution for implementing Integrated Water Resources Management (IWRM) using the markets, but in a limited way.

This chapter shows that although there have been important legal and institutional reforms since 1788, there are still more that may need to be undertaken in relation to water trading. There is still a lack of agreement on the balance between social economic and environmental outcomes and rent seeking behavior, insufficient transparency in some market processes (inside information), and environmental impacts, in spite of government intervention to buy back water for the environment. At the time of the early markets (between 1983 and 1992), environmental concerns were already raised in some regions. 1992 in fact heralded the era of the environmental concerns being paramount and the first consideration (see section on the epochs of water law; Harding and Fisher 1999).

All of the above and federal policy shifted toward a solution of locally based local water allocation plans by the state governments around 2000. These plans are state-based but set the consumptive pool of water and this reflects environmental concerns first. There has been considerable angst and lack of community consensus about the limits on water available to be used and traded in many places. In several places, the consumptive pool has been reduced by up to 50 percent by a legal process enforcing the decision of the Minister to protect the environment (McKay 2007, 2010, 2011a, b). Hence these instruments: the regional water plans are the key policy and legal instrument, and determine the amount of water left to be traded.

The argument of this chapter is that plans are the superior instrument and the markets need to be subordinate. This assessment is shared with some other authors (Heaney *et al.* 2006). Local water plans need to be the framework for water trading to be an instrument for IWRM. They must build community consensus on the amount of the consumptive pool and this must reflect longer term goals for inter-generational equity and intra generational equity by protecting the environment and some compatible economic uses. This process is a crucial first step. Then the markets can be used to make adjustments and to allow a respectful exit from farming.

Early and current water market issues

There have always been seven different types of water markets in Australia reflecting the interests of political power groups, old laws and institutional arrangements. Recently, some aspects of these have been grouped together and called a national water market. The process in each of the water markets was beset in the early days by legal issues concerning security for mortgages after the unbundling of water from land. The private sector banks and other lenders eventually took security over the water and the land assets. Other early issues were the lack of confidence in the market processes, lack of mechanisms to match buyers and sellers, the mobilization of previously un-used water and hence environmental issues with more use in a drought period and reluctance to permanently sell water. The reluctance to permanently sell water still remains.

There were issues earlier on, in some irrigation cooperative ventures, about water being traded out of the area leaving fewer growers to fund maintenance of irrigation infrastructure. A spectacular example was Coleambally Township, an irrigation town (McKay *et al.* 2010; Keremane and McKay 2011) which offered itself for sale when the legal water trading rules in New South Wales denied the co-op the right to limit trades out of the area. Clearly with water markets there were and are social justice issues and environmental issues.

In 2006, Brennan reported that there was limited development of markets into some sections of catchments and this reflected political and infrastructure barriers as most trading rules were tied to hydrological constraints reflecting geophysical limitations in transferring water (Brennan 2006). Despite this under the NWI in the next year, the State governments agreed to make further commitments to:

- prepare water plans with provision for the environment;
- deal with over-allocated or stressed water systems;
- introduce registers of water rights and standards for water accounting;
- expand the trade in water;
- improve pricing for water storage and delivery; and
- meet and manage urban water demands.

(Intergovernmental Agreement on National Water 2007)

To overcome the issues of matching buyers with sellers, Watermove was set up. Over the period to the present there have been many private sector water brokers set up to match buyers and seller, with a key one being Waterfind. Waterfind provides detailed information to water market participants on a demand basis and details of highest and lowest sale prices and reports on water market policy initiatives at all levels of governments State and Federal (see www.waterfind.com.au). Of late, again, there have been further issues of conflicts of interests in the water supply businesses and governments who may have been misusing information.

In this respect the Productivity Commission has noted that:

> possible conflicts of interest resulting from the multiple functions performed by infrastructure operators and government water agencies. In some cases, those conflicts were perceived to be having an adverse impact on water markets – either directly through breaches of competitive neutrality or indirectly through reduced confidence in market institutions.
>
> (Productivity Commission 2011)

Commonwealth intervention in the water market to buy water for the environment is a part of the overall macro Basin Plan, and this has been met with angst by some stakeholders (*Northern Times* 2012; ABC Rural News 2011). Indeed, the removal of Commonwealth purchasing demand in the southern Murray–Darling Basin has been estimated to result in a fall in permanent entitlement process of between 15 and 25 percent (Waterfind 2011). The Commonwealth had purchased 1,173,837 Ml of entitlements at a cost of AU$1,793,470,754 through the Commonwealth Environmental Water Holder (CEWH). As at 31 October 2011, CEWH holds 1,195,026 Ml of entitlements. The market value as of 6 December was AU$1,512,879,131, which is AU$280 million less than the cost of purchasing. The ways to use the Environmental water purchased under State and the National scheme, has been an issue in all parts of the Murray–Darling Basin. The Ministerial Forum requested the Basin plan to use the existing State frameworks and community networks to manage environmental Water (Murray–Darling Basin Plan 2011).

Each watering year, proposals for the use of environmental water can be put forward by any group operating in the Basin. Proposals are assessed against published criteria in accordance with the CEWH's framework for determining commonwealth environmental watering actions (CEWH 2009).

The criteria considered are :

- the ecological significance of the proposed watering site;
- the expected outcomes of the watering;
- potential risks;
- the long-term sustainability of the site (including whether the management arrangements are appropriate); and
- whether the watering proposal is cost-effective and operationally feasible.

Clearly, these criteria leave scope for disputes. The Commonwealth criteria works independently of any criteria developed in the States. There is a long-standing scheme in South Australia (Hughes and McKay 2009) applying to the River Murray.

For example, the Victorian Bill establishes the Victorian Environmental Water Holder as a new, independent, statutory body responsible for making decisions on environmental water use and managing environmental water holdings in Victoria (Victorian Government 2011). The Victorian Environmental Water Holder will consist of at least three commissioners, appointed by the Governor in Council. The administration of the environmental water holdings and in particular the timing of purchases by the governments involved has and is likely to generate issues about conflicts of interest. The issues raised relate to the selling of water to off shore buyers and the inflationary effect of government purchases.

The epochs of Australian water law: evolution of water rights and trading, and future governance challenges

The issues of the laws and institutions in each state will continue to have an impact in water market functioning in the future. There are several local issues in each state that show the importance of idiosyncratic rules and laws of a particular state.

The ten epochs in the development of Australian water law, and specifically of irrigation water supply for broad-scale irrigated agriculture, are set out below. All but the first two will have water markets as part of the legal arrangements. The last two epochs contain predictions of issues for the immediate future that may complicate the processes of water planning and the operation of water markets in some regions.

Epoch 1: 1788–1901 – state colonial law

Highly introspective in each State inheriting the Common law riparian doctrine and groundwater doctrines from England. These rules tied water access to land. The riparian doctrine was changed to allow non-riparian landowners access to river water during this epoch. The several states did this many ways and expanded water access, sometimes with government-funded infrastructure to deliver water in open channels. The English groundwater doctrine allowed the landowner to exploit groundwater. This was modified at various times by the states with licensing regulation (McKay 2007).

Epoch 2: 1901–1983 – fiscal federalism

Section 96 and section 100 of the Constitution limited the Commonwealth powers over water as an essential aspect to federation (McKay 2010). Exploitation thus remained with the states who were aggressive in development of both water and land. Each State allocated water under different rules to growers and most over-allocated, as the science was lacking. The legacy of different State laws was felt in relation to water trading where in the Southern Connected River Murray System spanning northern Victoria, southern New South Wales and part of South Australia there were over 180 categories of irrigation water entitlements (Shi 2005). These had, and still have, different levels of reliability, tenure periods, protection of the interest in the water license, and nomenclature.

Epoch 3: 1983–1994 – treaty power

Treaties' power enhanced federal intervention power over water, and also the Tasmanian dams case, ecologically sustainable development (ESD) principles were described as national policy, semi-formal water markets in some states. South Adelaide (SA) started permanent water trading in 1982 with NSW and Queensland in 1989 and Victoria in 1991. These give individual growers individual rights over water and entrench individualist aspirations. In many places the early water markets caused more water to be used as the unused water had a value so farmers sold it (McKay and Bjornlund 2007).

Epoch 4: 1994–2007 – weak national protocols funded by section 96, but strong state enforcement

Council of Australian Governments reforms required water markets, ESD and competition law reforms; hence, enhanced section 51 trade practices powers, state litigation on state water plans where the consumptive pool was reduced, regional delivery through state governments and other groups (McKay and Bjornlund 2007), National Water Initiative (NWI) detailed protocol of 80 state actions, during this period a subsidiary market opened up for peri-urban water re-use for intensive horticulture, and also for non-potable use in cities (see Hurlimann and McKay 2006; Keremane and McKay 2007; Wu *et al.* 2012).

Epoch 5: 2007 – the Water Act federal legislation asserting power over State water plans in Murray–Darling Basin

This used section 51 powers in the federal constitution and concurrent referred State powers under section 51(37). Water markets are subsidiary part of water management plans, which set the consumptive pool for a region. The Water Act provides national regulation of contents of plans in Murray–Darling Basin in the national interest, and potentially gives a broad power to alter State assessments of consumptive pool.

Epoch 6: 2007 – the National Water Initiative

By 2007, the National Water Initiative proposed that compatible institutional and regulatory arrangements for trade should be put in place, including principles for trading rules (paragraph 60 and Schedule G).

Epoch 7: 2007–2012 – food security issues and mining

There has been an expansion in this realm of the conflicts between water users and the social, cultural and economic aspects to water use and preservation of communities. The first has been and will continue to be a set of issues around rights of growers to historical water use regimes and this has been tied up with arguments about food security. In some regions, bio fuels demanded water and were able to purchase it from growers. While this moved water to higher-value uses, social justice issues were raised in some places.

In this epoch another issue has been the conflict between mining and agriculture where the mining processes lower a water table or impacts the land available for growing. There are examples in Queensland at the present time.

Planning is the framework for water resources management. The Water Act provides for a two-stage planning process. First a water resource plan is developed. This plan states the strategic goals of the catchment, states how much water is available and sets the principles for sharing the water amongst competing interests. Implementation is via a second plan, a resource operations plan, which includes the rules for trading of water allocations as well as rules for how operators of water supply schemes must operate their schemes and share the resource available at any point in time. Once approved, plans are valid for 10 years, at which time a new water resource plan (and resource operations plan) must be completed. The water planning process is the mechanism for implementing water trading and for granting tradable water rights. On completion of the resource operations plan existing water entitlements (which attach to land and are not tradable) are typically replaced with "water allocations," which are separate from land and can be traded. However, not all water entitlements are converted to tradable water allocations. Queensland has strengthened its Water Act to regulate underground water impacts associated with the extraction of petroleum and gas. Queensland declared a cumulative management area (CMA) where the impact of tow of more tenure holders overlaps. The Surat CMA was the first in March 2011; this is funded by a levy paid by Petroleum tenure holders. In other States Mining legislation may be used for this purpose and this gets away from the IWRM ideal.

Finally, in recognition of all of this the federal government is creating a water stewardship program. This involves:

- developing stakeholder endorsed principles and standards for responsible water use by commercial water users;
- establishing a verification process; and
- building a brand and logo that allows customers to recognise and reward responsible water users.

Also a National water plan report card (NWC 2012), a new interactive web resource has been launched by the National Water Commission to help people easily access information on the status of water planning across Australia.

This online web application, based on the inaugural report *National Water Planning Report Card 2011*, provides a summary of the status of water plans across Australia, including the quality of existing water plans, their implementation, and areas for future improvement.

Epoch 8: emerging and future

Spiritual and customary and economic water rights under human rights treaties

This aspect, which is still nascent in early 2012, is incorporating social, spiritual, and customary objectives into water plans taking account of an expanded concept of native title rights to water, water allocated to native title holders and certainly indigenous representation in water planning. There are some commentators talking about market based systems interacting with customary rights systems and the inevitable clashes especially if the native title systems morph into commercial rights to support economic aspirations. (Nikolakis 2007). These factors may push for increased centralization of control over water, which is also in evidence in National Water Initiative publications. There are real potential clashes in the institutional types desired by indigenous a communal system to run the land and water as opposed to the growers who prefer the present market system with individual rights (Jackson 2005). Queensland has proposed to establish in 2011 indigenous reserves with specified volumes of surface or groundwater available through a license. See Water and Other Legislation Amendment Bill 2011, amending the Queensland Wild Rivers Act 2005.

Epoch 9: foreign investment in the water market as an area of future regulation

The main issue with foreign investment in water markets is that there may need to be oversight from the Foreign Investment Review Board. The federal treasury interest on water licenses when a foreign player is buying an agribusiness worth more than AU$231 million.

Farmers' groups are worried that big players could corner areas of the market by buying up permanent rights in whole valleys, or being able to dictate what food is grown where by controlling water. "We don't have a problem with investment, or indeed, speculation in the water market,"' said Mr Gregson of the Irrigators Council. "We are concerned about market dominance. It's a recently developed, relatively fragile market" (Snow and Jopson 2010).

Epoch 10: climate change

Any adjustment to the consumptive pool in water plans (because of natural events such as climate change) after 2014 will need to be shared according to a risk formula if no other formula is agreed to (paragraphs 46–51 of the National Water Initiative). If adjustments are made to the consumptive pool in water plans because of new environmental objectives, then governments will bear the risks. However, no proportion was given, and it is an assumption that this refers to the state.

The water plans have often reduced water able to be used in order to protect the environment. This reduces the water able to be traded. The plans (there are over 157 of them at present) have done this in several ways, and an example will be given below. The policy accepted so far is that the first 3 percent, risk will be borne by the entitlement holder, from 36 percent to be shared between states and the Commonwealth in a 1 : 3 proportion; and greater than 6 percent shared equally between States and the Commonwealth. It is unclear whether the base total entitlement is that at June 2005 or whether it is a shrinking base (i.e. readjusted each year).

As scientific information improves on the effects of climate change on available water resources some areas are likely to have reduced amounts in consumptive pools. This will raise social justice issues and the conflict between the present generations and the need to make a living and the future generations. It is possible that this regime for risk sharing may be altered.

Current nascent issues: monopolistic behavior and ecologically sustainable development

Monopolistic behavior and restrictions of trade

There are several provisions in the general legislation in Australia which could cover any misuse of monopolistic power by an owner or group of owners of water. Another related issue is the restriction on trade outside of Irrigation areas. The issue for the Irrigation business was that with fewer growers it had to still cover infrastructure maintenance costs. However, this was declared an illegal restraint.

In 2006, the Australian Competition and Consumer Commission (ACCC) was invited (ACCC 2006) by three States to prepare a report to advice on the use of access and exit fees to irrigation infrastructure for Irrigation water suppliers in New South Wales, Victoria and South Australia. This was to assist them in complying with the NWI initiative, which required consistent inter-jurisdictional frameworks. Any water supply business imposing an access or exit fees was required under competition laws to be competitively neutral and to facilitate interstate and regional water markets in the southern Murray–Darling Basin.

This report was drafted in the context of promoting inter-regional and inter-jurisdictional trading of water trading to "facilitate the efficient use of water, both through making the opportunity cost of using water transparent and providing water to move from lower to higher value uses" (ACCC 2006).

The issue was that in the past land ownership brought with it water ownership but since the unbundling it was possible for water to be owned separately and hence be sold beyond the region leaving a smaller sub set of persons to pay for ongoing channel maintenance. The solution was to suggest a further unbundling. The right to have water delivered should be unbundled from any water entitlement, and should be recognized through a separate entitlement, which should be clearly specified, including permissible extraction/supply. Rates specified times, locations circumstances and service levels. Delivery entitlement (and therefore any obligations associated with holding the delivery entitlement) should be tradable. It was also suggested a cap on the termination fees to be set by the ACCC. This was the beginning of federal law creating a protocol for termination fees and other water entitlements.

The National Water Initiative in 2004 has agreed to "have in place pathways by 2004, leading to full implementation by 2006, of compatible, publicly-accessible and reliable Water registers of all water access entitlements and trades (both permanent and Temporary) on a whole of basin or catchment basis" (paragraph 59). The Australian Bureau of Statistics expects these registers will assist in providing more comprehensive and accurate water market information (Australian Bureau of Statistics 2006). This process has received even greater support under the Water Act 2007.

Water markets and ecologically sustainable development: the future with the implementation of the Water Act 2007

In the Water Act of 2007, the Federal Government has seized power to create a basin plan and the federal minister will have power to accredit state plans in the region. This will make up a basin plan. The minister will evaluate state-based plans on the basis of the objective that the plan is in the *national interest* and also if it *implements international agreements.* Clearly, any local plan that devises a water allocation policy is now subject to having them changed if it is not in the national interest. This will add an element of insecurity to water markets, as a plan for a share of a consumptive pool could have that pool reduced, in the national interest.

The Water Act of 2007 was drafted by Malcolm Turnbull based on the 2007 National Plan for Water security, outlined on 25 January 2007. With the change in government, it was amended by Minister Wong (Minster for Climate Change and Water) in 2008, and referred to a new plan called the "Water for the Future Plan," which was announced on 29 April 2008. This new plan introduced the concept of "critical human water needs" and also gave more power to the Murray–Darling Basin Authority. The Water Amendment Act of 2008 is based on a combination of Commonwealth constitutional powers and a referral of certain powers from the Basin States to the Commonwealth under Section 51(37) of the Constitution. The Act passed through the Commonwealth Parliament following the passage of referring legislation through the Basin states – Queensland, New South Wales, Victoria, and South Australia. This occurred in late 2008 Revised Explanatory Memorandum (Water Amendment Bill 2008). In the Water Act, the ESD definition has been reduced in scope from the previous formulation and it is used in relation to the creation of the Basin plan. Only the first two principles are in common with the original seven, but the integration principle still applies.

Thus, the following principles are principles of ecologically sustainable development as used in the Water Act:

1. Decision-making processes should effectively integrate both long-term and short-term economic, environmental, social and equitable considerations.
2. If there are threats of serious or irreversible environmental damage, lack of full scientific certainty should not be used as a reason for postponing measures (the Water Act 2007 defines "measures" to include also strategies, plans and programs) to prevent environmental degradation.
3. The principle of inter-generational equity – that the present generation should ensure that the health, biodiversity and productivity of the environment is maintained or enhanced for the benefit of future generations.
4. The conservation of biodiversity and ecological integrity should be a fundamental consideration in decision-making. "Biodiversity" means the variability among living organisms from all sources (including terrestrial, marine and aquatic ecosystems and the ecological complexes of which they are a part) and includes: (i) diversity within species and between species, and (i) diversity of ecosystems.
5. Improved valuation, pricing, and incentive mechanisms should be promoted.

The new ESD section does not emphasize community engagement, nor the trade aspect of resource use, nor international agreements. However, the last are covered in the objects clause of the Act. The last significant change to the past is that the Water Act has the object to manage water in the national interest and to implement international agreements; both of these will create enormous changes. The national interest will obliterate the state-based introspection that has characterized water management in Australia, and could lead to a decision to not accredit state plans on the grounds that the plan does not consider the national interest which could be the interests of users or the environment in another state.

The question still, however, is what do the set of five principles mean as a whole and do these create a determinate set of rules? It will be the judiciary that will need to make these words determinate. Determinacy is a principal legal problem defined as judicial manageability from the procedural or adjudicative perspective (Orakhelashvili 2008). The key issue that will arise will concern the tensions inherent in reviewing the exercise of Commonwealth Ministerial discretions that are reposed in the act. There is of course the tension in any judicial review of decisions made by ministers.

Conclusions: lessons for the regulation of water markets in Australia

Water markets are a subordinate part of the water reform process in Australia, and are subject to the local assessment of the consumptive pool in local water plans and the overall basin plan in the Murray–Darling Basin area. The determination of the consumptive pool in a region has been the cause of many disputes, and this will continue. There are different views over the quantum of environmental water, the way it is to be managed, and the notion of governments entering the market.

The water markets in Australia have been positive where they have allowed some people to exit farming but stay in the region. For others the water market has been an extreme source of uncertainty. The primary driver of water planning in Australia is now the achievement of the uncertain concept of ESD in regions. Water markets are a useful tool to achieve that and allow individuals some flexibility.

As explained, there has been many institutional reforms to incorporate water trading in Australia, but there may be some way to go to insure that they are part of integrated water resources management. To do so it would be important to consider dealing with extreme social justice issues, and acknowledge that the environment and its needs must come first. This may imply current losses in order to preserve longer-term interests. It will be crucial to have formal mechanisms to deal with the conflicts of values, especially where, as in Australia, the consideration of the environment is a new first priority.

Some institutional and legal reforms that would be necessary include:

- Have uniform laws over the entire country which separate water from land both for groundwater and surface water. The knowledge system needs to look at watersheds and factor in dependencies between groundwater and surface water.
- Have regional factors in water markets considered and rules set up to reduce the impact of trade on the environment. This is where most conflict will appear as most extant processes support overuse.
- Install meters and require reports from users of water so actual use can be monitored. The reports must be accurate and under some penalty of law for understating the water used.
- Charge for the raw water taken from rivers or aquifers maybe not to make a profit but to acknowledge the funds needed to enforce the regulation needed and the collection of scientific data (see points above and below).
- Define the water access right and allow temporary and permanent trade and have a register of trades.
- Have proforma contracts (as for land).
- Define water market participants who can buy and sell and if required limits on non-local ownership, if desired.

- Make buyers demonstrate and adhere to a watering plan so as to not degrade the river or aquifer.
- Impose general law explicitly on water market participants (i.e. not to misuse information or manipulate the market).
- Set up a one-stop shop and embed this within the most relevant body.
- Allow governments to buy water to protect the environment, but provide for a mechanism to stop inflationary effects of this.

For the future, the issues are the impact of new water from mining on the market price where water is generated or losses of water due to mining activity: the impact of indigenous claims as a rights-based claim to use water for economic purposes, which will reduce the previous consumptive pool in some places; the uncertainty over the meaning of words in the *national interest* and to *implement relevant international agreements* in the Basin Plan and the impact of these on the consumptive pool allocated in local water plans.

In summary, the theory behind introducing water markets was to shift water to higher-value uses and to preserve rural communities The data on this are patchy and incomplete. Overall, the practices of the markets tell a rich story of policy implementation, and provide many lessons about coherence of laws and institutions, as well the power of local lobby groups.

References

ABC Rural News (2011) "ABC Fiery Community Meeting at Deniliquin over Basin plan ABC Rural News. Australian Broadcasting Commission." Available at www.abc.net.au/rural/news/content/201010/s3037350.htm?site=rural (accessed 13 July 2012).

ACCC (2006) "A Regime for the Calculation and Implementation of Exit Access and Termination Fees Charged by Irrigation Water Delivery Businesses in the Southern Murray–Darling Basin." Advice to the Australian, New South Wales, South Australian and Victorian Governments. 6 November. Australian Competition and Consumer Commission. Available at www.accc.gov.au (accessed 5 March 2012).

Australian Academy of Science and Technology (2006) "Water Value Adding." Available at www.urbanecology.org.au/topics/watervalueadding.html (accessed 6 March 2012).

Australian Bureau of Statistics (2006) "Rural Water Use and the Environment: The Role of Market Mechanisms." Productivity Commission Submission by the Australian Bureau of Statistics. February. Available at www.pc.gov.au/__data/assets/pdf_file/0020/15068/sub017.pdf (accessed 13 July 2012).

Brennan, D. (2006) "Water Policy Reform in Australia: Lessons from the Victorian Seasonal Water Market." *Australian Journal of Agricultural and Resource Economics* 50(3): 403–23.

CEWH (2009) "Framework for Determining Commonwealth Environmental Watering Actions." Discussion paper. Available at www.environment.gov.au/water/publications/action/cewh-framework.html (accessed 5 March 2012).

Council of Australian Governments (1994) "Attachment A. Water Resource Policy." Council of Australian Governments' Communiqué, 25 February. Available at www.coag.gov.au/coag_meeting_outcomes/1994-02-25/docs/attachment_a.cfm (accessed 19 July 2011).

Davis, C. and Swinton, B. (eds) *Securing Australia's Water Future.* Syudney: Australia Focus Publishing.

Department of Water Land and Bio diversity Conservation (2012) Water Licences and permits. Available at www.sa.gov.au/subject/Water (accessed 2 March 2012; see also www.waterforgood.sa.gov.au/rivers-reservoirs-aquifers/river-murray).

Harding, R. and Fisher, E. (eds) (1999) *Perspectives on the Precautionary Principle.* Leichhardt, Australia: Federation Press.

Heaney, A., Gavan, D., Beare, S., Peterson, D. and Pechey, L. (2006) "Third-Party Effects of Water Trading and Potential Policy Responses." *Australian Journal of Agricultural and Resource Economics* 50(3): 277–93.

Hughes, S. and McKay, J. (2009) "The Contribution of Actors to Achieving Sustainability in Australia through Water Policy Transitions." In D. Huitema and S. Meijerink (eds), *Water Policy Entrepreneurs,* 175–92. Cheltenham, UK: Edward Elgar.

Hurlimann, A. C. and McKay, J. (2006) "What Attributes of Recycled Water Make it Fit for Residential Purposes? The Mawson Lakes Experience." *Desalination* 187(1–3): 167–77.

Intergovernmental Agreement on a National Water Initiative (2007) "Intergovernmental Agreement on a National Water Initiative." Available at www.nwc.gov.au/home/water-governancearrangements-in-australia/national-arrangements/Intergovernmental-agreement-on-a-National-Water-Initiative (accessed 3 March 2012).

Jackson, S. (2005) "Indigenous Values and Water Resources Management: A Case Study from Northern Territory." *Australian Journal of Environmental Management* 12(3): 136–46.

Keremane, G. and McKay, J. (2007) "Water Transitions and Urban Wastewater Reuse in Australia: Lessons for Developing Countries." *International Journal of Environment and Development,* 4(2): 249–66.

Keremane, G. and McKay, J. (2011) "Using Photostory to Capture Irrigators' Emotions about Water Policy and Sustainable Development Objectives: A Case Study in Rural Australia." *Action Research Journal* 9(4): 405–25.

McKay, J. (2007) "The Quest for Environmentally Sustainable Water Use and Constitutional Issues for Federal, State and Local Governments." *Reform, Journal of the Australian Law Reform Commission.*

McKay, J. (2010) 'Some Australian Examples of the Integration of Environmental, Economic and Social Considerations into Decision Making – The Jurisprudence of Facts and Context." In D. French (ed.), *Global Justice and Sustainable Development.* Leiden, The Netherlands: Brill.

McKay, J. (2011a) "Water Markets and Trading." In C. Davis and B. Swinton (eds), *Securing Australia's Water Future.* Sydney, Australia: Focus Publishing.

McKay, J. (2011b) "Australian Water Allocation Plans and the Sustainability Objective-Conflicts and Conflict-Resolution Measures." *Hydrological Sciences Journal* 56(4): 615–29.

McKay, J. and Bjornlund, H. (2001) "Recent Australian Water Market Mechanisms as a Component of an Environmental Policy that can Make Choices between

Sustainability and Social Justice." *Social Justice Research* 14(4): 375–402.
McKay, J., Keremane, G. and Gray, A. (2010) *Picturing Fresh water Justice in Rural Australia CRC Irrigation Futures.* With reflection by Judge Christine Trenorden, Chris Oldfield, Kieran O'Keeffe Hon Justice Preston. Mount Barker, Australia: CRC for Irrigation Futures.
Murray–Darling Basin Plan (2011) "Commonwealth Environmental Water." Available at www.mdba.gov.au/draft-basin-plan/supporting-documents/ewp/ewp_ch7/ewp_ch7_4 (accessed 13 July 2012).
National Water Market (2012) "National Water Market System." Available at www.nationalwatermarket.gov.au/index (accessed 3 March 2012).
Nikolakis, W. (2007) "Providing for Social Equity in Water Markets: The Case for Indigenous Reserve in Northern Australia." Available at www.nailsma.org.ao.
Northern Times (2012) *The Northern Times* 28 February.
NWC (2012) "National Water Planning Report Card." National Water Commission. Available at www.nwc.gov.au/reform/assessing/continuing/report-card (accessed 13 July 2012).
Orakhelashvili, A. (2008) *Interpretation of Acts and Rules in Public International Law.* Oxford Monographs in International Law. Oxford, UK: Oxford University Press.
Productivity Commission (2011) "Australia's Urban Water Sector." Available at www:pc.gov.au.
Shi, T. (2006) "Simplifying Complexity: Rationalising Water Entitlements in the Southern Connected River Murray System, Australia." *Agricultural Water Management* 86(3): 229–39.
Snow, D. and Jopson, D. (2010) "Thirsty Foreigners Soak Up Scarce Water Rights." *Sydney Morning Herald* 4 September. Available at www.smh.com.au/environment/water-issues/thirsty-foreigners-soak-up-scarce-water-rights-20100903-14uev.html (accessed 13 July 2012).
Victorian Government (2011) *Victorian CEC Competitive Neutrality Complaint Investigation Final Report.* Melbourne, Victoria: Victorian Government.
Water Amendment Bill (2008) "Water Amendment Bill 2008." Available at www.austlii.edu.au/au/legis/cth/bill_em/wab2008173/memo_2.html (accessed 13 July 2012).
Waterfind (2011) *Waterfind Newsletter* 6 December.
Wu, Z., McKay, J. and Keremane, G. (2012) "Governance of Urban Freshwater Some Views of Three Urban Communities." *Water* 39(1): 88–92.

17 Legal reforms that facilitate trading of water use rights in Spain

Antonio Embid Irujo

Introduction

The development of the legal framework for "Spanish water markets" has responded to some specific circumstances. These have been vital to its acceptability and have determined some of its legal characteristics. They are the following:

- *The need for water resources to be reallocated.* This means reallocating water use rights that had previously been allocated. This need is due to insufficient water availability to be able to supply water to the different water demands generated by a dynamic society such as the Spanish one at the beginning of the twenty-first century. There are no further resources to be allocated in the majority of the river basins. They have all been allocated already. Any new demands can only be attended to by reallocating resources that have already been allocated to other uses.
- *The introduction of economic incentives in the management of water and other natural resources.* Since the 1980s there have been proposals by economists specializing in environmental resources for introducing economic incentives in water management to avoid wasteful usage or deterioration of water resources. Prices of water in transactions would be one of the valid economic signals that would lead to improvement in water use and management. What is indicated here is not exclusively for water, but instead would also serve for other natural resources. The best-known example is that of the commerce around greenhouse gases – subscribed to by various instruments of international and community law – and which, to date, has yet to show any appreciable improvements in air quality.
- *The example of other countries.* The debate on trading that took place at the beginning of the 1990s was on the knowledge of actions occurring in other countries. Books, conferences, and interventions by public authorities focused on the experience in some areas of the United States of America – notably including the example of the Californian drought bank that operated in 1991. Likewise, the situations in Chile

and Australia were examined. From analyzing these experiences it was clear to legal reformers that there were different market models: some were more interventionist and others were more inclined towards deregulation and private initiative and even speculation (Chile). What needs highlighting in reference to the debate in Spain – as is explained later – is the common characteristics of "comparative right" and how it is constituted.

- *The agrarian recession.* With the volatility of crop prices the agricultural recession has been a constant over the last few decades, in the sense of a decline in the number of employees in the sector, in income, and in the scale of the share of agriculture in GDP. Under these circumstances, proposals naturally emerge for reallocating water resources used in the agricultural sector to other sectors (population and industrial supply). Agriculture would be the main candidate for reallocation given the large quantities of water resources used in this sector compared to the total (with variations from area to area, this percentage is undoubtedly around 70%). Farmers are expected to benefit from this as the introduction of trading also has the intention of offering economic resources to farmers in exchange for a reduction in their use of water.
- *The ineffectiveness of water reallocation implemented by the public ruling bodies.* The methods for water reallocation established in the "traditional" legal framework either do not work or are inefficient. The existing methods in the law are the water use rights reviews, the expiry of rights, the compulsory purchase of water use rights, and the administrative reallocation. All these were (and still are) contemplated by the law, before its reform, as means by which resources could be freed for other uses. Compulsory purchase and legal reallocations have a compensatory aspect that could also be encountered in the water use right review when this is required by water plans and causes damage to right holders (cf. art. 65 AWLT). Only the declaration of expiry of concessions contains no possibility of compensation (cf. art. 66 AWLT).

There has been a lack of interest on the part of the administration in pursuing massive reallocation of the resource through these methods contemplated in the existing traditional legal context, the main obstacle being the need to have financial resources for compensating established right holders. Rights which are to be revised, expropriated or reallocated have not been especially inclined to facilitate the intentions of the Administration and have tended to resist through legal instruments (recourse to courts) or political ones (organized social pressure).

Given all this, it is almost natural that trading was proposed. This would have the virtue of not being obligatory; it is the very action by private persons which voluntarily leads to reallocation. This is coherent with a conception of the relationship between the state and society of the conservative government that developed the reform of the Water Law 46/1999, dated 13 December.

Reallocation of water through trading could be defined as "indirect" reallocation, while that which happens through the traditional instruments (water use right review, expiry, compulsory purchase, legal reallocation) would be "direct" reassignment – a clear product of public action (normally by the water administration, but sometimes by the legislator) that aims to obtain these results by first freeing the resources and later reallocating them.

Transmission of water use rights and contrasts with water trading mechanisms

The meaning of the principle of water rights linked to land

It must be noted that under Spanish law, it has always been possible "to pass on" water use rights like any other real rights. These are transmissions (called concessions) under public control: those interested in the process must make the public authorities aware that it is to be undertaken. In some cases, the water use rights can be for a "public service" (e.g. for municipal drinking water supply) and these can only be passed on with prior authorization from the administration. In other cases, this transmission is always possible, as long as there is communication of the data regarding the new title-holder to the administration so these can be incorporated into the Water Register. The current AWLT continues to reflect this circumstance (cf. art. 63).

Behind this transmission possibility is the principle of "the link between water and land." This expression means that it is impossible to sell irrigation land, for example, without the parallel transmission of the water use right (or other rights) which makes this irrigation possible. Vice versa, it is not possible to sell the water use right without selling the land to which it is linked. This system was created in response to a deep-seated reason and to be coherent with the principles behind agricultural rights.

The deep-seated reason for this design by the legislator was to eliminate and prevent in the future the emergence of a social category of "water guardians" whose power would be based in the possession of water use rights, and who speculate with them – auctioning the effective use of these rights to the highest bidder from among the owners of land without water – to the point of eliminating the ability of farmers to be able to generate profits and invest. This shows the way in which the earlier legal reforms reacted to the specific physical and social conditions in the country, to improve the social conditions of farmers and facilitate productive investments and economic development.

From a practical perspective, linking water and land rights was a way to facilitate the identification of the area of the land with water rights so that there would not be future changes by the title-holder or otherwise this would not be legal.

Clearly, it is not possible to establish any system of trading of water use rights if the link between water and land rights is maintained. Therefore, one of the legal changes that were undertaken in the legal reform of Law 46/1999, dated 13 December, was that of unbundling water and land entitlements. The reform specified that this would be the case only if this was needed for registering contracts of trading water use rights. As regards everything else, the principle continues to be a solid descriptive marker within Spanish water law (cf. art. 61.2 AWLT).

Clarifications of the concepts introduced in relation to trading by law 46/1999

Given the discussion above, it is possible to clarify some of the characteristics of the trading formulas that were established by the Water Law Reform (Law 46/1999, dated 13 December and later regulations in the AWLT):

- *We are not truly faced by a "water market."* In Spanish law, water is neither bought nor sold because it is a good that lies within public dominion (cf. AWLT art. 2a), and these goods are inalienable (art. 132 in the Spanish Constitution). It is a "market" of water use rights (titles). What can be acquired (and ceded) are water use titles and not water.
- *There is trading of rights of use of public water.* What can be acquired and ceded are rights which the title-holders have for the use of public water and which they have acquired through different forms: concession, law, historical titles.

It is useful to draw attention to the latter characteristic, which could appear surprising. In fact, there are "private" waters in Spain that lie on the fringes of the existing water use rights system given the terms of Law 29/1985, dated 2 August, on Water (transitory disposition for second and third parties) which allows – in accordance with the option chosen by title-holders – the permanence of "private" ownership of water for some title-holders. This is clearly indicated by these dispositions and rubber-stamped by the CC in its 227/1988 ruling. Water would continue to be private property if the users maintain the legal situation that they had on 1 January 1986 – the date upon which Law 29/1985 became enforceable. If there is any change in the manner of usage (i.e. that derived from its transmission to a third party), then these would become public waters that could only be used for a beneficial use through a concession granted *ex lege*. Trading of water use rights, as referred to in Law 46/1999, can only operate with on rights over public water. This is so since to be able to engage in trading remaining private water owners must apply for this water to be registered as public domain.

Trading is currently regulated by articles 67 onwards in the AWLT. There are two types of trading formulas:

- Contracts for ceding w*ater use rights*. These are contracts that can occur out of the free will of individuals, title-holders of the water, who register themselves with one as the acquirer and the other as the seller in a contract of this type. These contracts have to respect "the public interest" and are subject to public control as explained later.
- *The acquisitions and sales of the centers for the exchange of water use rights.* The Law 46/1999, of 13 December, authorizes the creation of these centers as part of the existing river basin authorities. These centers can make public tenders for the acquisition of water use rights from those title-holders who are interested to do so. The centers also make public tenders to sell (make concessions) to other title-holders. All this needs to happen in a framework of transparency of procedures that applies to all the procedures of public contracting. The model for these centers is the Water Bank created in California during the drought of 1991. One of the causes under which these centers for the exchange of water use rights may be constituted is the existence of an exceptional drought situation.

Contracts for ceding water use rights

Law and regulations: the question of public control

The Water Law Reform established the characteristics of these contracts that allowed trading to take place. The articles 67ff. from AWLT and 343ff. from Regulation of Public Domain Water (RPDW) establish the following:

- *These are contracts between "users" of water use rights.* The Law requires that both the one that cedes the right and the one that acquires it is a water user. This means that there is no chance that, through these contracts, water can be acquired by anyone who does not have a water title – independently of the quantity that is being traded. The RPDW review, of May 2003, specified what must be understood as a "user." These are title-holders of water use rights of public water or those that are using "temporarily" private waters – being this the legal status of the former title-holders of private water who chose to go under the public regime (this was possible to be done from 1 January 1986 to 31 December 1988). As explained earlier, private owners of water cannot engage in this type of contract unless they change their status. Agreements among members of the same water user association lie also outside the purview of this regulation as they are internal actions within the community, which can be freely undertaken if their ordinances permit them.

- *Water use rights become unbundled from property of land so as to facilitate trading.* The contracts between agricultural users must identify which land will cease to be irrigated or will be irrigated with less water (by the ceding party) and which will be irrigated (by the acquirer).
- *The AWLT specifically forbids ceding water from non-consumptive to consumptive uses.* For example, a hydroelectric water use title-holder cannot cede the rights to an agricultural title-holder or a municipal one, the aim being to avoid allowing trading to increase consumptive uses.
- *These contracts can only apply to preferential uses as defined in the River Basin Plans or, by default as per article 60 of the AWLT (normally urban and irrigation users).* There are possible exceptions in extremis (with water rights being sold to industrial uses where this have a lower preference, for example), but these contracts need to be approved by the Ministry of Environmental, Rural and Marine Affairs and not only by the River Basin Authority.
- *The regulations establish the volume of water that can be ceded.* According to the AWLT and the RPDW review of 2003 the volume cannot be all that is established in the title of the right holder that is ceding water – to avoid the sale of "paper water." It can only subject to trading the water that has been "really used by the seller" – as per practice during the previous five years, bearing in mind the possibility of amendments in this quantity that may have been established in the River Basin Water Plans.
- *The duration of the contract has not been definitively clarified in the regulations.* The Law establishes that these are "temporary" contracts – referring to water use rights also as "temporary." This means that if the maximum duration of the existing water use rights is of 75 years, the maximum duration of a contract for water use rights could only be the time left from those 75 years.
- *Regarding the compensation to be received by the one that is ceding water use rights there are no clear decisions either in the AWLT or in the RPDW.* The AWLT refers us to the regulation and the RPDW to the decision of the relevant ministry as regards the maximum amount to be paid. In any case, it can be seen that "compensation" is mentioned but price is not.
- *Contracts can be accompanied by other contracts to make them possible.* For example, a contract with the relevant public authority can be made for the use of public infrastructures to transport water: this entails paying of tariffs that have been created for the use of these infrastructures.

The Law also establishes the specific form of *public control over these contracts:*

- *Once the contract has been registered by both parties, it is valid but not effective. The effectiveness depends upon the explicit authorization from the administration.* In order to obtain this authorization, the regulation indicates that a copy of the contract must be sent to the respective user associations (of the seller and of the acquirer) as well as to the basin authority. If the contract involves agricultural uses the Autonomous Governments – that hold exclusive responsibilities over agricultural matters – need to issue a report. In addition a process of public information regarding the contract must be in place so that those potentially affected can respond in the manner they deem most expedient.

- *Authorization for the contracts lies with the basin bodies, who have to issue it within two months.* If this period passes with no specific decision, the contracts will be understood to have been authorized by positive administrative silence.[1]
- *Authorization will specifically fix the volume to be ceded and the obligation of installing a counter.*
- *The criteria for authorizing the contract is regulated.* If the authorities refuse to give permission the reasons for this need to be given. These criteria relate to concerns on the contracts negatively affecting:
  - the pattern of resource exploitation in the basin;
  - the rights of third parties or the environmental flows;
  - the state or the conservation of the water ecosystems.

  Beyond these general refusal criteria, the regulations indicates that authorization will also be refused whenever any other legal requirement has not been fulfilled.
- *During the period in which the basin authority has to grant authorization,* this authority can also opt to exercise a preferential right of acquisition through the payment of a monetary compensation which has been agreed upon.
- *When transfers involve users from different river basins,* contracts need to be authorized by the Ministry of the Environment, Rural and Marine Affairs, and need to be done by Law or to have it included in the National Water Plan.

Finally, the contracts must be registered in the Water Register and can be accompanied by the authorization to use public infrastructures and the authorization to construct infrastructures, if required.[2]

The centers for exchange of water use rights

These centers are run by the public powers. The centers can issue public tenders for the acquisition of water use rights and also issue offers of those rights previously acquired. Private individuals can access water use rights by bidding for them to the public organism and not as a result of a private agreement between two parties.

The AWLT established the regulations that rule the creation and functioning of the Centers. The Cabinet of Ministers is the body that authorizes the River Basin Authorities to create them at times of extreme drought, over-exploitation of aquifers, and so on. This resembles the model of the

Water Bank created in California during the drought of 1991. Centers have been authorized in the Júcar, Segura, Guadiana, and Guadalquivir basins.

The centers are created within the River Basin Authority. This means that they are just an administrative body with no independent legal standing. The legal capacity lies with the River Basin Authority itself. The centers are then subject to the public sector contracting legislation (Law 30/2007, dated 30 October). The acquisitions can be both temporary and definitive and the criteria to be used by centers to decide on competing bids must be made public at the time of issuing the call for tenders.

The original regulation of Law 46/1999, established that the centers could not withhold any of the rights acquired: all of those obtained had to be offered to bidders who also had to be users. Royal Decree-Law 9/2006, made some changes to this. Now the centers are authorized to maintain rights for environmental use and for the Autonomous Regions to implement their own water policies.

Legal developments facilitating trading during droughts

Experiences and considerations

In the Spanish legislation the existence of droughts has been the main justification for allowing voluntary private contacts and for the establishment of public centers for the exchange of water use rights.

During the drought that occurred in Spain from 2004 to 2008 some Decree-Laws were approved to facilitate the implementation of trading. They did so by excepting or adding some principles to those already known. Here we examine these texts and their contents.

- *Royal Decree-Law 15/2005, dated 16 December* established urgent measures to regulate the rights of water use in transactions. The Royal Decree-Law incorporated some additional regulations to those already existing. These were:
  - The first of them deals with the titles that the water users must hold so as to be able to engage in contracts. The Decree-Law widens the potential right holders that can engage in contracts. Previously it was only possible for title-holders to do so. The Royal Decree specifies that owners of water use rights in "public initiative" irrigated areas whose maximum volume of water use has been reflected on the River Basin Water Plan can also be considered as users and can then engage in contracts. The reasoning for this change is very simple. Irrigation farmers in irrigation areas that were developed by the state do not hold water use rights. If these were not allowed to engage in trading then large amounts of water for irrigation would not be able to be traded. The Royal Decree-Law established that these title-holders would be considered users and their titles would

be registered in the Water Register. To do so the Royal Decree established some specific rules to facilitate their inclusion (art. 2 of Royal Decree-Law 15/2005).
  - The authorization for contracts for trading water between river basins would be given by the Central Government Water Director (previously this was a matter for the Minister).
  - For the duration of the Royal Decree-Law, it will not be necessary to transfer water from one basin to another as per the contracts as these transfers are covered in a Law or National Water Plan (second final provision).
  - The duration of this Royal Decree-Law is temporary and would last until 30 November 2006.
  - There are other provisions for the use of infrastructures, such as those of the Tagus–Segura Transfer.
- The second legal initiative during the drought was the *Royal Decree-Law 9/2006, dated 15 September* through which urgent measures were adopted to palliate the effects of drought on villages and irrigated agrarian zones in specific river basins. What stands out in this text is the following:
  - There is an extension of the application of the previous Royal Decree-Law 15/2005 until 30 November 2007.
  - It authorizes the Centers to undertake tenders so as to temporarily or definitively acquire water use rights and use them to "create reserves with a purely environmental objective, both temporarily and definitively" and to cede them "to the Autonomous Regions, with an agreement in advance to regulate the cession and later use of the water. The cession must be registered in the Water Register in the basin." This is an important provision given that in the original AWLT regulations (and which are still in force in its article 71) all the rights acquired by the Centers must inexorably be passed on to users who wish to acquire them.
- *Royal Decree-Law 9/2007, dated 5 October,* proceeded to grant another extension to Royal Decree-Law 15/2005 until 30 November 2008.
- *Royal Decree-Law 3/2008, dated 21 April,* established exceptional and urgent measures to guarantee the water supply to areas affected by the drought in the province of Barcelona. This is an unusual Royal Decree-Law,[3] and even more curious is the way in which it can come to an end as this depends on the drought finishing or a desalination plant in Barcelona becoming operational. Since there was great rainfall before the end of April, this led the Catalonian and Spanish governments to declare the end of the drought situation and, therefore, this regulation was never enforced.[4]

In this case, it is worth mentioning that article 3 of the Royal Decree-Law foreseen that one of the ways to obtain the volume of water that was needed to supply water to the city of Barcelona would be through allowing the contracts ceding water from the irrigators of the Ebro basin who belong to their users associations. This measure caused political, social and media furor and the irrigators of the Ebro, via their Federation refused to register any kind of contract of this type, promising, however, to implement water saving measures to be able to provide water to Barcelona.

Final assessment

Regulations allowing water trading in Spain are detailed, technically well constructed, and do not facilitate speculative operations or contrary to the general interest. Despite the occasional problem detected in the manner of configuring public control (positive administrative silence), the administration has sufficient tools to be able to control the development of this market as to align operations with the general interest. In addition, the lack of specification in the length of the temporary contracts is a matter which needs to be addressed in a regulatory reform – given that it seems necessary to reconcile the length of these contracts with what can be deduced from current water planning. A lack of specificity does not seem to be the best way to achieve this.

Observation of the transactions that have occurred over the last few years show that trading has been particularly active in extreme situations – drought – and that these transactions contributed to palliating some of the negative effects (contracts between a Community of Irrigators in Castilla la Mancha and the Union of the Tagus–Segura transfer, for example). The activities of the centers for the exchange of water use rights seem to be, in the case of the Special Plan for Alto Guadiana, one of the main ways by which it would be possible to end of the situation of environmental non-sustainability which exists in this area due to the overexploitation of groundwater, threatening the survival of the National Park at Tablas de Daimiel.

The final judgment on the functioning of water trading in Spain still requires a detailed study of the transactions undertaken and their effects (economic, environmental, social, etc.). Here, a positive critical assessment of existing regulations from a legal point of view has been made. It would be especially relevant to analyze the "prices" for the transactions undertaken since under the legal ordinance this would need to equate to that of "compensation." This seems to indicate that there is no intention that exchanges take place at "market" prices. Rather, this should be a simple "compensation" from which the real elements of a market will be absent as regards to price formation (the classic supply–demand relationship).

In any case, legislative developments seem to show that Spain does not consider one single formula for water allocation and management. Rather, these trading formulas are forms of indirectly reallocating resources. They will be quantities of water of little significance compared to the total volume being managed. Although this does not mean that this insignificance might not considered vital in critical situations – in droughts, which is when these transactions have taken place thus far.

Lastly, it is the belief of this author that these indirect reallocation tools must remain within our water rights legislation. The important thing is to have configured legal regulations with sufficient scope and flexibility so that different parts can be applicable and effective at different times according to the water situation being experienced. The current situation is so complex and the society to which we belong is so dynamic that it is necessary to ask for – and provide – a legal framework that provides flexibility, and that offers opportunities (and never obstacles) for water managers and for society. To date, the legal regulation summarized in this work has almost all the characteristics required to constitute an opportunity.

Notes

1. The Government of Aragon went to court on the grounds of unconstitutionality – still unresolved – because it understood this prescription to be contrary to the constitutional regime of public domain. There would be no possibility, claimed the appellant, to gain rights over public domain by means of administrative silence, as, in a similar way, that silence is considered negative in the granting of water use rights, pollution emission permits, etc.
2. In this case, a deadline of four months will be established for authorization via positive administrative silence as well. This timescale seems very limited if it is borne in mind that on many occasions a works project must be formulated, or perhaps it needs to be submitted to an environmental impact study.
3. See a complete reading of the content and meaning of the distinct legal problems in the Report from the Government of Aragon's Legal Assessment Commission – 81/2008, dated 12 May.
4. The Royal Decree-Law was revoked according to the terms in its final third provision. However, there was a formal confirmation of the ending of the state of emergency through the Cabinet Agreement of 6 June 2008, which declared the concurrence of the reason for the end of the validity of Royal Decree-Law 3/2008, dated 21 April, on exceptional and urgent measures to guarantee the water supply for the populated zones affected by the drought in the province of Barcelona, published through the Resolution of 6 June 2008, from the Secretary of State for Rural and Water Affairs.

References

Embid Irujo, A. (ed.) (1996) *Precios y Mercados del Agua.* Madrid, Spain: Civitas.

Embid Irujo, A. (2000) "Una Nueva Forma de Asignación de Recursos: El Mercado del Agua." *Revista de Obras Públicas* 50.

Esteve Pardo, J. (2003) "El Mercado de Títulos Administrativos: Asignación Objetiva, Reasignación Transparente." In *Estudios de Derecho Público Económico. Libro Homenaje Al Prof. Dr. D. Sebastián Martín-Retortillo,* 743ff. Madrid, Spain: Civitas.

Molina Giménez, A. (1998) "Administrativización vs. Liberalización: Los Mercados del Agua en España." *Revista de Administración Pública* 146: 357ff.

Molina Giménez, A. (2007) "Contrato de Cesión de Derechos de Uso de Agua." In A. Embid Irujo (ed.), *Diccionario De Derecho De Aguas*, 464ff. Madrid, Spain: Iustel.
Molina Giménez, A. (2007) "Centros de Intercambio de Derechos del Agua." In Embid Irujo, A. (ed.), *Diccionario de Derecho de Aguas*, 322ff. Madrid, Spain: Iustel.
Navarro Caballero, T. M. (2007) *Los Instrumentos de Gestión del Dominio Público Hidráulico. Estudio Especial del Contrato de Cesión de Derechos al Uso Privativo de las Aguas y de los Bancos Públicos De Agua.* Valencia, Spain: Tirant Lo Blanch.
Navarro Caballero, T. M. (2007) "Representa un Verdadero 'Mercado del Agua' el Contrato de Cesión de Derechos al Uso Privativo del Agua?" *Revista Aranzadi de Derecho Ambiental* 11: 203–11.
Navarro Caballero, T. M. (2008) "Las Transacciones de Derechos al Uso del Agua y su Transferencia a las Cuencas Receptoras del Tajo–Segura." *Revista Aranzadi de Derecho Ambiental* 14: 103–21.
Setuain Mendia, B. (2004) "Aspectos Normativos de los Mercados de Aguas: Últimas Aportaciones Desde la Reforma del Reglamento del Dominio Público Hidráulico." *Revista de Administración Pública* 163: 349–87.

18 Drought management, uncertainty and option contracts

Almudena Gómez-Ramos

Introduction

The increase of uncertainty in the availability of water resources is perhaps the most significant challenge for water resource allocation today. The system of allocation of scarce and uncertain water resources under scenarios of increasing demands need to change for economies to prosper. Indeed, there is evidence showing that many of the social problems related to water allocation today are not trivial.

Water management and allocation systems may not be able to cope today, as they were designed under different conditions. During the past century, the prevailing model in the majority of developed countries was essentially based on state initiatives allocating increasing water supplies resulting from building new water infrastructures. Today, this model may not work. It may be inefficient in improving water accessibility and does not discriminate quality and quantity. Furthermore, it may also be inefficient in securing reliability of water supplies.

One of the causes of the unsatisfactory progress in water management may be the way in which water problems have been conceived and understood. Dealing with the growing conflicts over water requires a broader perspective. This requires analyzing not only the efficiency of the allocation of resources between competing uses, but also the allocation of the involved risk. One of the new paradigms created under the auspice of the "new culture of water" is that of managing water risk, caused both by climate instability and human intervention, in the different phases of the hydrological cycle. In the Spanish context, for example, the aggravation of water stress in most of the river basins is becoming increasingly evident, and is caused not only by climatic variability, but also by the increasing demands and changes in the water uses. Most studies agree that uncertainty in water availability will rise, especially in the Mediterranean basins.

This situation poses the need to seek new allocation mechanisms that allow access to water but incorporating risk into the allocation process. Allocation of water according to the distribution of water use-related benefits would need to be formulated on the basis of a more complex conception of water rights, which would allow an efficient allocation of risk between different agents and consumers. In order to design such as system, it is necessary to consider the level of risk that every user is willing to assume in a context of water supply uncertainty.

This chapter examines different aspects related to allocation of water resources in a context of uncertainty. To do so it examines the case of the Spanish legal framework to determine how, in this case, the allocation of risk or the reliability of access to water could be conceived in such a way as to allow the present licensing/concession system to transfer risk between participating agents. The proposal is that option contracts can be introduced. This paper expects to highlight the value of this mechanism and its potential for conflict resolution. An empirical application will be shown containing concrete potential results from the implementations of this mechanism. The empirical application of the option contracts is shown in the case of the water supply of the city of Seville, through the transaction between the irrigation district of El Viar and EMASESA, the public company responsible for fresh water provision in Seville.

The relevance of risk in water resource allocation

Risk is an inherent element in the management of water resources in water scarce countries such as those in the Mediterranean basin, which makes the problem of allocation more complex. Here, the dynamic of allocation is periodically altered by drought periods, the beginning, duration and intensity of which are unpredictable. Risk management is a key factor in the entire process of water administration. There can be two situations. One is when a risk is accepted and classified by social agents and institutions as a known risk. The second is when uncertain episodes of great severity are completely unknown. For Cubillo (2002), in the latter case it is necessary to provide the system with prediction mechanisms that help anticipate the impact and the responses to these events. Risk becomes an element to be considered and minimized in each water management decision taken.

This new approach to water management challenges the traditional approach to water management based on building infrastructures, allocating water made available by them and establishing priorities of use. Experience has proven that traditional policies for increasing water supply are not able to assign water and risk, as they are too rigid for situations of uncertainty. Thus, it is necessary to explore new mechanisms that essentially aim to deal with it and provide a better and more efficient allocation of water resources. Market-based risk instruments appear to be potentially useful tools, as they are able to be very flexible in assigning risk. These instruments are created via the design of appropriate responses to different scenarios of scarcity that can happen, with an associated probability.

Legal and institutional framework for the transfer of water use rights in a context of uncertainty

Water use rights are bundles of attributes including different conditions for access such as seniority of water rights, conditions of quality and quantity of water, location of the capture or exclusivity of the use.

Efficient risk-sharing arrangements are based on the premise that each user's willingness to pay for each water attribute differs, because the marginal use value of water is different for every user. They have a different ability to endure periods of scarcity, to postpone consumption and to endure lower water quality. An efficient water allocation under stressful conditions requires the capacity to disentangle the different aspects of the bundle of rights, some of which could be tradable among competing users. For this aim, it is necessary to design long-term, secure, well-defined water rights, and to ensure that land and water rights are separated (Bjornlund and Rossini 2005). Efficient risk-sharing mechanisms also require greater flexibility in the transfer of water rights, allowing only some of the risk-related aspects of water rights to be transferred. Under conditions of uncertain water availability, elements such as security of access under pre-set conditions or timely access to scarce resources, are essential attributes for planning demands and available resources.

In the case of Spain, institutional and legal frameworks are limited in terms of ability to manage water reallocation and risks. In this case it is not possible to implement some alternative possibilities for achieving water allocation in a context of uncertainty. Water rights are conceived in the Spanish legislation in a compact way. This conception has lost validity nowadays not only for legal reasons, but also for economic and technological reasons. The reform of the Spanish Water Law in 1999 opened up a real range of trading options and the possibility for beneficial agreements between water rights holders. However, the system of water rights in the Law has been limited in terms of its application and potential for managing risks. Technological advances that enable changes to differentiate different characteristics of the resource in terms of quality and quantity are not recognized by the current system of water rights in Spain, which is not flexible enough to assign water according to quality criteria, for example.

In the Spanish legislation, hydrological planning assumes risk management functions in situations of scarcity through drought plans. It would be important that drought plans have clear conditions of reliability and clarity of property rights under drought situations, so that specific risk management instruments can be implemented in an active way. That is, the owners of the rights do not need to be limited to passive acceptance of the decisions taken by the water authority during droughts on reductions of water allocations, but they should be able to have mechanisms to react.

The allocation of resources has to consider that users not only demand legal certainty of their rights, but also reliability or availability of water resources, and it has to consider that these are not demanded or valued in the same way by the different users. Therefore, by segmenting water rights into their most distinguishable and essential attributes and providing reliability by assigning a certain volume of water under pre-established conditions, it is likely that two users will find more possibilities for exchange of risk to be mutually advantageous, and obviously more advantageous than those offered by the existing legal system, which depends on decisions taken by the Water Authority at a specific time.

Water markets and risk

Water markets have been the economic instrument par excellence in the so-called mature water economies (California, New Mexico, Australia, Spain, Chile, etc.). But it has also been recognized that the effectiveness of water trading is explicitly influenced by various uncertainties existing in water use systems (Luo *et al.* 2007; Calatrava and Garrido 2005b). When transactions are carried out in situations of uncertainty there can be a disparity between the expected water transferred at the time the decision was taken and the real water available at the time when the transfer takes place. This disparity can affect the positions of the parties participating in the transaction (Calatrava and Garrido 2005a) and as a result, the final decision may be influenced more by the level of risk that each party is prepared to assume than by their real water needs.

Also, it is necessary to consider that the effectiveness of the market is very sensitive to transaction costs, and exchanges fail when costs are very high (McCann and Easter 2004; Luo *et al.* 2007). It is evident that negotiation costs are directly related to the uncertainty of water availability, so the greater the uncertainty and scarcer the available information, the lower the probability of reaching an agreement. However, market success depends on the transparency and symmetry of the available information needed to create the conditions of perfect competence. Past experience of the application of temporary leases of rights shows the need for bodies or agencies to implement formal rules to control the settlement of transactions and, at the same time, to monitor and assess their effects. A system for transferring water use rights, regulated in a stable manner and adapted to a particular context of uncertainty, will improve the reliability of the system, making it less vulnerable.

In contexts of greater frequency of droughts, trading of water use rights may need other water policy requirements to ensure the effective transfer of water to higher value users during drought periods. For Bjornlund and Rossini (2005), more sophisticated markets and instruments need to be developed to ensure that these constant redistributions of entitlements and seasonal allocations can take place quickly and with low transaction costs. These mechanisms will allow flexibility in exchanges, as they will be adapted to the contexts of uncertainty which take place. In this respect, in

Australia, there is a proposal to implement a suitable legal framework that would be able to activate transfer contracts that include climatic conditions clauses or specific levels of transfers adjusted to water availabilities for a specific period of time and at an established price. This is, in essence, as will be seen later, an option contract.

This more flexible kind of instrument would be clear and proactive in anticipating risks. It should be able to start working automatically if it is linked to specific thresholds or objective data. However, it requires the design of a contract that considers a complex context that takes into account all the possible activation stages of the mechanism and the negative externalities that the activation entails. Formal risk-sharing tools require the formulation of agreements that minimize ambiguities or problems of enforcement. But this rigidity enables the contracting parties to plan ahead and evaluate the resulting risks more rigorously. Most treaties for managing transboundary water resources have these types of risk-sharing components. In Spain, the Tagus–Segura transfer is run with reference to the storage levels of key reservoirs that dictate when and how much can be transferred at any given time. But very often, political pressure is put on decision makers to allow for the use of reserves when they are low, as occurred in Spain from 1993 to 1995 (Giansante *et al.* 2002).

Problems linked to environmental and social externalities derived from the liberalization of allocation mechanisms can be overcome by centralized instruments such as water banks. After a period of experience and learning, water banks can be organized to behave like real risk management tools, activating responses automatically under pre-set water scarcity conditions established in the plans. Water banks would allow basin/water authorities to allocate scarce resources, creating awareness of scarcity costs and, ultimately, reducing the economic and social impacts of the drought. The main advantage of the banks is the ability to adjust, between supply and demand, in time and place, thereby facilitating exchanges. Water banks are easy to control and more transparent than other modalities for water reallocation and are likely to be widely accepted (Garrido and Gómez-Ramos 2009).

The success of this market mechanism depends on the level of integration of these transfers for the adjustments to supply and demand, in hydrological plans. Legal and environmental considerations and the effects on third parties are basic elements to be considered. This means that it is necessary to design closed, taking into account the need to preserve environmental flows rates and protect the rights of downstream users. It is also important institutional intervention to facilitate sharing experiences and information, to allow potential buyers and sellers to access the market on equal terms.

In the case of Spain, water banks have been developed as a result of the Royal Decree-Laws on Urgent Measures in 2005, 2006, and 2007 (15/2005, 9/2006, 9/2007). The extent of success of the way in which water banks are operated is not clear, as few exchanges have been made, and an enormous amount of specific administrative and technical expertise was required to create the infrastructure of the banks. This expertise was no doubt required to provide minimum guarantees for the correct functioning of the banks. Nonetheless, as these initiatives have not succeeded in creating a stable framework of exchanges, nor permanent offices have been set up, we cannot state that the Spanish experience of water banks demonstrates them to be tools that are able to manage risk automatically (Garrido and Gómez-Ramos 2008).

The option contract as a risk-sharing mechanism

The main feature which characterizes the option contract is the existence of two elements linked to it: uncertainty and time. Uncertainty, in turn, has two basic sources: uncertainty with respect to the future availability of water, and uncertainty with respect to the value of water in the future, given a certain level of water supply. In the short term, the value and the availability of water are definitely correlated, but in the long term there are several elements that can distort this relationship, due to the fact that the average rise in the value of water is influenced by the increasing demand for water and the reallocation of the water supply for different uses.

In the medium term, temporary and permanent reallocation of water use rights present serious disadvantages in terms of planning supply for different users. In terms of the asymmetrical distribution of risk between sellers and buyers, the option contract appears to be at a halfway stage between temporary and permanent trading of water use rights. In the case of temporary trading, it is the seller who possesses complete information about the amount of water available before the price of the transaction is specified. The buyer is worse positioned in relation to the seller because of it has inelastic demand, as its position incorporates a combination of uncertainty over the future water supply and the resulting price. These are disincentives to their participation in the market.

The uncertainty associated to final water availability and its price could be overcome through securing water supply in the long term. Buying permanent water rights, the buyer will reduce its uncertainty in relation to future price of water supplies thereby risking only the natural variation in the water supply. The sale of permanent water use rights involves transferring more of the risk to the seller, who is committed to selling the water at a set price in advance, without considering the variability of the supply, which could affect the price.

The option contract is a natural evolution of temporary markets and permanent transfers of water use rights, which avoids the costly and politically controversial nature of permanent transfer of rights. It also solves temporary markets by reducing uncertainty caused by a random supply of water.

This situation can be avoided by implementation of an option contracts by which water right holders agrees in advance to transfer water to another user in periods of scarcity, at a certain price. The seller anticipates the risk of being in a scarcity condition. As regards the effects of uncertainty over the conditions of water provision, the advantage of an option contract with respect to a temporary market is that the former allows greater security in terms of provision, since the option leads to an increase in the average supply in the medium term (Michelsen and Young 1993; Howitt 1998). Option contracts can be implemented through a water bank, as the stored water in the bank is always available for use (Israel and Lund 1995).

Option contracts are able to spread the risk associated with the supply and the price between the current user of the water and potential buyers, thereby stimulating an active and efficient market. In addition, option contracts have the advantage of reducing the effects on third parties of other modalities of rights transfers. In this respect, option contracts are able to lessen the economic effects of the transfer in the area where the affected parties are concentrated, for example farmers in the case of transfers from agricultural uses. This is the case because firstly, the water is used by the irrigators most of the year. If they agree to the possible relinquishment of the available water during a drought, they can be compensated every year (whether there is a drought or not) by means of an annual premium. Second, as farmers remain active in the region (since the sale option is not executed every year), the profits from the premium are invested directly in the area. And lastly, the negotiations between parties on the level of the premium in the medium to long term provide the opportunity to appropriately assess the effects on third parties.

Compensation to irrigators who have signed an option contract should not only compensate for the relinquishment of a certain amount of water, but should also compensate for the risk assumed by sellers. This risk is linked to the uncertainty of how much, how often and when that water should be transferred. It is evident that, in a context of random water inflows, an option contract that requires the transfer of water, providing that certain pre-set conditions are met, requires certain changes to the management of the reservoirs that provide water to those involved in the water transfer. These changes would entail a loss of flexibility in relation to farmers' crop decisions, and would therefore increase the economic risk for farmers and reduce their expected profits.

The water authorities should observe the rights of the intervening parties regardless of whether the option is executed. That is the most essential requirement for the implementation of the option contract. What this involves in terms of the framework of current water use rights in Spain should not be underestimated. In a situation in which the conditions of the option are exercised, the water authorities should never be able to prevent it because it would render the contract completely invalid and, as a result, the whole premise of using the options contact would be not credible. The intervening parties' confidence in the validity of the contract should be absolute, especially if it is a long-term contract.

Options contracts and drought management in Spain

The hydrological drought in most of the Spanish territory in the years 2004–5 has been the direct cause of trading of water use rights mostly in the southern basins. The 1999 Water Law establishes trading of water use rights through two procedures: Exchange Centers and Cession Contracts. The aim was to make the water transfer regime more flexible for responding to exceptional situations of water scarcity. The ultimate aim was to activate and boost reallocation, but always in the event of the exceptional nature of the aforementioned drought situation.

Without entering into detail on the environmental and economic impact of transactions carried out in this period, it would be appropriate to consider their validity as a risk management tool, given that they have been authorized in situations that are of an exceptional nature and therefore represent a higher risk. It is precisely this view of the water situation as being exceptional that is responsible for the considerably reactive nature of the response to drought, as it is almost a palliative response. Thus, trading has been activated by law for the moment in which the exceptional situation is declared, and immediately following on from that, contracts have been implemented between interested parties. They are very short timescales for activating and controlling complex mechanisms step-by-step, a task which is carried out by the river basin authorities who are authorized by law to do so. As a result, the process has lacked the dynamism and the flexibility that a reactive response requires in order to mitigate the effects of the drought. Also, the lack of time has prevented greater transparency and prior economic assessment of the possible social and environmental impacts of water transactions.

The aforementioned considerations may lead one to think that option contracts would be a better instrument for solving most of the limitations detected in water trading in countries such as Spain:

1. First, option contracts are ideal for managing droughts using a proactive approach, or in other words one that anticipates risks. This type of contract could fit well into the drought plans. Thus, the plans would be able to foresee the existence of pre-established Exchange Centers allowing them to launch purchase options that would logically be activated in drought situations and in the event of specific conditions of hydrological deficit. Thus, a framework would be created that would be stable in the long term, but that would also be flexible enough to adapt the call for options to risk situations (see Garrido and Gómez-Ramos 2009).
2. The economic, social and environmental assessments in these option contracts, via the established risk premium, presents an approach that is more cost-effective than the other alternatives. Logically this assessment should be revised periodically (including the signing of a new contract) due to the variable nature of water supply and demand.

3. The implementation of long-term agreements through option contracts which can be included in drought plans or guidance, have the capacity to be protected from ever-changing political conjunctures and controversies.

In order to prove the validity of the option contract, an empirical application has been simulated in the context of the Guadalquivir River Basin, by means of developing a transfer contract between the irrigation area of the Viar, which is supplied with water from the "El Pintado" dam, and EMASESA, the company in charge of the water supply of Seville City and the metropolitan area (Gómez-Ramos and Garrido 2004).

Table 18.1 shows the value of the premium for different types of contract (established for a four-year period) and for different volumes of water to be transferred.

Table 18.1 Price of option contract for each contract modality (for four years)

Volume of water (m^3 millions)	*Contract A (€/ha)*	*Contract B (€/ha)*	*Contract C (€/ha)*
0	0	0	0
30	102	102	102
60	164	162	157
90	412	387	338
120	1,068	958	735
150	1,709	1,501	1,094
180	1,918	1,644	1,105

Source: Gómez-Ramos and Garrido (2004)

Table 18.2 shows a comparison of costs between the traditional measures of the EMASESA drought plan and the cost of the options contracts. The table also shows the expected average social costs linked to water deficit endured by citizens and the financial costs associated with the two alternative strategies – the reduction of water allocations in the drought plan and the option contracts.

Table 18.2 Costs incurred during drought period 1992–5 with option contracts and the EMASESA drought plan

Costs	*Option contract*	*EMASESA drought plan*
Social cost (€ millions)	480,173	325,818
Financial cost (€ millions)	3,882	28,790
Total cost (€ millions)	484,055	354,608

Source: Gómez-Ramos and Garrido (2004)

The main negative consequence of this kind of contract is that farmers in El Viar that would have agreed to trade their water rights during the drought would lose part of the productive potential of their irrigated land. They can minimize that because the conditions of the cession would give them enough time to anticipate the consequence of the reduction in water availability. This would allow them to cultivate crops that consume less water, thus adapting to the restrictions imposed by the contract. For EMASESA the contract would entail the addition of a new virtual dam that would provide 90 hm^3 annually if required, which would involve a cost over four years of approximately 4 million euros. Figure 18.1 simulates the evolutions of stock level of the El Pintado dam considering three scenarios: no policy scenario, which implies no actions against drought; EMASESA handbook, which considers the measures included in the drought management handbook of Seville water supply company; and "options contract," which includes the implementation of the mechanism here proposed. In this last case the results are satisfactory, as the reserve levels of the supply system of Seville would improve substantially. That obviously supposes lower social and financial costs during a drought period.

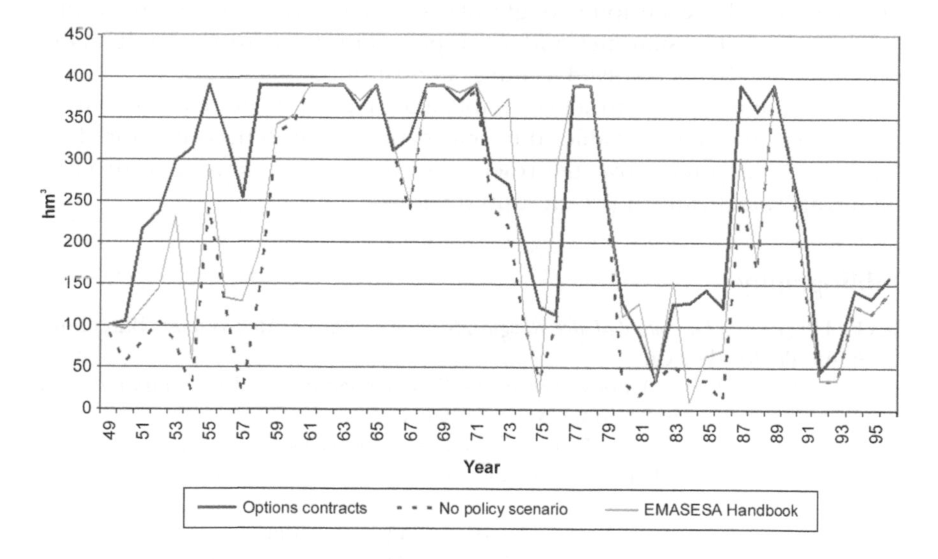

Figure 18.1 Simulation of the levels of water stock of the water supply system of Seville in different scenarios
Source: EMASESA and own elaboration

Conclusions

Water management in contexts of water economy has entered a stage of maturity characterized by a high and increasing demand for water, an inelastic and limited water supply, and an exhaustion of traditional water policies based on water supply augmentation, due to the high environmental, social and economic costs, and demand for new tools that are able to respond to challenges faced in this phase of water management. The search for new water management mechanisms in these contexts must inevitably include risk as an integral element in the integrated water management principles. The ever-present uncertainty in the "risk society" in which we live, demands the exploration of new management systems that must be able to anticipate and respond to these situations, contributing to the provision of equal access to water, not only in terms of the amount and quality of water, but also in terms of guaranteed provision.

The inclusion of risk in management demands a new conception of water use rights. These should be defined separately on the basis of the different aspects of the water use, such as the quality, guarantee, volume, or seasonality. The inclusion of risk involves the possibility of transferring the guarantee of use instead of the right in compact form. This would facilitate transactions and allow greater flexibility when adapting to the different contexts of uncertainty.

Option contracts behave like risk management tools that can be built into action and prevention drought plans as proactive measures, providing legal certainty, transparency and a compensation framework that includes the different effects involved in integrated resource management.

Today, the option contract is a viable alternative for the management of exchange centers that enables the concession system to become more flexible, as it guarantees strategic reserves within the framework of drought planning, specifically in pre-alert predictions.

Bibliography

Beck, U. (1986) *La Sociedad del Riesgo: Hacia una Nueva Modernidad.* Barcelona, Spain: Paídos Ibérica.

Bjornlund, H. (2006) *A Framework for the Next Generation of Water Market Policies.* Canberra, Australia: National Water Initiative.

Bjornlund, H. and Rossini, P. (2005) "Fundamentals Determining Process and Activities in the Market for Water Allocations." *Water Resources Development* 21(2): 355–69.

Calatrava, J. and Garrido, A. (2005a) "Modeling Water Markets under Uncertain Water Supply." *European Review of Agricultural Economics* 32(2): 119–42.

Calatrava, J. and Garrido, A. (2005b) "Spot Markets and Risk in Water Supply." *Agricultural Economics* 133: 131–43.

Calatrava, J. and Gómez-Ramos, A. (2009) "El Papel de los Mercados de Agua Como Instrumento de Asignación de Recursos Hídricos en el Regadío Español." In J. A. Gómez-Limón *et al. La Economía del Agua en el Regadío Español*, 295–319. Almeria, Spain: Fundación Cajamar.

Cui, J. and Schereider, S. (2009) "Modeling of Pricing and Markets Impacts for Water Options." *Journal of Hydrology* 374(1–4): 31–41.

Cubillo, F. (2002) "Gestión de la Demanda y Garantía de Abastecimiento." Paper presdented at *III Congreso Ibérico de Gestión y Planificación de Aguas*, Seville, 13–17 November.

Garrido, A. and Gómez-Ramos, A. (2008) "Risk Sharing Mechanisms Supporting Planning and Policy." In A. Iglesias, A. Cancelliere, F. Cubillo, L. Garrote, and D. Wilhite (eds), *Coping with Drought Risk in Agriculture and Water Supply* Systems, 133–53. Berlin, Germany: Springer.

Garrido, A and Gómez-Ramos, A. (2009) "Propuesta Para la Implementación de un Centro de Intercambio Basado en Contratos de Opción." In J. A. Gómez-Limón *et al.*, *La Economía del Agua en el Regadío Español*, 321–40. Almeria, Spain: Fundación Cajamar.

Garrido, A. and del Moral, L. (2000) *Drought Management in the Lower Guadalquivir River Basin.* SIRCH Project Final Report. Seville/Madrid, Spain: Universidad de Sevilla/Universidad Politécnica de Madrid

Giansante, C., Babiano, M., Aguilar, M., Garrido, A., Gómez, A., Iglesias, E., Lise, W. and Moral, L. (2002) "Institutional Adaptation to Changing Risk of Water Scarcity in the Lower Guadalquivir Basin." *Natural Resources Journal* 42(3): 521–63.

Gómez-Ramos, A. (2009) "La Nueva Economía del Agua." In MMARM, *Crónicas del Agua: La Importancia del Agua en Nuestra Cultura*, 85–97. Madrid, Spain: Ministerio de Medio Ambiente Rural y Marino.

Gómez-Ramos, A. and Garrido, A. (2004) "Formal Risk-Sharing Mechanism for Allocating Water Resources. The Case of Option Contracts." *Water Resources Research* 40: W12302 (doi:101029/2004WR003340).

Howitt, R. E. (1998) "Spot Prices, Option Prices, and Water Markets: An analysis of Emerging Markets in California." In K. W. Easter, M. W. Rosegrant, and A. Dinar (eds), *Markets for Water: Potential and Performance*, 119–40. Norwell, MA: Kluwer Academic Publishers.

Israel, M. and Lund, J. R. (1995) "Recent California Water Transfer: Implication for Water Management." *Natural Resources Journal* 35: 1–32.

Luo, B., Huang, G. H., Zou, Y. and Yin, Y. (2007) "Towards Quantifying the Effectiveness of Water Trading under Uncertainty." *Journal of Environmental Management* 83: 181–90.

McCann, L. and Easter, W. L. (2004) "A Framework for Estimating the Transaction Costs of Alternative Mechanisms for Water Exchange and Allocation." *Water Resources Research* 40: W09S09 (doi:101029/2003WR002830).

Michelsen, A. M. and Young, R. A. (1993) "Optioning Agriculture Water Rights for Urban Water Supplies During Drought." *American Journal of Agricultural Economics* 75: 1010–20.

Rogers, P. (1994) "Assessing the Socioeconomics Consequences of Climate Change on Water Resources." *Climate Change* 28: 179–208.

Tomkins, C. D. and Weber, A. W. (2010) "Option Contracting in the California Water Market." *Journal of Regulations Economics* 37: 107–41.

Tsur, Y. and Zemel, A. (1998) "On Event Uncertainty and Renewable Resource Management." In D. D. Parker and Y. Tsur (eds), *Decentralization and Coordination of Water Resource Management.* Norwell, MA: Kluwer Academic Publishers.

Vose, D. (2000) *Risk Analysis: A Quantitative Guide.* Chichester, UK: John Wiley.

Young, M. D. and McColl, J. C. (2002) *Robust Separation: A Search for a Generic Framework to Simplify Registration and Trading of Interest in Natural Resources.* Australia: CSIRO.

Part IV

Incentives and prices in water trading

Introduction: incentives and prices in water trading

Carlos Mario Gómez

By tackling the many existing opportunities to enhance the efficiency with which water is allocated and used in the economy, water trading can make a real contribution both to economic progress and to the preservation of the critical water resources on which the entire economy depends.

In contrast to command and control, water trading belongs to the general category of economic policy instruments, and then its outcome relies on spontaneous and voluntary individual decisions rather than on prescribed or norm-driven behavior. For this reason its failure of success depends critically on the ability of the chosen instrument, for example water trading, to align the individual decisions to buy or sell water with the more desired courses of action at the scale of the river basin or of the entire economy. This implies that, particularly in the case of water, the final decisions on what is the desired status of conservation of water resources and what kind of pressures are acceptable cannot be left to the market as its belongs to the political process. Water trading is then a flexible instrument to find the best way to adapt the functioning of the economy to the ability of water ecosystems to provide valuable economic services within the limits previously decided in the decision making process.

This section of the book explores two particular dimensions in which the particular incentives provided by water markets can be judged as superior to other alternative economic and command and control instruments. The first, explored by Franklin M. Fisher in Chapter 19, is the capacity of water trading to reveal the less visible opportunity costs of providing water services, and then to make the case in favor of adopting least social cost alternatives. The second dimension, explored by Jay R. Lund in Chapter 20, is the ability of water trading to alleviate the important information burden implied by public decisions on the best way to allocate water across space and among its many alternative uses. After considering the best decision support models available to inform public decisions on water resource allocation (presented by Lund in his chapter), one may conclude that, when information is scattered and imperfect, markets may perform better than the best command and control alternative.

According to Fisher, in order to reach efficient decisions, market prices need to be equal to all the costs of producing any particular good or service, including the opportunity costs involved. This is not generally the case of water services. In a market economy utilities can indeed translate to the water tariffs the capital and operational financial costs implied in the in the necessary processes of water abstraction and impoundment, transport, treatment, delivery, and so forth. But without public regulations, water tariffs will hardly translate to final water users the external cost associated with the impaired capacity of water ecosystems to provide in stream environmental services. In the same sense, when water cannot be reallocated between users, water tariffs are uninformative on the opportunity cost of using water for one purpose instead of its best available alternative.

In his chapter, Fisher connects both the environmental and the opportunity cost of providing water services. When the non-commercial opportunity costs are ignored an economy can go along with traditional options exacerbating both the water crisis and the political fight for the control of the critical water sources.

Considering the overall opportunity cost may help society in finding out many situations where enhancing the conservation of natural systems may lead to significant welfare gains. Comparing the best available alternatives assessed in the WAS model presented in his chapter Fisher uses the case of desalination to provide a ceiling to the value of water. Desalination is not the efficient answer to the world's water problems but the example is just a way of showing that one can put a finite value on water. According to that, the water services provided by the mountain aquifer and the Golan Heights in the disputed territories of the Middle East are not worth a war. Pushing this logic ahead, Fisher shows how finding a bargaining solution to the water problem can be an essential step for conflict resolution in the area as far as there are wide scope for mutually beneficial solutions among Israel, Jordan, and Palestine.

Jay Lund's chapter explores the informational problems implied in finding the best way to allocate water resources. In this respect, the main advantage of water trading is that its effectiveness does not require having in advance detailed information on the value of water on any place and for any alternative use. Engaging in trade is a voluntary decision, thus the market is itself a preference revelation mechanism. No particular prior information on people's preferences is then required in advance for trading to find mutually beneficial deals. By definition, when a transaction takes place, the maximum willingness to pay of those buying water is higher than the minimum compensation required by the sellers of water property rights and the deal improves the welfare of those engaged in it. Allowing for water trading thus reduces the information burden of water authorities as they do not need detailed information beyond that required to reach a decision on the overall constraints on the water use in the entire economy.

This particular informational advantage of water trading becomes more relevant when the many disadvantages of the alternative – for instance, using the normative models available to inform a centralized decisions on water allocation – is considered. Following this approach, Lund presents an ordered assessment of the optimization and simulation models available to support water allocation decisions stressing on its usefulness to inform practical decisions. In Lund's words, applied optimization models are "too smart" in having perfect hydrological foresight, seeing floods and droughts in advance, and being unrealistically ready for them. Simulation models are, in contrast, "too dumb" and not able to take advantage of changing water supplies. Although predicting the future remains an impossible task markets can provide information on willingness to pay to prevent future losses by, for example, improving water storage, paying a prime risk by the reduction of current water uses and buying insurance to protect themselves for future losses.

Rather than relying on imperfect data and models, allowing people to take their own decisions in a bargaining environment might be a more promising way to obtain the critical information required for a better response to current and future water challenges.

19 Models for optimal water management and conflict resolution

Franklin M. Fisher

Introduction: Fishelson's example

Water is necessary for life. Water is very important. But such importance does not exempt water from the laws of economics. In this paper, I show how somewhat unfamiliar ways of thinking about water lead to some surprising (and surprisingly useful) results, both in terms of water management and in terms of conflict resolution.

I begin with an example due to the late Gidon Fishelson of Tel Aviv University. As with other examples in this paper, it is drawn from the Middle East, where I have been the Chair of the Water Economics Project (WEP), a joint effort of Israeli, Jordanian, Palestinian, Dutch, and American experts.[1] However, it is applicable around the world. No matter how much one values water, one cannot rationally value it at more than it would cost to replace. Hence, the possibility of seawater desalination places an upper bound on the value of water on the coast. In the case of Israel and Palestine, the cost of desalination on the Mediterranean Coast is roughly US$0.60 per cubic meter.[2] Hence water in Tel Aviv or Gaza (see Figure 19.1) is not worth more than US$0.60 per cubic meter.

But the water sources in the region are mostly not on the Mediterranean Coast. In particular, a good deal of the water in dispute between Israel and Palestine is underground in the (ineptly named) Mountain Aquifer. To extract it and convey it to the coast would cost roughly US$0.40 per m^3. Hence ownership of that water is worth no more than about US$0.20 per m^3 (US$0.60 – US$0.40).

Now, 100 million m^3 per year is a large amount of water in the dispute. But 100 million m^3 per year is worth no more than US$20 million per year (and, in fact, is often worth considerably less). Such a sum is not worth fighting about. More dramatically put, it is not worth the cost of a single fighter plane. Israel's GDP exceeds US$200 billion. Water is not worth war.

The point of this example is not that desalination is the efficient answer to the world's water problems. It isn't. The point is rather that thinking about the value rather than only the quantity of water can produce useful results – in particular, the result that water is not beyond price and can be traded off for other things.

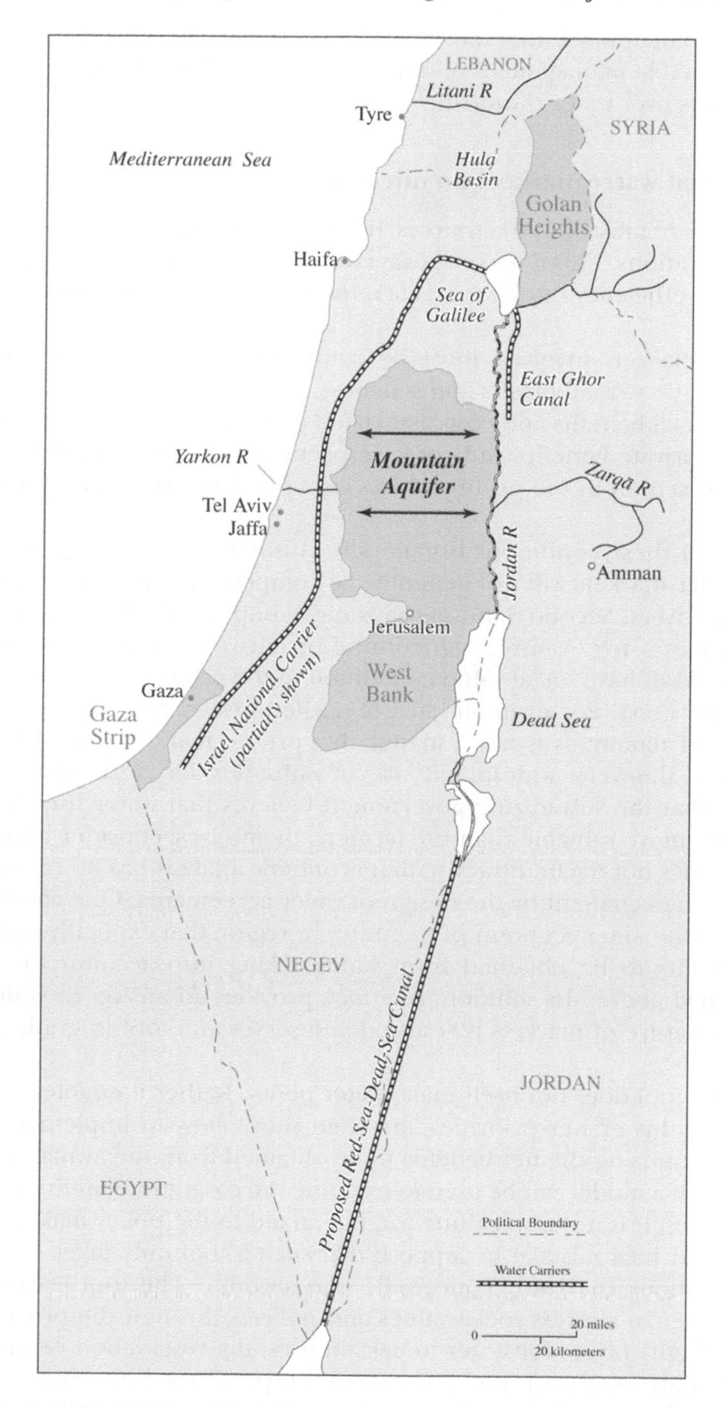

Figure 19.1 Regional map with water systems
Source: Figure P-1 in Fisher *et al.* (2005: xv); adapted from Wolf (1994: 175)

As we shall now see, that way of regarding water can both serve to inform the sustainable management of water resources and also for the resolution of conflicts over water ownership.

Why actual water markets do not work

In the case of most scarce resources, free markets can be used to secure efficient allocations. This does not always work, however; the important results about the efficiency of free markets require the following conditions:

- The markets involved must be competitive consisting only of very many, very small buyers and sellers.
- All social benefits and costs associated with the resource must coincide with private benefits and costs, respectively, so that they will be taken into account in the profit-and-loss calculus of market participants.

Neither of these conditions is generally satisfied when it comes to water. First, water markets will not generally be competitive with many small sellers and buyers. Second (and perhaps more important), because water in certain uses – for example, agricultural or environmental uses – is often considered to have social value in addition to the private value placed on it by private users. For example, in Chile, the preservation of the wetlands habitats of flamingos is not a matter that private markets consider. More generally, the very widespread use of subsidies for agricultural water implies that the subsidizing government believes that water used by agriculture is more valuable than the farmers, themselves, consider it to be.

This does not mean, however, that economic analysis has no role to play in water management or the design of water agreements. One can build a model of the water economy of a country or region that explicitly optimizes the benefits to be obtained from water, taking into account the issues mentioned above.[3] Its solution, in effect, provides an answer in which the optimal nature of markets is restored and serves as a tool to guide policymakers.

Such a tool does not itself make water policy. Rather it enables the user to express his or her priorities and then shows how to implement them while maximizing the net benefits to be obtained from the available water. While such a model can be used to examine the costs and benefits of different policies, it is not a substitute for, but an aid to the policy-maker.

It would be a mistake to suppose that such a tool only takes economic considerations (narrowly conceived) into account. The tool leaves room for the user to express social values and policies through the provision of low (or high) prices for water in certain uses, the reservation of water for certain purposes, and the assessment of penalties for environmental damage. These are, in fact, the ways that social values are usually expressed in the real world.

I first briefly describe the theory behind such tools applied to decisions within a single country. I then consider the implications for water negotiations and the structure of water agreements.

The Water Allocation System tool

The tool that I shall principally discuss is called WAS, for "Water Allocation System." It is a single-year, or steady-state annual model, although the conditions of the year can be varied and different situations evaluated. A related but more powerful tool – MYWAS, for "Multi-Year Water Allocation System" – permits consideration of a sequence of years or seasons. I discuss WAS first.

The country or region to be studied is divided into districts. Within each district, demand curves for water are defined for household, industrial, and agricultural use of water. To assure sustainability, extraction from each natural water source is limited to the annual average renewable amount. Allowance is made for treatment and reuse of wastewater and for inter-district conveyance. This procedure is followed using actual data for a recent year and projections for future years.

Environmental issues are handled in several ways. As stated, water extraction is restricted to annual renewable amounts; an effluent charge can be imposed; the use of treated wastewater can be restricted; and water can be set aside for environmental (or other) purposes. Other environmental restrictions can also be introduced.

The WAS tool permits experimentation with different assumptions as to future infrastructure. For example, the user can install wastewater treatment plants, expand or install conveyance systems, and create seawater desalination plants.

Finally, the user specifies policies toward water. Such policies can include: specifying particular price structures for particular users; reserving water for certain uses; imposing ecological or environmental restrictions, and so forth. This is where social values that are not simply private values come in.

Given the choices made by the user, the model allocates the available water so as to maximize total net benefits from water. These are defined as the total amount that consumers are willing to pay for the amount of water provided less the cost of providing it.[4]

Shadow values and scarcity rents

It is an important theorem that, under very general conditions, when an objective function is maximized under constraints, the solution also generates a set of non-negative numbers, usually called "shadow prices," but here called "shadow values" to emphasize that these are not necessarily the prices to be charged to water users. Such shadow values (which are the

Lagrange multipliers corresponding to the various constraints) have the property that they show the amount by which the value of the thing being maximized would increase if the corresponding constraints were to be relaxed a little.

In the case of the WAS model, the shadow value associated with a particular constraint shows the extent by which the net benefits from water would increase if that constraint were loosened by one unit. For example, where a pipeline is limited in capacity, the associated shadow value shows the amount by which benefits would increase per unit of pipeline capacity if that capacity were slightly increased. This is the amount that those benefiting would just be willing to pay for more capacity.

The central shadow values in the WAS model, however, are those of water itself. The shadow value of water at a given location corresponds to the constraint that the quantity of water consumed in that location cannot exceed the quantity produced there plus the quantity imported less the quantity exported. That shadow value is thus the amount by which the benefits to water users (in the system as a whole) would increase were there an additional cubic meter per year available free at that location. It is also the price that the buyers at that location who value additional water the most would just be willing to pay to obtain an additional cubic meter per year, given the net benefit maximizing water flows of the model solution.[5]

It is important to note that the shadow value of water in a given location does not generally equal the direct cost of providing it there. Consider a limited water source whose pumping costs are zero. If demand for water from that source is sufficiently high, the shadow value of that water will not be zero; benefits to water users would be increased if the capacity of the source were greater. Equivalently, buyers will be willing to pay a non-zero price for water in short supply, even though its direct costs are zero.

A proper view of costs accommodates this phenomenon. When demand at the source exceeds capacity, it is not costless to provide a particular user with an additional unit of water. That water can only be provided by depriving some other user of the benefits of the water; that loss of benefits represents an opportunity cost. In other words, scarce resources have positive values and positive prices even if their direct cost of production is zero. Such a positive value – the shadow value of the water *in situ* – is called a "scarcity rent."

Where direct costs are zero, the shadow value of the resource involved consists entirely of scarcity rent. More generally, the scarcity rent of water at a particular location equals the shadow value at that location less the direct marginal cost of providing the water there.[6] Just as in a competitive market, a positive scarcity rent is a signal that more water from that source would be beneficial were it to be available.

An important use of shadow values is illustrated in Figure 19.2, where water in a lake (L) is conveyed to locations a, b, and c. It is assumed that the only direct costs are conveyance costs and that the capacity of the

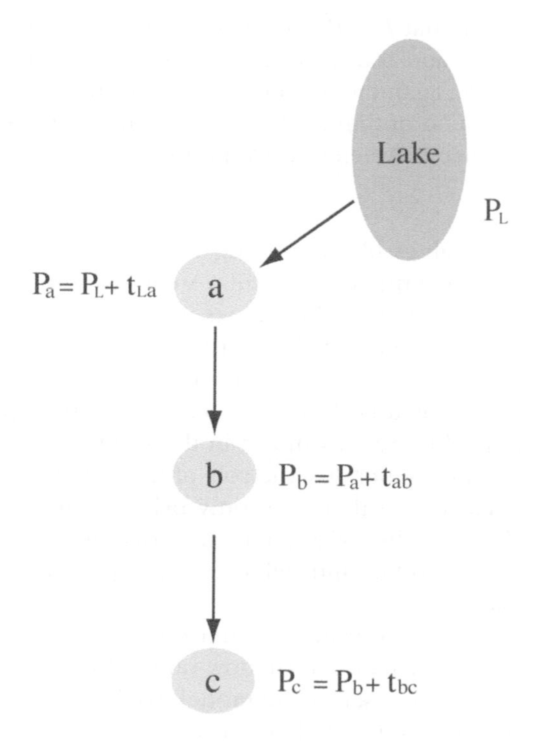

Figure 19.2 A use of shadow values
Source: Fisher *et al.* (2005: 17, Figure 1.5.1)

conveyance lines is not a binding constraint.[7] The marginal conveyance cost from the lake to *a* is denoted t_{La}; similarly, the marginal operating conveyance cost from a to b is denoted t_{ab}; and that from b to c is denoted t_{bc}. The shadow values at the four locations are denoted P_L, P_a, P_b, and P_c, respectively.

To see that the equations in Figure 19.2 must hold, begin by assuming that $P_a > P_L + t_{La}$ and that there is extra conveyance capacity from *L* to *a* at the optimal solution. Then transferring one more cubic meter of water from *L* to *a* would have the following effects. First, since there would be one cubic meter less at *L*, net benefits would decline by P_L, the shadow value of water at *L*. (That is what shadow values measure.) Second, since conveyance costs of t_{La} would be incurred, there would be a further decline in net benefits of that amount. Finally, however, an additional cubic meter at *a* would produce an increase in net benefits of P_a, the shadow value of water at *a*. Since $P_a > P_L + t_{La}$ by assumption, the proposed transfer would increase net benefits; hence, we cannot be at an optimum.

Similarly, assume that $P_a < P_L + t_{La}$. Then too much water has been transferred from L to a, and transferring one less cubic meter would increase net benefits. Hence, again, we cannot be at an optimum.

It follows that, at an optimum, $P_a = P_L + t_{La}$, and a similar demonstration holds for conveyance between any two points.

Q.E.D.

Note that shadow values play a guiding role in the same way that actual market prices do in competitive markets. An activity that is profitable at the margin when evaluated at shadow values is one that should be increased. An activity that loses money at the margin when so evaluated is one that should be decreased. In the solution to the net benefit maximizing problem, any activity that is used has such shadow marginal profits zero, and, indeed, shadow profits are maximized in the solution.

That shadow values generalize the role of market prices can also be seen from the fact that, where there are only private values involved, at each location, the shadow value of water is the price at which buyers of water would be just willing to buy and sellers of water just willing to sell an additional unit of water.

Of course, where social values do not coincide with private ones, this need not hold. In particular, the shadow value of water at a given location is the price at which the user of the model would just be willing to buy or sell an additional unit of water there. That payment is calculated in terms of net benefits measured according to the user's own standards and values.

This immediately implies how the water in question should be valued. Water *in situ* should be valued at its scarcity rent. That value is the price at which additional water is valued at any location at which it is used, less the direct costs involved in conveying it there.

Note that such water shadow values take full account of the fact that using or processing water in one activity can reduce the amount of water available for other activities and thus has opportunity costs. The shadow values include such opportunity costs, taking into account system-wide effects. (This is particularly important in using WAS for cost–benefit analysis.)

One should not be confused by the use of marginal valuation in all this (the value of an additional unit of water). The fact that people would be willing to pay much larger amounts for the amount of water necessary for human life is important. It is taken into account in the optimizing model by assigning correspondingly large benefits to the first relatively small quantities of water allocated. But the fact that the benefits derived from the first units are greater than the marginal value does not distinguish water from any other economic good. It merely reflects the fact that water would be (even) more valuable if it were scarcer.

It is the scarcity of water and not merely its importance for existence that gives it its value. Where water is not scarce, it is not valuable.

Cost–benefit analysis of infrastructure

Before proceeding, it is useful to understand how WAS can be used in the cost-benefit analysis of proposed infrastructure projects and how it handles capital costs (which can be quite substantial).

Consider the discussion of the lake and the conveyance line (Figure 19.2). Suppose that there were no existing conveyance line to carry water from the lake to city *a*. Suppose further that, if the WAS model were run without such a conveyance line, the resulting shadow values would be such that $P_a < P_L + t_{La}$, where t_{La} is the per-cubic-meter conveyance operating cost that would be incurred were such a conveyance line in place. Such a result would show that the conveyance line in question should not be built, because it would not be used even if it were. On the other hand, if the inequality were reversed, so that $P_a > P_L + t_{La}$, then that conveyance line might well be worth building – but whether it should be built would depend on the capital costs involved.

There are two ways of incorporating capital costs into the analysis of WAS results. One option is to impute an appropriate charge for capital costs to each cubic meter of water processed by the proposed facility.

I illustrate this for the case of desalination plants in Israel, using projections made in the mid-1990s. Figure 19.3 shows the shadow values obtained for 2010 both in a situation of normal availability of natural resources ("normal hydrology") – the upper numbers – and in a severe drought when that availability is reduced by 30 percent – the lower numbers. Israel's price policy ("fixed-price policies") of 1995 are assumed to remain in effect. These policies charge each user class the same amount in all districts and heavily subsidize water for agriculture while charging higher prices to household and industrial users. Note that Israel's practice of reducing the quantity of subsidized agricultural water in times of drought has not been modeled, so the results are more favorable to the need for desalination than would be the case in practice.[8]

The important result with which to start can be seen in the upper shadow values for the coastal districts: Acco, Hadera, Raanana, Rehovot, and Lachish. The highest shadow value is at Acco (near Haifa) and is only US\$0.319/m^3 – well below the cost of desalination. This means that desalination plants would not be needed in years of normal hydrology.

On the other hand, such plants would be desirable in severe drought years. In the lower numbers in Figure 19.3, desalination plants operate in all the coastal districts at an assumed cost of US\$0.60/m^3. The required sizes of such plants can be obtained by running WAS without restricting plant capacity and observing the resulting plant output.

Results for 2020 are similar, although, as one should expect, it does not take so severe a drought to make desalination efficient, and the required plant sizes in each district are larger.

Of course, much of the cost of desalination consists of capital costs –

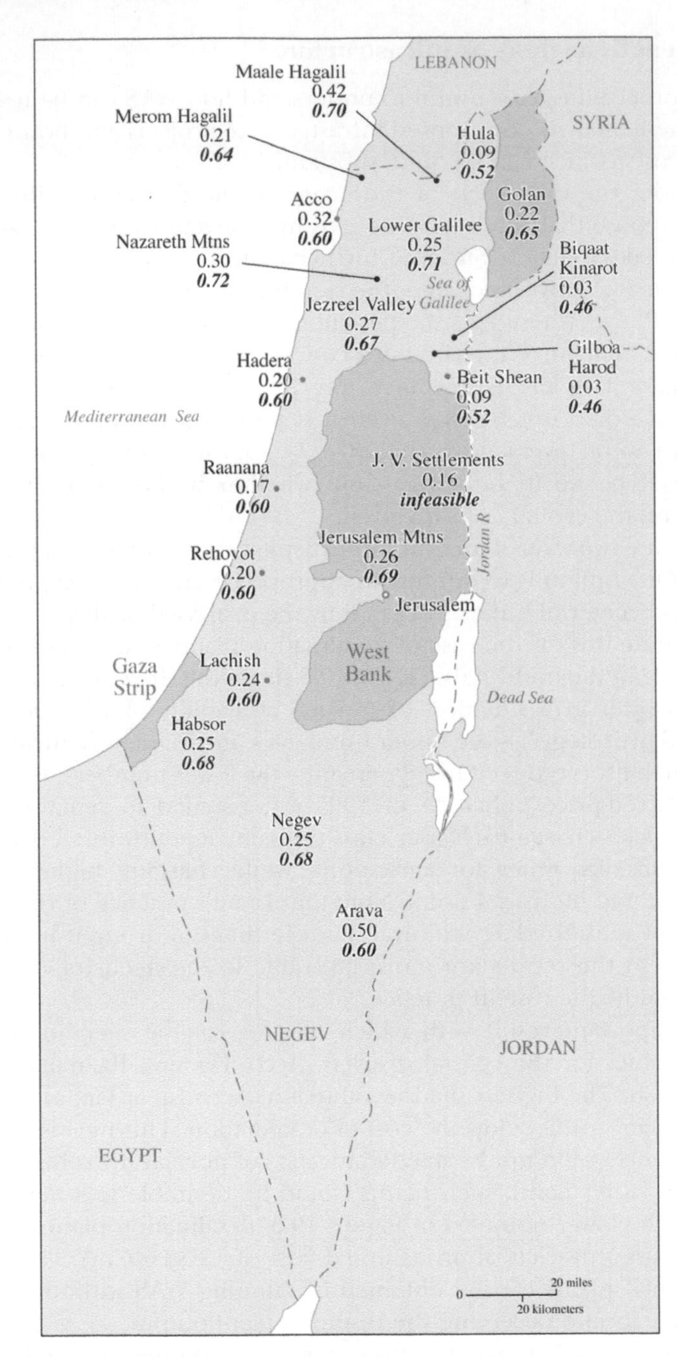

Figure 19.3 2010 shadow values with desalination: normal hydrology vs. 30 percent reduction in naturally occurring fresh water sources (fixed-price policies in effect)
Source: Fisher *et al.* (2005: 102, Figure 5.3.1)

here included in the price (or target price) per m^3. Such costs are largely incurred when the plant is constructed. After that, the plants would be used in normal years unless the operating costs were above the upper shadow values in Figure 19.3 (highest US$0.319/m^3). Israel therefore needed to consider whether the insurance for drought years provided by building desalination plants is worth the excess capital costs. Note that the system of fixed-price policies contributes substantially to the need for desalination; without such policies, the plants required for severe drought would be far smaller and some would not be required at all. In fact, WAS can be used to estimate the plant capacities required.

The second option is more direct. One runs the WAS model with and without the proposed infrastructure.[9] This generates an estimate of the annual increase in benefits that would result from having such infrastructure in place. Given the estimated life of the projected infrastructure, one can repeat the exercise for the expected conditions of the various years of that life. Then, choosing a discount rate, one calculates the present value of such benefits and compares that with the capital costs.

Multi-Year Water Allocation System

MYWAS offers a much more convenient treatment and moves away from the single-year, steady-state treatment of WAS. MYWAS deals readily and directly with problems over time by maximizing the present value of net benefits over a number of future years or time periods, using a discount rate specified by the user. Capital costs are treated as cash outflows when they occur.

Here is a (presumably partial) list of applications. In all of them, as in all WAS applications, system-wide effects and opportunity costs are automatically dealt with, and the user's own decisions and values are implemented.

- **The timing, order, and capacity of infrastructure projects:** MYWAS allows the user to specify a menu of possible infrastructure projects, their capital and operating costs and their useful life. The program then yields the optimal infrastructure plan, specifying which projects should be built, in what order, and to what capacity.[10] This is a major advance.
- **Storage management:** Most obviously, it is now easy to deal with storage issues, in particular the decisions as to how much water should be stored or released from reservoirs. The decisions involved can be for inter-year or for inter-seasonal storage.
- **Aquifer management:** Man-made storage is not the only kind. Water can also be transferred between time periods by increasing aquifer pumping when water is relatively abundant and reducing it when water is relatively scarce. This means that the use of aquifers and other natural water sources no longer needs to be restricted to the average yearly

renewable amount in the model (with that average adjustable by the user). Rather, by specifying the effects of withdrawal on the state of the aquifer, the user can obtain a guide to the optimal pattern of aquifer use over time, including guidance as to aquifer recharge.

- **Fossil aquifers:** The rate at which a fossil aquifer should be pumped can also be determined endogenously through the use of MYWAS rather than being specified exogenously by the user. That rate will generally vary over time as conditions change.
- **Climatic uncertainty:** Of course, optimal planning over time will depend on the climate, and climate – especially rainfall – is variable and uncertain. MYWAS enables the systematic study of the effects of such uncertainty on optimal planning by providing the means to examine optimal decisions as a non-linear function of climate variables. Other uncertainties, such as those involved in population forecasts, can also be dealt with.
- **Global warming:** Of course, the multiyear nature of MYWAS makes it suitable for examining the effect of different global warming scenarios on optimal infrastructure.
- **Water quality:** If desired, MYWAS (or WAS) can be adapted to permit a more sophisticated treatment of multi-dimensional aspects of water quality than available in the original version of WAS.
- **Effect of discount rate:** Obviously, MYWAS can be used to examine the effects of the choice of discount rate on all aspects of the optimal solution.

Using WAS for policy examination

It is important to understand, however, that infrastructure issues are not the only ones that can be examined with the use of WAS (or MYWAS). WAS can also be used to evaluate contemplated water policies. Such policies, directly or indirectly, will typically involve water reallocation, and this will have both costs and benefits. As with infrastructure, because WAS looks at the entire water system of a country, those costs and benefits can be estimated and analyzed. Moreover, the distribution of the effects – both geographically and by water user type – can be measured, so that the policy-makers can find out who gains and who loses, and can consider whether taxes or subsidies should be used to mitigate undesirable effects.

While WAS does not make or require specific water policies, it does provide a powerful tool that can inform and assist decision-makers in the formulation of rational policies. A country contemplating serious changes in its water system or water policies would do well to construct a WAS model.

Conflict resolution: negotiations and the gains from trade in water permits

Optimal infrastructure planning is not the only use of WAS, however. By using WAS (or MYWAS) to value water in dispute, water disputes can be monetized, and this may be of some assistance in resolving them.

Consider bilateral negotiations between two countries, A and B. Each of the two countries can use its WAS tool to investigate the consequences to it (and, if data permit, to the other) of each proposed water allocation. This should help in deciding on what terms to settle, possibly trading off water for other, non-water concessions. Indeed, if, at a particular proposed allocation, A would value additional water more highly than B, then both countries could benefit by having A get more water and B getting other things which it values more. Note that this does not mean that the richer country gets more water. That only happens if it is to the poorer country's benefit to agree.[11]

Of course, the positions of the parties will be expressed in terms of ownership rights and international law, often using different and conflicting principles to justify their respective claims. The use of the methods here described in no way limits such positions. Indeed, the point is not that the model can be used to help decide how allocations of property rights should be made. Rather the point is that water can be traded off for non-water concessions, with the trade-offs measured by WAS.

In addition to monetizing water disputes, WAS can facilitate water negotiations by permitting each party, using its own WAS model, to evaluate the effects on it of different proposed water arrangements. As we now exemplify, this can show that the trade-offs just discussed need not be large.

Water on the Golan Heights (see Figure 19.1) is sometimes said to be a major problem in negotiations between Israel and Syria, because the Banias River that flows from the mountains of the Golan is one of the three principal sources of the Jordan River.[12] By running the Israeli WAS model with different amounts of water, we have evaluated this question.

In 2010, the loss of an amount of water roughly equivalent to the entire flow of the Banias springs (125 million cubic meters annually) would be worth no more than US$5 million per year to Israel in a year of normal water supply and less than US$40 million per year in the event of a reduction of thirty percent in naturally occurring water sources. These results take into account Israeli fixed-price policies towards agriculture.

Note that it is *not* suggested that giving up so large an amount of water is an appropriate negotiating outcome, but water is not an issue that should hold up a peace agreement. These are trivial sums compared to the Israeli GDP (gross domestic product) of more than US$200 billion per year or to the cost of fighter planes.

Similarly, a few years ago, Lebanon announced plans to pump water from the Hasbani River – another source of the Jordan. Israel called this a *casus belli* and international efforts to resolve the dispute were undertaken. But whatever one thinks about Lebanon's right to take such an action, it should be understood that our results for the Banias apply equally well to the Hasbani. The effects on Israel would be fairly trivial.[13]

Water is not worth war!

Monetization of water disputes, however, is neither the only nor, perhaps, the most powerful way in which the use of WAS can promote agreement. Indeed, WAS can assist in guiding water cooperation in such a way that all parties gain.

The simple allocation of water quantities after which each party then uses what it "owns" is not an optimal design for a water agreement. Suppose that property rights issues have been resolved. Since the question of water ownership and the question of water usage are analytically independent, it will generally not be the case that it is optimal for each party simply to use its own water.

Instead, consider a system of trade in water "permits" – short term licenses to use each other's water. The purchase and sale of such permits would be in quantities and at prices (shadow values) given by an agreed-on version of the WAS model run jointly for the two (or more) countries together. (The fact that such trades would take place at WAS-produced prices would prevent monopolistic exploitation.). There would be mutual advantages from such a system, and the economic gains would be a natural source of funding for water-related infrastructure.

Both parties would gain from such a voluntary trade. The seller would receive money it values more than the water given up (else, it would not agree); the buyer would receive water it values more than the money paid (else, it would not pay it). While one party might gain more than the other, such a trade would not be a zero-sum game but a win–win opportunity.

The gains from WAS-guided cooperation: Israel, Jordan, and Palestine

I now present results for Israel, Jordan, and Palestine, illustrating the gains from trilateral cooperation.[14]

I concentrate on two sources of water that are the subjects of conflicting claims. These are the Jordan River and the so-called Mountain Aquifer (see Figure 19.1). Both of these are (very roughly) of equal size, each yielding about 650 million cubic meters a year. The Jordan River is claimed by all three countries, while the Mountain Aquifer is claimed only by Israel and Palestine. Since the gains from cooperation are a function of the water ownership assumptions made, we obtain results for selected varying assumptions about such ownership. *It must be emphasized that such assumptions are not meant as a political statement. They are illustrative only.*

For the Jordan River, we examine ownership cases as follows:

A. Israel 92 percent, Jordan 8 percent; Palestine 0. (This is approximately the existing situation.)
B. Israel 66 percent; Jordan 17 percent; Palestine 17 percent.
C. Israel 33.3 percent; Jordan 33.3 percent; Palestine 33.3 percent.

For the Mountain Aquifer, we examine ownership cases varying from Israel 80 percent/Palestine 20 percent (close to the existing situation) to Israel 20 percent/Palestine 80 percent, by shifts of 20 percent at a time.[15]

It is assumed that, both for Israel and for Jordan, the fixed-price policies of the late 1990's are in place. For both countries, this means subsidies for agriculture and, for Israel, higher fixed prices for the other sectors. The Palestinian water price in each district is assumed to equal the corresponding shadow value.

Figure 19.4 shows the results for 2020. In it, Israel is represented in black, Jordan in white, and Palestine in white with diagonal black stripes. Each of the four panels corresponds to a different ownership of the Mountain Aquifer, while the three positions on the horizontal axis (A, B, and C) correspond to the three different assumptions as to Jordan River ownership. The heights of the bars show the gains from WAS-guided cooperation in millions of (1995) US dollars.

The first thing to notice is that, in general, the smallest gains are Jordanian. Not surprisingly, generally, the less water Israel owns, the more it has to gain from cooperation. On the other hand, Jordan gains more from cooperation the *more* water it owns – selling permits to Israel.

Like Jordan, Palestine presents a mixed picture. It also tends to benefit more from cooperation the larger is its share of the Jordan, selling permits to use river water to Israel. On the other hand (Figure 19.4, panel I), when Palestine owns relatively little Mountain Aquifer water, it also benefits as a buyer.

It should be emphasized that Israel is not always a seller in the cases portrayed. Nor is it invariably a buyer. Further, it is not always the case that the country owning the least water has the most to gain from cooperation. Sellers benefit also.

The most important general conclusion from all these cases should be clear. WAS-guided cooperation in water would benefit all parties – Israel and Palestine the most. As we shall now see, the gains from such cooperation generally exceed those that would be obtained from moderately large ownership shifts. This is particularly true under cooperation.

The gains from ownership shifts are measured as follows. Holding constant the distribution of ownership in one of the two water sources being studied, we look at the change in benefits that accrue to each of the parties as a result of moving from one of the ownership cases examined above to the next. (For example, in the case of the Mountain Aquifer, we

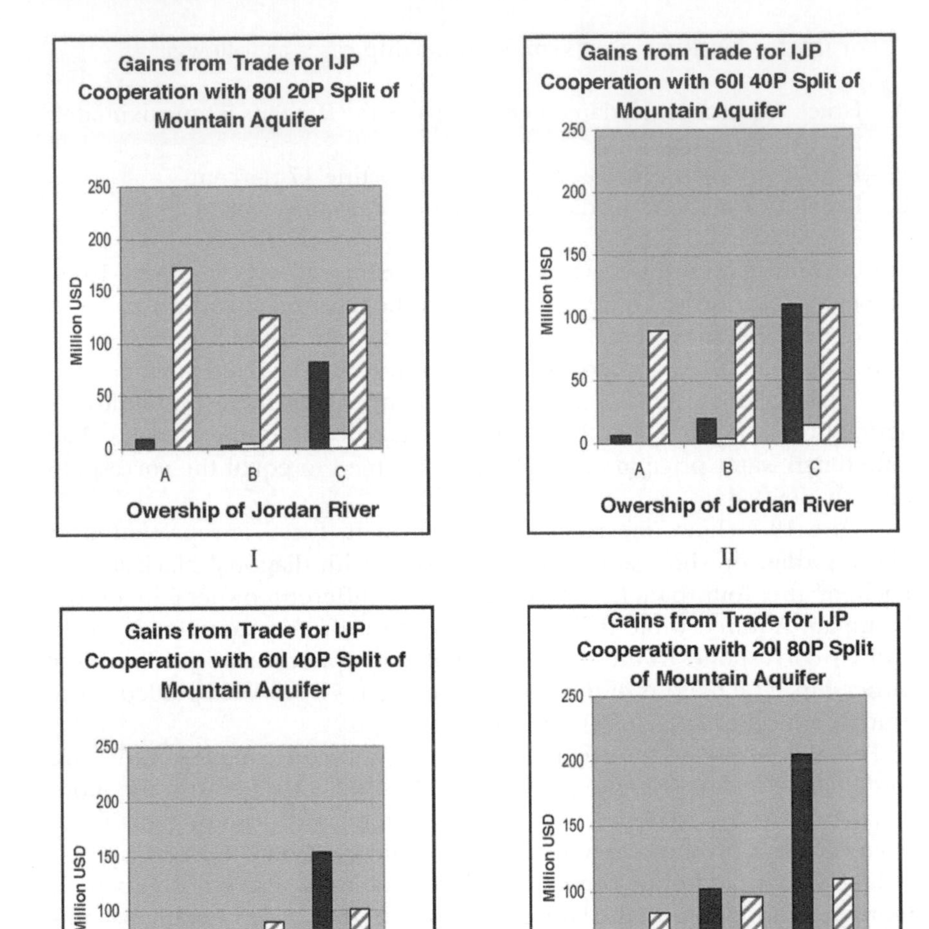

Figure 19.4 Gains from trilateral cooperation in 2020
Source: Fisher and Huber-Lee (2008–9: 108, Figure 5)

hold ownership in the Jordan River constant and examine the gains – or losses – to Israel and Palestine from an ownership shift of the aquifer from Israel 80 percent/Palestine 20 percent to Israel 60 percent/Palestine 40 percent, and from there to Israel 40 percent/Palestine 60 percent, and so on, repeating the exercise for each case of ownership of the Jordan River.)

We then normalize the results by expressing them as the gain from a 10 percent ownership shift.

The gains from such shifts under trilateral cooperation are constant for each of the two resources. The reason for this is that, under cooperation, the optimal water flows in the WAS solution are independent of the ownership assumptions. Hence, the only gains from changes in ownership are the changes in the money that ownership represents. But all water permit trades take place at the shadow values for the optimal solution. These are the scarcity rents of the water resources involved and also do not depend on ownership. Hence, the value of a 10 percent shift in the ownership of a given resource is independent of the initial ownership assumptions.

Under trilateral cooperation, the gains from such shifts in 2010 would be only about US$5 million per year for a shift in ownership of 10 percent of the Mountain Aquifer and about US$7.5 million per year for 10 percent of the Jordan River. In 2020, where the gains from cooperation are also larger, the corresponding gains from 10 percent ownership shifts would be about US$15 million per year for the Mountain Aquifer and US$25 million per year for the Jordan River.

It should come as no surprise, however, that the value of ownership shifts would be considerably different (and usually higher) when there is no cooperation. Moreover, in that case, the value would be substantially different for different parties (reflecting the fact that there are gains to be had from trading in water permits) and also widely different for different ownership circumstances.

Note, then, that one value of WAS-guided cooperation is that it reduces the value of ownership shifts, making them easier to negotiate.

Note again that it is *not* the case that the gains from cooperation are high only when the party receiving those gains has little water and the value of ownership shifts are high to it. That phenomenon naturally tends to occur when the big gainer is a *buyer* of water permits. But large gains also occur when the party receiving those gains has a large amount of water and the value of ownership shifts are low. In such cases, the big gainer is a *seller* of water permits. In fact, the largest Palestinian gains from trilateral cooperation occur both when Palestine has very little water and when it has a good deal.

Moving onward, the gains from WAS-guided cooperation would be greater in other ways than are shown above. In particular, as populations and other factors change, a quantity agreement that is adequate when signed can easily become out of date and a source of new tension. WAS-guided cooperation provides a flexible means of readjusting water usage in a way that all parties benefit.[16]

In addition, our results show clearly that Israel and Palestine would both benefit from the creation of a sewage treatment plant in Gaza, with the treated effluent sold to Israel for use in agriculture in the Negev. This means that Israel has a positive economic incentive to assist in the construction of such a plant. That would be a confidence-building measure that does not impinge on the core values of either party.

Concluding remarks

Of course, there are issues to be settled in all this – security among them – and space does not permit me to deal with them here. But the main points should be clear:

- While actual markets will not work efficiently, the use of economic models such as WAS or MYWAS can handle the issues that lead to the failure of actual markets. Such models do not only deal narrowly with money but also permit the user to express social values for water that are not private ones. They can be a powerful tool for water management and infrastructure planning.
- While WAS does not make or require specific water policies, it does provide a powerful tool that can inform and assist decision-makers in the formulation of rational policies, showing the costs and benefits from contemplated policies, both direct and indirect through water reallocation. A country contemplating serious changes in its water system or water policies would do well to construct a WAS model.
- Such models also point the way to the transformation of water conflicts into "win-win" situations. They show that disagreement over water ownership can be made a minor issue and the benefits from cooperation in water *usage* can be considerably larger than those from shifts in water ownership.

Acknowledgments

This paper draws heavily on the work of a number of colleagues (see Fisher *et al.* 2005), and especially on work done jointly with Annette Huber-Lee. For other papers on the topic, see Fisher (2007, 2008), and Fisher and Huber-Lee (2006, 2008–9, 2010, 2011a, and especially 2011b.)

Notes

1. The most complete discussion of the work and history of the WEP is Fisher *et al.* (2005). See also Fisher and Huber-Lee (2006).
2. All prices are in 1995 US dollars. This ignores the environmental consequences of desalination, but the same principles would apply were those consequences to be included.
3. The pioneering version of such a model (although one that does not explicitly perform maximization of net benefits) is that of Eckstein *et al.* (1994).
4. The total amount that consumers are willing to pay for an amount of water, Q^*, is measured by the area up to Q^* under their aggregate demand curve for water. Note that "willingness to pay" includes ability to pay. The provision of water to consumers that are very poor is taken to be a matter for government

policy embodied in the pricing decisions made by the user of WAS.

5. If the user of the model – for example the government of a country – would value additional water in a particular location more than would private buyers, then the shadow value reflects that valuation.
6. If this calculation gives a negative figure, then the scarcity rent is zero, and water is not scarce at the given location.
7. This is the simplest case, but results similar to that given in the text hold in any case.
8. The infeasibility listed for the Jordan Valley Settlements in the drought case reflects the fact that the full amount of subsidized water required for the supply of agriculture there cannot be delivered.
9. A similar method can be used to analyze different proposed water policies.
10. This problem was encountered in Fisher *et al.* (2005: ch. 7), in its analysis of Jordanian issues.
11. If trading off ownership rights considered sovereign is unacceptable, the parties can agree to trade short-term permits to use each other's water.
12. The others are the Hasbani, which rises in southern Lebanon, and the Dan, which rises in pre-1967 Israel.
13. Of course, the question naturally arises as to what the effects on Syria and Lebanon, respectively would be in these two situations. Without a WAS model for those two countries, we cannot answer that question. Both countries would surely profit from such a model.
14. Results for bilateral cooperation have also been obtained, but there is no room to present them here.
15. The Mountain Aquifer in fact consists of several sub-aquifers. I have made no attempt to divide ownership except in the arbitrary manner described in the text.
16. Note that this applies even if the initial ownership allocation just happens to be that of the WAS optimizing solution – an unlikely event.

References

Eckstein, Z., Zackay, D., Nachtom Y. and Fishelson, G. (1994) "The Allocation of Water Resources between Israel, the West Bank and Gaza – An Economic Analysis." *Economic Quarterly* 41: 331–69.

Fisher, F. M. (2007) "Water Management, Infrastructure, Negotiations and Cooperation: Use of the WAS Model." In H. Shuval and H. Dweik (eds), *Water Resources in the Middle East: Israel–Palestinian Water Issues – From Conflict to Cooperation*, 119–32. Berlin, Germany: Springer.

Fisher, F. M. (2008) "Water Value, Water Management, and Water Conflict: A Systematic Approach." In E. Wiegandt (ed), *Mountains: Sources of Water, Sources of Knowledge*, 123–48. Advances in Global Change Research, vol. 31. Dordrecht, The Netherlands: Springer.

Fisher, F. M. and Huber-Lee, A. (2006) "Economics, Water Management, and Conflict Resolution in the Middle East and Beyond." *Environment: Science and Policy for Sustainable Development* 48(3, April): 26–41.

Fisher, F. M. and Huber-Lee, A. (2008–9) "WAS-Guided Cooperation in Water: the Grand Coalition and Sub-Coalitions." *Environment and Development Economics* 14(Special Issue 1, February): 85–115.

Fisher, F. M. and Huber-Lee, A. (2010) "Water for Peace: a Game-Changer for Israel, Palestine, and the Middle East." *WaterBiz* 2(3): 112–14.

Fisher, F. M. and Huber-Lee, A. (2011a) "Sustainability, Efficient Management, and Conflict Resolution in Water." *The Whitehead Journal of Diplomacy and International Relations* 12(1): 63–78.

Fisher, F. M. and Huber-Lee, A. (2011b) "The Value of Water: Optimizing Models for Sustainable Management, Infrastructure Planning and Conflict Resolution." *Desalination and Water Treatment* 31(1): 1–23.

Fisher, F. M., Huber-Lee, A., Arlosoroff, S., Eckstein, Z., Haddadin, M., Hamati, S. G., Jarrar, A., Jayyousi, A., Shamir, U. and Wesseling, H. (2005) *Liquid Assets: An Economic Approach for Water Management and Conflict Resolution in the Middle East and Beyond.* Washington, DC: RFF Press.

Wolf, A. (1994) "A Hydropolitical History of the Nile, Jordan, and Euphrates River Basins." In A. S. Biswas (ed.), *International Waters of the Middle East from Euphrates–Tigris to Nile*, 152–99. Oxford, UK: Oxford University Press.

20 Optimization modeling in water resource systems and markets

Jay R. Lund

Introduction

There is a long history of formal optimization modeling in water resources engineering and management. Early mathematical optimization dates back to Varlet (1923) for flood control and water supply, and Massé (1946) for hydropower operations, with numerical optimization methods dating back at least to Little (1955) in his application of stochastic dynamic programming to hydropower. Many applications have sought optimal operating rules for reservoirs or optimal sizing and operation of reservoir systems (Yeh 1985; Labadie 2004). While engineering applications of optimization have largely focused on the operation of reservoirs, economists applying optimization often focus on finding optimal prices.

Complex societies and economies have evolved and grown in many regions, with water resource infrastructure having been significantly developed. More complex water management problems have arisen that go beyond the optimal operation of a single component or class of water system components (reservoirs or prices). Contemporary water management problems call for a more integrated and comprehensive optimization of water management within a larger and more diverse economy, and within an institutional structure which is far more decentralized than is traditionally assumed.

Contemporary water management in the developed has by many potentially conflicting water users and uses, many forms of infrastructure with many water management options, and many water managers. California's water system has hundreds of dams, many dozens of aquifers, tens of thousands of kilometers of aqueducts, and highly diverse water demands ranging from high-valued permanent crops, to low valued annual crops, to residential, commercial, and industrial urban demands, to a host of environmental demands for instream flows, wetlands, and maintaining cool or warm water temperatures depending on the local species of concern (Hanak *et al.* 2011). This complex system is managed by a network of interacting institutions, including dozens of federal and state agencies with either project management authority or regulatory authority and about

3,000 local water districts and suppliers with locally-elected leadership and thousands of water contracts among these entities to coordinate water operations. Indeed, water demands are perhaps the most important single component of any functional water system – and these decisions are typically made by millions of sovereign water users each day.

This paper reviews application of optimization modeling to such contemporary water management problems. We begin by briefly summarizing early applications of optimization, mostly for simple single-purpose problems. We then explore the use of more extensive optimization models to help guide thinking, operations, and policies for larger and more diverse problems. We also pay particular attention to the role of water marketing in such systems and uses of optimization modeling for developing policies, plans, and operations which work with water markets.

Applications of optimization in water management

Over more than 50 years, optimization has been applied to a wide variety of water problems. This review will be brief, with a few examples of applications for real-time operations, operations planning, facility planning, and policy-making.

Real-time operations

Real-time operation applications involve the use of optimization to make water management decisions in real operational time-frames, ranging from decisions on which turbines or pumps to employ in the coming minutes to scheduling water releases, diversions, or pumping over hourly, daily, and perhaps weekly time frames. Real-time applications of optimization tend to be mostly for single-purpose operations such as hydropower scheduling or the operation of pumps and storage in water distribution systems. Over these short time-frames, the optimization problem is well-specified and fairly tractable. Many large hydropower systems employ optimization to aid unit scheduling in the face of external price and load fluctuations (Jacobs *et al.* 1995). Most use linear programming methods, often making constant-head assumptions for short periods (BC Hydro; Shawwash *et al.* 1999). Some use is made of dynamic programming for very short term unit scheduling (which turbines to employ) or longer-term operations (Quebec Hydro). Single-purpose real-time optimization benefits from having a well-defined objective for optimization, typically maximizing hydropower revenues or minimizing energy costs for distribution system operations.

Operations planning

Operations planning can have two distinct directions, seasonal operations planning and system re-operation studies. Seasonal operations planning seeks to make operational schedules for the coming season, based on forecasts of water availability and water demands, so operations can respond to wet or dry conditions or anticipated changes in demands that might accompany changes in crops prices or environmental conditions. System re-operation studies are longer-term affairs where the operating policies and rules for a system are re-examined and updated to respond to changes in regulatory conditions, regional water demands, and perhaps perceived changes in climate. Applied seasonal operations planning studies have been done by the Corps of Engineers, but simulation remains the most common tool for such problems (Murk and Lund 1996), outside of hydropower systems. Water purchase decisions are often made in a seasonal operations planning time frame. Several papers develop methods for optimizing water purchases for environmental and urban water purchases (Hollinshead and Lund 2006).

Optimization has found greater use for system re-operation studies (Lund and Ferreira 1996; USACE 1993; Tilmant *et al.* 2008), although simulation modeling remains central to this activity as well. Re-operation studies also have become more important with increasing need to coordinate activities among diverse elements of large systems, often controlled by different institutions. Several papers discuss optimization of various water market forms in planning long-term operation strategies (Jenkins and Lund 2000; Lund and Israel 1995).

Facility planning

Many early applications of optimization in water resources were for identifying optimal sizes and configurations of reservoir systems. These studies were at a time towards the end of the historical period of major reservoir construction (Luss 1982; Butcher *et al.* 1969). However, some facility planning aspects remain in contemporary water management, as existing systems are adapted to changes in water demand and supply conditions.

Policy

Optimization models can be particularly useful for examining long-term policies which must operate under sometimes very different conditions from the present, when existing simulation models are likely to work well. Examples include how a water system might operate differently if operations changed to reflect the use of water markets, changes in water allocation policies, changes in climate, integrated water management, dam removal, or major investments (Null and Lund 2006; Tanaka *et al.* 2006, 2011; Medellín-Azuara *et al.* 2007, 2008; Pulido-Velázquez *et al.* 2004; Jenkins *et al.* 2004; Rosenberg *et al.* 2008; Harou and Lund 2008; Harou *et al.* 2010).

Markets and modeling

Optimization modeling has several advantages and uses in studying water markets. These uses vary with the water marketing problem and the maturity of water market development in a region. Water markets fundamentally change the political and financial relationships among water management institutions in a decentralized regional system (Pulido-Velázquez *et al.* 2004). Early in the introduction of markets, great policy uncertainty exists regarding the effects of markets on the stability of water systems and the groups which benefit from them (Vaux and Howitt 1984). Planning concerns naturally exist on how operations with markets will affect demands on other water supplies, infrastructure operations, and water deliveries under various hydrological conditions, particularly droughts (Jenkins *et al.* 2004). Water demand decisions also can change when water markets are introduced (Cary and Zilberman 2002). Ultimately, optimization models can be applied in an operations planning time frame to help determine market participation decisions and how much of which water products available in a market should be purchased under different conditions (Hollinshead and Lund 2006). The major advantage of optimization models in the study of water markets is that they can better represent the more distributed nature of decision-making with markets; their greatest disadvantage is that they often must represent other policy and physical processes in a more simplified way than simulation models.

Simulation and optimization

Most major water systems worldwide have at least one system model. Often a basin will have several models, with different time scales (hourly, daily, or monthly) and different spatial resolutions to aid with different operational or planning decisions. Usually these are simulation models. There are frequent discussions in the literature of the complementarity of simulation and optimization modeling, particularly for planning and policy purposes (Jacoby and Loucks 1972). For such problems, regional water systems have immense numbers of water management options from which decision-makers can choose. The numbers of alternative combinations of options make examination of most alternatives combinatorially impossible by simulation modeling alone. Even with only 100 binary options, which can be either selected or not, there would be $2^{100} \approx 1.3 \times 10^{30}$ potential alternatives to examine by simulation.

Optimization models, which usually require a simpler representation of the system to be mathematically tractable, commonly handle tens of thousands of decision variables (even millions on occasion). This allows optimization models to examine the broad decision-space of integrated water management solutions for those which appear to be most promising. These identified promising solutions can them be tested and refined using more detailed simulation modeling (Lund and Ferreira 1996).

Strategies for model development and use

Various approaches are available for developing models that incorporate water markets as part of more complex water management systems. I would like to highlight modeling which has disintegrated components, modeling with distributed components, and integrated modeling, and discuss the roles of databases in these processes.

It is common to have a variety of models which are loosely linked to provide simulation capability for a basin. These models are often the result of a history of model development in the basin by different interests and agencies. The models are often only loosely coordinated, and could be characterized as disintegrated components of a regional modeling system. Humans are needed to make these models work together; this is not a smooth process, but can inject additional human thought (and expense) into any resulting analysis. Optimization cannot be done in this manner; it is too laborious and slow. But many of the most complex systems can only be simulated in this way, examining very few alternative policies.

The modular development of distributed model components can help rationalize the modeling of complex systems, reducing the need for humans at the interface of modeled processes. Ideally, the distributed component models have well-designed input and output variable sets which allow rational and rapid communications between modules, allowing simulation to be much more easily done, and some forms of optimization (perhaps evolutionary algorithms). The modularization allows model development to occur while the system is being used, with components being improved incrementally to upgrade the modeling system. Alas, this approach is rare, since a river basin rarely has the ability to develop the overarching technical framework and institution needed to design and develop a distributed model system.

Integrated modeling is then all processes to be modeled are combined into a single integrated model. The CALVIN model of California's intertied water system is an example (Jenkins *et al.* 2004). Traditionally, this type of model has been wonderful for its initially developed task, but has been cumbersome to adapt to the ever-changing analytical needs of real river basins.

The increasing scope and detail of system models of water problems and the rising speed of computing has led us into a new era where models and modeling algorithms are less of a problem than managing the input and output data. Models cover much larger areas in ever greater detail, so models increasingly represent our integrated knowledge and understanding of water systems in far greater detail than an individual can possess. Databases should both store inputs data sets for modeling and contain the documentation, explanation, and origins of these numerical values are important to make these models, and the understanding they represent, something more than a "black box." Databases and models increasingly represent the experience of veterans who have worked with the system and the training-ground of new professionals to a system. Modeling and databases are the textbook for technical people to more rapidly and completely document a system, in much the same way that Frontinus (97 AD) used his treatise on the water supply of ancient Rome to better understand Rome's water supply, as well as to communicate it to others.

Modeling water markets in California using optimization

The CALVIN model of California's intertied water supply system was developed between 1998 and 2001 to examine integrated water management and water markets for California; see Figure 20.1 (Draper *et al.* 2003; Jenkins *et al.* 2004). It is an economic-engineering optimization model based on generalized network flow programming using the HEC-PRM reservoir system optimization software. Urban, agricultural, and hydropower water demands are economic, based on users' willingness to pay for water delivery and use. Operating costs are also included for pumping, treatment, and urban water quality.

The model contains over a million decision variables for optimizing a wide range of surface water, ground water, pump, treatment plant, recharge, reservoir operation, water conservation, water reuse, and water allocation decisions. The model is usually optimized for monthly decisions over a 72-year period, allowing a range of wet and dry conditions to be considered.

The model has been applied to a wide range of studies of water marketing, integrated water management, climate change, dam removal, conjunctive use, environmental policy, and other issues (Jenkins *et al.* 2001, 2004; Newlin *et al.* 2002; Draper *et al.* 2003; Tanaka and Lund 2003; Tanaka *et al.* 2006, 2011; Pulido-Velázquez *et al.* 2004; Null and Lund 2006; Lund *et al.* 2007; Medellín-Azuara *et al.* 2008; Harou and Lund 2008; Rosenberg *et al.* 2008). As an optimization model, it can adapt operations economically to major changes in many aspects of the statewide water system. While this presumes a somewhat unrealistic level of coordination statewide, California's decentralized system does have a fairly good record of cooperation, particularly with the advent of water markets.

Some tricky issues

Optimization modeling of large regional water resource systems faces many tricky issues. These issues do not mark the difficulty or futility of regional modeling, but rather mark that the approach has been successful enough to struggle with details. Most major water systems now rely significantly on simulation modeling capability and there is increasing investigation and use of optimization modeling. Some of these tricky issues often require the modeler to make trade-offs regarding the purposes of the model and

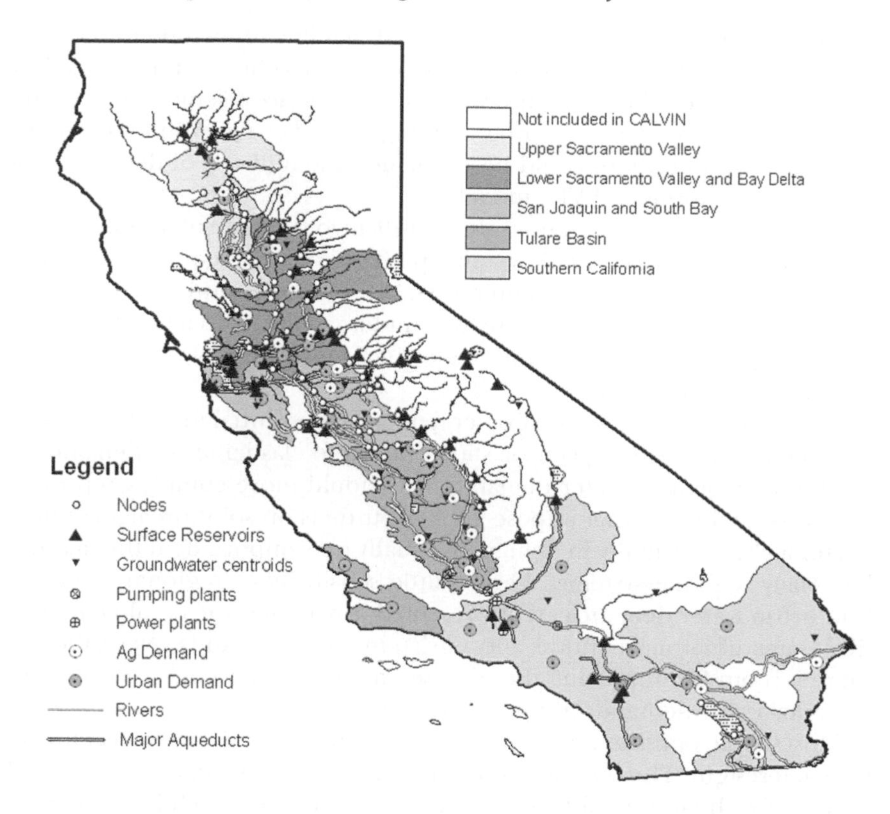

Figure 20.1 Schematic of CALVIN model superimposed on a map of California

available data and representations. These trade-offs then require interpretations and cautions when the model is used outside of its strict and formal range of application, as will be required practically. "All models are wrong, but some are useful," as George Box once wrote, and an imperfect modeling capability is often superior to our unaided intuition when there is no time to build modeling capability better suited to a particular question.

Several of these tricky issues are briefly described in the following paragraphs. First, should we model many details of spatial and temporal representation with a simple (fast) mathematical formulation or a rougher representation of the system using a more complex (but slower) mathematical formulation? The CALVIN model includes a relatively detailed representation of California's water system on a monthly time-step, with a wide range of water management infrastructure, operations, and water demand management options – all represented as a simple generalized network flow optimization formulation. With the fast, cheap, and public domain solver comes the limitations of this very simple formulation – perfect hydrological foresight, inability to include non-convexities, fixed (non-head-dependent) pumping costs, and inability to include in the optimization system-wide hydropower production or other non-network-flow variable of interest. However, models which can address some of these limitations, solved by dynamic programming, genetic algorithms, or other non-linear optimization, could never include the millions of decision variables implied in the CALVIN model.

Groundwater representation always will involve trade-offs. Solution of the detailed groundwater flow (and perhaps quality) equations implies great data and computation requirements. It is simpler and faster to represent groundwater as a surface reservoir underground, with a fixed unit cost for pumping withdrawals and neglecting detailed flows and surface-groundwater interactions.

Economic representation of water demands is essential for water marketing studies. Should long-run or short-run price elasticities of demand be used in representing water demands? Or should more complex representations be included (that impose greater burdens on solution algorithms). Fortunately, uncertainty in demand is usually only important at the margin. For many regional systems, there is little sensitivity of regional results to how urban water demands are represented, since they are so valuable relative to marginal agricultural and environmental demands. Nevertheless, there is room for improving the representation of water conservations and marginal water demands in the urban sector.

Most simple optimization models are "too smart" in having perfect hydrological foresight. The optimization can see floods and droughts coming and prepare for them unrealistically in advance. While too much is often made of this limitation, it is nonetheless real and sometimes very significant. Simulation models, in contrast, are often "too dumb" in not taking advantage of opportunities that smart water system operators would likely see and take advantage of. There is an inherent trade-off in selecting models which are "too smart" or "too dumb" in terms of hydrological knowledge, as well as, for water marketing, knowledge of water demands (prices).

Another contrast between simulation and optimization modeling is that optimization models often assume zero institutional viscosity, whereas simulation models often assume zero institutional flexibility. Neither approach is likely to be the case during real droughts or other times of challenge.

Alas, we must always think about model results.

Mathematics of markets and negotiated consensus

For large water resource systems with many interested parties, it is sometimes said that negotiated or otherwise political solutions are better than market-driven water re-operations. This might sometimes be true, but negotiated and political solutions are often difficult for large decentralized systems, with many interested parties who have an ability to stall or otherwise prevent timely changes to water management to accompany improvements in the understanding of the water system or external changes such as changes in climate change or water demands.

If n parties must agree to a negotiated solution, and each party has a probability P_a of being agreeable to a given solution, then the probability of all parties agreeing to the solution is P_a^n. Even parties whose best interests are served by an agreement can have internal political, personal, or organizational reasons for being disagreeable (Madani and Lund 2012). The mathematics of the probability of an overall agreement under these circumstances appear in Figure 20.2. The prospects of a purely negotiated solution seem bleak with many required parties, even for high internal abilities of parties to be agreeable.

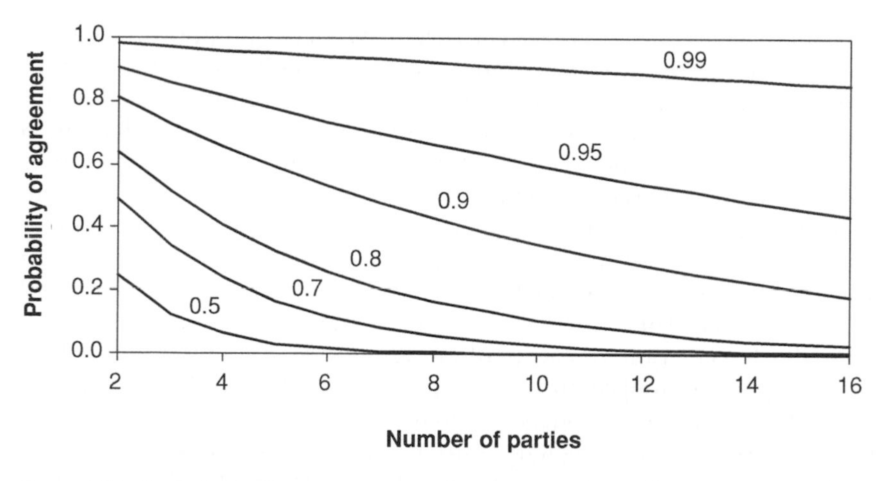

Figure 20.2 Probability that agreement consensus can be achieved with number of parties and probability of individual party agreement

While markets have many imperfections, they are fairly good at giving many parties an interest in coming to an agreement. Without markets, senior and junior right-holders are in a legal struggle; with functioning markets, newer users with higher valued uses can come to reasonably quick agreement with more senior right-holders, who themselves now have incentives to come to agreement. While not solving all problems of negotiated problem-solving, such markets sometimes can resolve some problems. Where rapid or sustained flexibility is needed with individual parties having incentives to cooperate, markets can be a useful policy instrument.

Mature versus immature markets

Most controversies regarding water markets are where they have not become accepted. This sounds a bit tautological, but is also true in the sense that in regions where water markets have developed, they have become part of the fabric of water management and the operation and regulation of the water system has largely adjusted to and profited from their existence.

California has gone significantly through this type of transition, and optimization models have had a changing role in the course of this development. Vaux and Howitt (1984) were among the first to employ optimization to assess the potential value of water markets in California to reduce the cost of water provision and improve its economic efficiency. Early in the history of major water markets, such analysis was important for policy-making.

However, well-developed water markets are not monolithic, but can take various forms, including permanent sales, long-term sales, spot market sales, and various forms of options (particularly dry-year options). As water markets have become more accepted, modeling has become more sophisticated to include greater representation of infrastructure capacities and operations and how water markets function (Newlin *et al.* 2002; Jenkins *et al.* 2004; Pulido-Velázquez *et al.* 2004). In such studies, we can see the larger potential of water markets to help coordinate the many otherwise disparate interests involved in managing most large regional and inter-regional water systems. In many cases, markets change the nature of water management (in a game-theoretic sense) such that agencies which previously competed for water now have incentives to cooperate and share water and water operations. Such cooperative behavior often can be seen in economically-driven regional system optimization models (Jenkins *et al.* 2004).

Finally, optimization models can have a role in aiding potential water buyers and sellers in making water purchase and sale decisions from among a wide range of potential water products. Lund and Israel (1995) present a two-stage optimization model to aid urban water purchasers in selecting amounts of water for permanent and spot-market purchase, as well as the purchase of options and the exercise of options. Hollinshead and Lund (2006) propose a more detailed three-stage linear program for an environmental water account to purchase and sell water over the course of a season as hydrologic, infrastructure, market, and environmental conditions become clearer.

Conclusions

Optimization modeling is becoming more common for water management for policy, planning, and operational decision-making. While still unusual in practice, it offers some distinct advantages when trying to explore novel solution approaches in complex systems or when trying to adapt operations quickly to changing conditions with well-defined objectives. For water marketing, optimization is particularly useful for its ability of better represent the mode decentralized nature of decision-making with markets.

The role of optimization in water market studies varies with the maturity of markets and local conditions. Early in the development of markets, optimization has been useful in policy studies to compare the potential economic value of markets with other water management and policy alternatives. As markets are being developed, optimization models help identify potential market transactions and desirable changes in infrastructure and operations to accompany the introduction of markets. Potential problems and undesirable impacts of markets can also be identified from such model runs. With mature markets, optimization results can help guide operational decisions on water purchases, sales, and water operations in the context of markets.

References

Butcher, W. S., Haimes, Y. Y. and Hall, W. A. (1969) "Dynamic Programming For The Optimal Sequencing Of Water Supply Projects." *Water Resources Research* 5(6): 1,196–1,204.

Carey, J. M. and Zilberman, D. (2002) "A Model of Investment under Uncertainty: Modern Irrigation Technology and Emerging Markets in Water." *American Journal of Agricultural Economics* 84(1): 171–83.

Draper, A. J., Jenkins, M. W., Kirby, K. W., Lund, J. R. and Howitt, R. E. (2003) "Economic-Engineering Optimization for California Water Management." *Journal of Water Resources Planning and Management* 129(3): 155–64.

Frontinus, S. J. (97AD) *The Water Supply of the City of Rome.* Trans. Clemens Herschel. Charleston, SC: Nabu Press [2012].

Hanak, E., Lund, J., Dinar, A., Gray, B., Howitt, R., Mount, J., Moyle, P. and Thompson, B. (2011) *Managing California's Water: From Conflict to Reconciliation.* San Francisco, CA: Public Policy Institute of California.

Harou, J. J. and Lund, J. R. (2008) "Ending Groundwater Overdraft in Hydrologic–Economic Systems." *Hydrogeology Journal* 16(6, September): 1,039–1,055.

Harou, J. J., Medellín-Azuara, J., Zhu, T., Tanaka, S. K., Lund, J. R., Stine, S., Olivares, M. A. and Jenkins, M. W. (2010) "Economic Consequences of Optimized Water Management for a Prolonged, Severe Drought in California." *Water Resources Research* 46: W05522 (doi:10.1029/2008WR007681).

Hollinshead, S. P. and Lund, J. R. (2006) "Optimization of Environmental Water Account Purchases with Uncertainty." *Water Resources Research* 42(8, August): W08403.

Jacobs, J., Freeman, G., Grygier, J., Morton, D., Schultz, G., Staschus, K. and Stedinger, J. (1995) "Socrates: A System for Scheduling Hydroelectric Generation under Uncertainty." *Annals of Operations Research* 59: 99–133.

Jacoby, H. D. and Loucks, D. P. (1972) "Combined Use of Optimization and Simulation Models in River Basin Planning." *Water Resources Research* 8(6): 1,401–14.

Jenkins, M. W. and Lund, J. R. (2000) "Integrated Yield and Shortage Management for Water Supply Planning." *Journal of Water Resources Planning and Management* 126(5, September/October): 288–97.

Jenkins, M. W., Draper, A. J., Lund, J. R., Howitt, R. E., Tanaka, S. K., Ritzema, R., Marques, G. F., Msangi, S. M., Newlin, B. D., Van Lienden, B. J., Davis, M. D. and Ward, K. B. (2001) *Improving California Water Management: Optimizing Value and Flexibility*. Center for Environmental and Water Resources Engineering Report No. 01-1. Davis, CA: Department of Civil and Environmental Engineering, University of California. Available at http://cee.engr.ucdavis.edu/faculty/ lund/CALVIN.

Jenkins, M. W., Lund, J. R., Howitt, R. E., Draper, A. J., Msangi, S. M., Tanaka, S. K., Ritzema, R. S. and Marques, G. F. (2004) "Optimization of California's Water System: Results and Insights." *Journal of Water Resources Planning and Management* 130(4, July): 271–80.

Labadie, J. W. (2004) "Optimal Operation of Multireservoir Systems: State-of-the-Art Review." *Journal of Water Resources Planning and Management* 130(2, March/April): 93–111.

Little, J. D. C. (1955) "The Use of Storage in a Hydroelectric System." *Operations Research* 3: 187–97.

Lund, J. R. and Ferreira, I. (1996) "Operating Rule Optimization for the Missouri River Reservoir System." *Journal of Water Resources Planning and Management* 122(4, July/August): 287–95.

Lund, J. R. and Israel, M. (1995) "Optimization of Transfers in Urban Water Supply Planning." *Journal of Water Resources Planning and Management* 121(1): 41–8.

Luss, H. (1982) "Operations Research and Capacity Expansion Problems: A Survey." *Operations Research* 30(5, September–October): 907–47.

Madani, K. and Lund, J. R. (2012) "California's Sacramento-San Joaquin Delta Conflict: from Cooperation to Chicken." *Journal of Water Resources Planning and Management* 137(2, March/April): 90–99.

Massé, P. (1946) *Les Reserves et la Regulation de l'Avenir Dans La Vie Economique*. Paris, France: Hermann.

Medellín-Azuara, J., Lund, J. R. and Howitt, R. E. (2007) "Water Supply Analysis for Restoring the Colorado River Delta, Mexico." *Journal of Water Resources Planning and Management, ASCE* 133(5, September/October): 462–71.

Medellín-Azuara, J., Harou, J. J., Olivares, M. A., Madani-Larijani, K., Lund, J. R., Howitt, R. E., Tanaka, S. K., Jenkins, M. W. and Zhu, T. (2008) "Adaptability and Adaptations of California's Water Supply System to Dry Climate Warming." *Climatic Change* 87(Sup. 1, March): S75–90.

Murk, N. and Lund, J. R. (1996) *Application of HEC-PRM for Seasonal Reservoir Operation of the Columbia River System*. Technical Report RD–43. Davis, CA: Hydrologic Engineering Center, US Army Corps of Engineers.

Newlin, B. D., Jenkins, M. W., Lund, J. R. and Howitt, R. E. (2002) "Southern California Water Markets: Potential and Limitations." *Journal of Water Resources Planning and Management* 128(1, January/February): 21–32.

Null, S. and Lund, J. R. (2006) "Re-assembling Hetch Hetchy: Water Supply Implications of Removing O'Shaughnessy Dam." *Journal of the American Water Resources Association* 42(4, April): 395–408.

Pulido-Velázquez, M., Jenkins, M. W. and Lund, J. R. (2004) "Economic Values for Conjunctive Use and Water Banking in Southern California." *Water Resources Research* 40(3, March): W03401.

Rosenberg, D., Howitt, R. and Lund, J. R. (2008) "Water Management with Water Conservation, Infrastructure Expansions, and Source Variability in Jordan." *Water Resources Research* 44: W11402 (doi:10.1029/2007WR006519).

Shawwash, Z. K., Siu, T. K. and Russell, S. O. (1999) "The B.C. Hydro Short Term Hydro Scheduling Optimization Model." *Power Industry Computer Applications, PICA '99.* Proceedings of the 21st 1999 IEEE International Conference.

Tanaka, S. K. and Lund, J. R. (2003) "Effects of Increased Delta Exports on Sacramento Valley's Economy and Water Management." *Journal of the American Water Resources Association* 39(6, December): 1509–19.

Tanaka, S. K., Zhu, T., Lund, J. R., Howitt, R. E., Jenkins, M. W., Pulido, M. A., Tauber, M., Ritzema, R. S. and Ferreira, I. C. (2006) "Climate Warming and Water Management Adaptation for California." *Climatic Change* 76(3–4, June): 361–87.

Tanaka, S. K., Buck, C., Madani, K., Medellín-Azuara, J., Lund, J. and Hanak, E. (2011) "Economic Costs and Adaptations for Alternative Regulations of California's Sacramento-San Joaquin Delta." *San Francisco Estuary and Watershed Science* 9(2, July): 28.

Tilmant, A., Pinte, D. and Goor, Q. (2008) Assessing marginal water values in multipurpose multireservoir systems via stochastic programming, *Water Resources Research* 44: W12431 (doi:10.1029/2008WR007024).

USACE (1993) *Columbia River Reservoir System Analysis: Phase II.* Technical Report no. PR–21. Davis, CA: Hydrologic Engineering Center, US Army Corps of Engineers.

Varlet, M. (1923) "Étude Graphique des Conditions d'Exploitation d'un Réservoir de Régularisation." *Annales des Ponts et Chaussées, Partie Technique* 93e année, 11e série-time 64, fasc. 4. tome 2, July–August.

Vaux, H. J. and Howitt, R. E. (1984) "Managing Water Scarcity: An Evaluation of Interregional Transfers." *Water Resources Research* 20(7, July): 785–92.

Yeh, W. W-G. (1985) "Reservoir Management and Operations Models: A State-of-the-Art Review." *Water Resources Research* 21(12): 1,797–1,818.

21 Conclusions and recommendations for implementing water trading

How water trading can be part of the solution

Josefina Maestu and Almudena Gómez-Ramos

There is important experience on trading today that allows us to draw some relevant conclusions and recommendations. This chapter draws the main messages from prior chapters on what are the challenges for water management and the new tools available, on the trading experiences, and on the concerns on water trading and how we are dealing with them. It does so in a positive manner, making recommendations on how trading should be considered and implemented.

Conclusions and recommendations made in this chapter are on water scarcity and drought, future scenarios, and institutional reforms; on improving efficiency in water allocation and the role of water trading; on how water trading can provide a mechanism to delay and possibly eliminate undertaking costly civil water works; on the price in water trading operations; on the role of public sector to regulate water trading performance; on the legal and institutional frameworks to facilitate reallocation through voluntary agreements; on improving water markets flexibility and transparency through water rights reform; on improving water quality by water quality trading; on water markets to cope with risk and uncertainty; and on detailed issues for improving implementation of water trading.

On water scarcity and drought, futures scenarios, and institutional reforms

Water resources may be affected as rainfall and evaporation patterns get modified due to climate change. In some regions, climate change will decrease water availability, increasing the opportunity cost of water and, therefore, decreasing the availability of water per capita. But at the same time, other climate-driven changes, for instance the rise in temperature, will increase water use. Whether climate change will cause future per capita water use to increase or decrease depends on the magnitude of changes in supply and demand. Equally, as well as a potential reduction in water resources, one of the most dramatic impacts for future water planning (and future water trading) is the increased frequency and scale of extreme events such as droughts and floods.

Climate change, as a global process, could potentially increase the incidence and outcomes of future water disputes and conflicts. Pressures over water could not only be directly influenced by climate-driven changes in water availability, but also by people's behavioral response to climate change, impacting water demand. If future water use increases, disputes could increase as well, not only in areas where water is already scarce, but also in areas where water is currently abundant, but expected to become scarce due to climate change.

In order to tackle increasing social concerns and disputes over water, international funding and policy efforts are required to strengthen conflict resolution mechanisms and develop the right institutional frameworks for cooperation before the system is tested or put under pressure due to climate change and increased water scarcity. These efforts should target on countries with severe water scarcity, increased water demand, and where there is already a history of potential water-related environmental conflicts. A necessary condition to facilitate the negotiation is to develop a well-established and enforced water property rights system. It is important to consider the relative strengths and weaknesses of particular property rights systems, including open access and indigenous rights, for other water resource scenarios.

Increasing water stress and recent droughts, along with unfavorable climate change scenarios, have brought greater awareness over the need of more effective planning for water shortages around the world. Ultimately, tackling the inter-related issues on water scarcity, drought, and climate change adaptation will require a combination of local action, national direction, and regional and global coordination. There is also a place for private investment, as well as public oversight and funding. A balance must be struck between local autonomy and general standards for meeting basic human needs. The test will be to make the systems capable of adapting to change (i.e. resilient water right systems), which are flexible enough to cope with change, yet stable enough to provide security.

International frameworks and programs, such as the UNCCD and UNFCCC initiatives, the Millennium Development Goals, and the Hyogo Framework for Action, provide important coordinating mechanisms, and advocate the importance of risk reduction, through a more proactive planning and water resources management. Global institutions must coordinate the implementation of tasks, like the identification and implementation of alternative risk management strategies – critical for reducing the risks associated with water stress, drought, and climate change – and endorse the funding necessary to carry out these activities. However, in terms of appropriateness of scale, specific mitigation and preparedness actions should ideally implemented at the national and local level.

Climate change will also require innovation and experimentation on several areas, in particular related to new water allocation instruments, to ensure equitable access to resources, taking into account uncertainty management, by incorporating this in the planning process.

On improving efficiency in water allocation and the role of water trading

Historically, the allocation of water use rights has not been determined by efficiency criteria. In the new circumstances, characterized by increasing water scarcity and water conflicts, an efficient use becomes even more important to make economic growth sustainable and compatible with the limited availability of water resources. The reallocation of water use rights through voluntary agreements – with some form of compensation – may be useful to improve the allocation of water; yet it will not necessarily always lead to efficient water use. Societal debates and discussion will have to be opened up in seeking a balance between the most equitable and the most efficient water uses, which are not necessarily always compatible. Some difficult trade-offs might have to be made, which can only be done legitimately though a transparent and consultative process with all the relevant stakeholders, which would ideally include those with prior water use rights and other, newer uses, or uses impacted by re-allocation patterns.

Allocative efficiency on its own is not the only objective of water policy. Allocative efficiency would need to be considered along other objectives of public policy such as equity, spatial balance, preservation of rural activities and the environmental services provided by them, guaranteeing employment and rural incomes. This way water trading – rather than being an exclusive instrument for water policy – becomes an important support or complementary tool to make improved water efficiency compatible with other water policy objectives such as hydrological balance, employment and preservation of traditional activities. The ideal policy mix would normally include a mix of regulation, market based instruments and information or horizontal instruments. Furthermore, when water markets are introduced, there may be the need to design complementary measures to achieve the public objectives that are pursued as a whole by water policies.

When water trading becomes a real option, voluntary transactions could become an important instrument to improve efficiency in water use. By allowing potential selling of "options" to water use rights, water markets may provide the missing incentives to encourage both a more responsible water use and the implementation of water conservation measures to enhance, for example, water reutilization and investment in wastewater treatment techniques. In a long term arrangement, water trading may provide the financial incentives for research, development and innovation of advanced technologies for using and conserving water. Equally, there is a lot to be learnt potentially from informal water markets that develop spontaneously (such as those discussed in this book in India) to cope with water scarcity, as a bottom up approach and its potential advantages and limitations.

Water trading can provide a mechanism to delay and possibly undertaking costly civil water works

Water trading can be designed so that both water and risk are efficiently allocated. Trading can consider not only economic compensation to users but also risk compensation. This involves unbundling the attributes of water rights, and establishing trading mechanisms that allow for exchange of certain, not all, attributes. Water use rights can be defined as contingent to water supply to be able to cope with droughts.

Other contingency arrangements that involve trading can be implemented to smooth out or even out the negative consequences of short term water scarcity. "Spot markets" (active water transfers during drought periods) temporary transactions and agreements between water use right holders, and permanent sales of water entitlements among holders are some of the available alternatives. Further research is needed, however, to know under which conditions these different types might (or might not) be more suitable. Furthermore, water banks can take an active role in transactions buying and selling water rights directly, or holding the rights to improve environmental flows, or to achieve other social benefits.

However, the practical implementation of water trading schemes is still challenging in real life cases. Since the true economic value of water is dependent on time and location, the number of potential participants in any possible water transaction is less than the minimum required for an effective competition to exist. In most cases the water trading framework is characterized by a bilateral monopoly rather than a perfect framework for competition. Efficiency in water transactions depends on highly complex water market information, since both private and public agents act with imperfect information (water availability, external costs and water productivity are somehow uncertain) and the information available to all the participants is not complete (if there is asymmetric information about willingness to pay and accept payment). Finally, costs and benefits of any water transaction are not always internalized in the potentially attainable private benefits. Under certain circumstances, voluntary water rights trading are indeed the source of policy relevant information. One potential way to reduce transactions costs due to information asymmetry is to develop collective agreements with water users to share the information load and obtain true, real time data on water use and trading. This therefore, equally requires detailed knowledge on potential incentives for cooperation and information sharing.

By improving potential buyers' welfare without worsening the welfare of sellers of water rights, the voluntary agreements incentivized by water markets can potentially make room to mutually beneficial improvements; whilst at the same time reveal the scarcity rent of water to both society and policy-makers.

On the price in water trading operations

By revealing the willingness to pay of the potential buyers (closer to their marginal productivity of water) and the required compensation by potential sellers (higher than its marginal productivity of water because willingness to accept includes other factors like legal or cultural considerations), water prices may signal the water scarcity value. The price paid in a water transaction would be then a signal of the scarcity rent that would accrue when water is used by those users that can obtain more value for it. Water allocation must consider not only the opportunity cost of water use, but also the production costs. Trading prices should take into consideration the costs of management, monitoring and the costs of infrastructure conveyances. In this sense the cost recovery principle of water services should be applicable in all water trading operations.

Price may also depend on the relative power of the concerned parties and especially, the key role played by the regulator like water authorities and public agents, acting at the same time as promoters or facilitators of water transactions and assuring that private transactions preserve the public interest to prevent monopolistic prices.

On the role of public sector to regulate water trading performance

In spite of the efficiency advantages of water markets, public policy intervention is essential for a number of reasons. The reasons include: preservation of collective goods, including the environmental benefits associated with water ecosystems conservation, the control of market power of the different participating agents, insuring homogeneous information for the partners involved and transparency of operations.

An adaptive water planning processes sufficiently robust is necessary to ensure environmental and other public benefits. Water institutions should regulate water transaction contracts so that this takes into account environmental flows and downstream user interests. The river basin authorities also have a role in monitoring and evaluating the effects of transactions, to prevent unintended consequences.

In practical water trading situations, with a limited number of agents (each one acting with uncertain and limited information), and with potential external effects over collective goods and other third parties, it is the public sector that may need to set the playing ground where "market transactions" are made and "voluntary agreements" are reached. By setting the rules of the game – potentially traded amounts of water, the permits and prices of using public facilities to transfer water from one location to the other, accreditation of potential buyers or sellers, payments due to potential externalities caused by the transaction, and so forth – public intervention in water trading might define the outcome of any potential voluntary agreement in a significant way.

Public authorities must guarantee that all those involved in trading have actually legally effective water use licenses/rights covering the amounts of water actually traded. This is important to insure that trading does not create incentives for higher water use and uncontrolled water extractions.

Recent experiences of implementing short term transactions of water use rights demonstrate the need of water agencies to regulate and oversee transactions, including the establishment of formal rules for the operation of transactions. Formal trading rules not only take into account compensation issues but also consideration of what are the legally established water use priorities, the level of security of the different licenses, the degree to which general welfare may be affected and the environmental impact of the resulting proposal use. Water institutions must inform about water markets rules to ensure that potential buyers and sellers access to market at equal conditions. Also, public water institutions should consider public participation in order to protect for society's interests in environmental sustainability through processes of allocating water that take into account the public interest.

To avoid monopolistic prices, water markets need to develop instruments that ensure transparency and information to all parties potentially interested in entering into trade. The potential use and benefit of new technologies on information management have to be fully explored and utilized.

Government intervention is essential then to prevent monopolistic prices, avoid the "activation" of unused rights – unless they reflect conservation practices – and promote quick seasonal allocation with low transaction costs. Good performance of water markets demands synergies with others economic instruments like pricing for water services. Government intervention can determine the distributive results of trading, with the government acting as regulator.

The amount of water rights effectively used in the economy and the amount of water rights traded must be clearly known by public authorities and affected stakeholders, taking into consideration the desired level of water conservation and the status of all the water bodies providing these values to society. The level of water conservation guaranteed by the public sector can also be increased or relaxed by using market mechanisms – to increase environmental flows or to recover the level of aquifers – all this requires a clearly defined set of objectives for public water policy, including water use and water conservation, something that is still a challenge to most of the countries and regions where water markets are considered as an option.

On the legal and institutional framework to facilitate reallocation through voluntary agreements

For water trading to occur, it is necessary to have in place an institutional and legal system that takes into account the complex nature of water. The economic and cultural context must also be appropriate for the development of water trading. Social factors are an essential element in the development of any policy tool in compliance with the law. Broad social legitimacy in judging design and performance are a necessary, almost essential condition to guarantee or ensure successful implementation of trading mechanisms.

It is necessary to develop holistic legal reforms from the outset to facilitate water trading, promote economic efficiency and an adequate environmental allocation of water, as well as better manage supply risks and lessen environmental impacts in the short to medium term. It is important to modify water use rights so that there are differences consider the timing and peak load congestion issues and to incorporate water quality aspects in water rights. This holistic reform can imply adaptive statutory water planning processes which will ensure water for the environment and other public benefits – cultural uses, as well as spiritual sites for some communities – and define how much water is available for extractive and consumptive uses and how much water has to be kept in the system to assure its environmental sustainability and long term viability.

Water transactions must consider local interests. It is important to point out that there are countries in which the water management system is insufficiently mature for the implementation of water markets. In these cases, emphasis has to be placed on the implementation and improvement of appropriate systems for the allocation and registration of water rights. Voluntary instruments – education and training – as well as adequate regulatory frameworks, have to be in place before adopting approaches like water trading. Any approach needs to be tailored to local circumstances.

On improving water markets flexibility and transparency through water rights reform

For voluntary trading agreements to function properly the importance of a well-defined set of water property rights cannot be exaggerated. In many cultures and legal systems across the world (including indigenous) water itself as a resource is essentially perceived as a public asset and only the water use "rights" can be traded. At any moment, these options may be constrained by water supply, water use priorities, and a complex system of legal regulations. Overuse of water resources implies that in many cases the amount of allocated water rights actually may exceed the real capacity of the river basin to provide water services to the local or regional economy. This is a condition that may be worsening in many regions of the world, with increasing future uncertainty of available water resources, due to the effects of climate change. In this context, entering into the property rights market with a "legal" right (or so-called "paper rights") that is not grounded on an available physical resource must be avoided. A problem that will have to addressed in many areas of the world is the over allocation of water rights as well as the fact that an important share of water resources effectively being used in the economy proceed from illegal abstractions.

Water rights must be then well defined to facilitate water transactions. There should be a registry on the ownership of water entitlements and of the transaction agreements that is transparent and readily available to all at a low cost, and with the full backing of the law. Options include a publicly operated system similar to the one used for land titles, or a system similar to the public register of companies. These registries could be administered by a public agency or by a regulated private entity. However, the main cases where water right registries are working properly have also relied on a strong, positive collaboration with those that either own the rights or use the rights.

The term "water markets" hides the essence of the rights that effectively are – and must be allowed to be – traded. In practice what is effectively traded is not water but the temporary/or permanent option to use a certain quantity of water at some time under some very specific conditions. In order to achieve the objective of sharing risk efficiently a new understanding of water rights must be introduced. Water rights or water access licenses or concessions should be unbundled to identify and single out their attributes so that access entitlements, seasonal allocations and use approvals can be managed using separate instruments and independent processes. Unbundled rights allow for transferring also the right to access water most reliably. Water rights then must include different dimensions such as: volume, season, quality, time of use, and so on. Water transfer contracts could consider trading only some of these attributes to facilitate flexibility in water transactions. Water users have to be directly involved in defining these contracts,

Improved seasonal allocation of water through water markets need to be based in well-defined contracts. These kinds of mechanisms can allow water trading with more flexibility and will be better adapted to the uncertainty context they are designed to respond to. Water right systems regulated in a stable way and adapted to the contexts, can serve to increase the guarantee of water supplies and make them less vulnerable. Both third party effects and environmental externalities can be incorporated in the contracts in a concrete way.

Future forecasting on potential future uses and scenario planning should also be undertaken, particularly in relation to the economic return of the water traded and whether there are any limitations for future trading.

On improving water quality by water quality trading

Water trading can also be an instrument to improve water quality. The range of potential economic uses that can be supplied by a water source depends on its specific quality. For this reason, improving chemical quality becomes a way to increase the number of potential buyers and the financial value of the tradable water rights. Water markets may create incentives for the use of reused or regenerated water. Trading of regenerated water can provide the financial revenues to compensate for the extended treatment required to make wasted water adequate for reuse. By reducing both water abstractions and pollution loads the trading of regenerated water can help improving the quality of water ecosystems and reduce water scarcity.

Quality trading should imply significant pollution reductions and encourage innovation. Water quality trading instruments offer a means to establish more robust water rights. However, water must be a certain minimum quality before it is discharged back to a river or aquifer. There should be a strict rule on water quality because of externalities.

The development of efficient instruments will benefit from prior reforms that establish clear water quality goals, targets and catchments based nutrient load (or mass) budgets. An effective policy of water quality trade (WQT) should include stringent pollution caps, measurement of pollution emissions, strong and swift enforcement, and may imply reform of existing prescriptive or "command and control" policies.

One of the most important limitations to WQT is not of a regulatory nature, but a cultural one. Policy-makers and planners need to factor in a transitional period, from an administrative perspective, and ensure that the main potential traders are on board.

Point sources will be more easily included if they are subject to performance-based regulation – that is, the use of load-based emission discharge limits, rather than continuing with concentration-based standards – in an attempt to control both local acute issues as well as cumulative regional loads. With respect to spatial distribution of pollution, or "hot spots," good monitoring of trading and of environmental conditions is essential in that it provides added confidence about the capacity to change course if needed. The new mixed approach adopted by the current EU water framework directive could be a useful testing case.

On water markets to cope with risk and uncertainty

Water markets can be a tool to manage risk in an uncertain context. Changing climate requires continued innovation in water markets to develop tools to cope with greater uncertainty. Uncertainty scenarios and risk must be included into water planning.

Unbundled rights would allow users to directly manage risks commensurate with their production/consumption trade-offs, which is becoming even more of an issue given the impacts of climate change. It is important to implement more sophisticated mechanisms to be able to consider risk and uncertainty: more efficient markets, more secure water entitlements, and an unbundling of the rights traditionally embedded in the water entitlements. Option markets, futures and conditional contracts are the main new mechanisms to improve risk allocation and risk sharing between buyers and sellers. Option contracts are one of the means to harmonize competing users that demand different access reliability, ensuring transparent risk transfer mechanisms.

On improving implementation of water trading

Monitoring water trading

Governments should not issue more entitlements than there is capacity in the system, given environmental objectives, and should always keep the sum of all entitlements in line with system capacity, making clear to all what will happen when unused/unlicensed water is activated. Reversing current situations where legal property rights exceed river basin capacities remains a challenge in most water-scarce countries – a challenge that needs to be addressed with greater uncertainty over available water resources – due to climate change.

Water markets implementation requires adequate human and regulatory means and measures to monitor transactions in order to avoid non-legal re-allocations and discharges.

In order to control the volume of water effectively transferred, installation of meters and conversion from area based licenses to a volumetric management system is needed. Once meters are in place volumetric rights make conversion into unbundled rights easier. On occasions, other monitoring techniques like satellite information, real-time data and GIS could also be incorporated into the monitoring framework

Another element of value would be to institute various screening routes to review trading requests based on the potential disturbance, third-party or externality effects. This would allow for prioritizing resources in the most potentially dangerous exchange, while streamlining those that are potentially more innocuous.

Annual assessment reports (as well as ideally term progress reports) about foreseeable effects of water trading and a post monitoring of final results and impacts should be carried out. Reports should contain analysis of legal, economic, environmental, and social impacts of water trading.

All the above-mentioned monitoring and enforcement issues lead to the conclusion that the administrative cost of water markets is high enough to be considered an important issue. Given economies of scale, these transaction costs might not be justified if transactions are few and scattered. Its importance decreases as water transactions are more common, repeated, and cover more geographical areas, or if costs are shared with water users themselves, either directly through an administrative fee charged by the public authority or charging for the administrative cost to users themselves, while the administration remains as the arbiter, or if it only steps in when needed.

Managing droughts

Water markets could be an instrument to manage drought periods and shortage situations. This way allocation of water use rights become contingent on water supply and, given prior water rights distribution, provides enough flexibility to mitigate the negative economic impact of the reduction of the guarantee of water supplies by establishing a proper compensation to low value added water users. With this aim, the priorities of use must be well defined and included in drought plans in order to guarantee urban supply and achieve environmental objectives.

There are cases where water rights are not being used because of the lack of profitable productive alternatives in a particular location. If water markets serve to activate unused water rights, instead of enhancing water conservation and sustainability, they will in fact contribute to increase present and future water scarcity.

Water banks could be an instrument to manage droughts. Water banks could foresee pre-alert scenarios in which some mechanisms of acquisitions of water volumes would be activated. This option need to demonstrate to be more cost-effective than other alternatives (considering also other objectives of public policy).

Option contracts between urban and agricultural users would be suitable instruments to be incorporated in drought planning as a proactive measure that helps increase security, while avoiding building expensive additional infrastructures.

Establishing market exchanges when water shortages are critical should perhaps be avoided, unless experience has been accumulated in similar, albeit less critical circumstances.

Preserving environmental flows

Externalities are more efficiently managed using separate instruments or mechanisms (such as pricing or environmental standards) in a manner that provides an incentive for people to avoid creating them. The preservation and maintenance of the water flow regimes needs to be more secure than that used to allocate water for other purposes. This could be "owned" by governments, as for the Murray–Darling Basin Ministerial Council acting as environmental water holder in Australia.

Improving transparency and access to market

Equity and fairness objectives require careful attention about the way that trading opportunities, allocation decisions and policy changes are announced and set. Water trading is more effective when information about the opportunities for trading and the prices being paid and offered is made available to all participants at the right time.

Water trading contracts should be made public and participation of stakeholders during the whole process ensures the required transparency. This could include access to available information through a public register containing all water transfers and to all previous transactions and monitoring reports.

The public regulator should make public studies about the value of water in the different uses and the real cost of water trading based on the opportunity cost, costs of conveyance, environmental and third party effects.

Selecting water market options

The suitability of each option depends on local circumstances, uncertainty scenarios, and the regulatory context. However, the options selected have to take into account priorities of allocation and the water demand of different users and their location, as well as the existence of adequate conveyance facilities.

When the objectives pursued are linked with structural changes (overexploitation of aquifers, environmental flow recovery, restructuring of the local sector, promoting new local activities), trading permanent water rights by recovering water use rights – including land acquisition – may be more appropriate.

If the purpose is reducing risk, this may be better achieved through annual spot markets. Annual spot water markets allow for reducing farmers' economic vulnerability derived from variability and uncertainty water availability.

The existence of water trading in agriculture has some degree of novelty, whose consequences go beyond the mere welfare increase. Annual trading with water entitlements allows farmers to sell surplus water or to complete their allocated water, but it also forces them to reconsider the profitability of farm activities. In addition, the establishment of green water credits could be explored to acknowledge the environmental value of rain-fed agriculture.

Effective and reliable water banks not only improve water allocation but also provide more efficient tactical responses to cope with supply uncertainty. The success of water banks depends on the integration of water trading with environmental management and with supply and demand management, and must be integrated in the strategy of water planning at a river basin scale.

Environmental, legal and third party considerations are important in the development and implementation of water banks. For this, the drafting and enforcement of binding contracts among various entities is required. In addition, there has to be an educational effort of the water authorities to inform potential buyers and sellers of water users' rights about the functions of water banks. Water banks reduce uncertainty because the water price reflects water scarcity and influences irrigators' production decisions.

Medium- or long-term option contracts allow urban water authorities to access additional water during periods of scarcity, avoiding high transaction and infrastructure cost that would need otherwise to be paid to insure water provision at critical times. The main drawback is that this kind of contracts must account for and detail any likely eventualities, and the valuation for parties involved may become quite complex. For example, external prerequisites associated with the fulfillment of ecological flows in sensitive river water bodies may be added to the contract's provisions to reduce third party or environmental effects. Common to these mechanisms is the recognition that a group of users holds rights over resources that can be used to offer service reliability to another party at a mutually beneficial price.

Index